Other Books in the Six Sigma Operational Methods Series

Michael Bremer • *Six Sigma Financial Tracking and Reporting*

Parveen S. Goel, Rajeev Jain, and Praveen Gupta • *Six Sigma for Transactions and Service*

Praveen Gupta • *The Six Sigma Performance Handbook*

Thomas McCarty, Lorraine Daniels, Michael Bremer, and Praveen Gupta • *The Six Sigma Black Belt Handbook*

Alastair Muir • *Lean Six Sigma Statistics*

Andrew Sleeper • *Design for Six Sigma*

Design for Six Sigma for Service

Kai Yang, Ph.D.
*Department of Industrial and
Manufacturing Engineering
Wayne State University
Detroit, Michigan*

McGraw-Hill

New York Chicago San Francisco Lisbon London Madrid
Mexico City Milan New Delhi San Juan Seoul
Singapore Sydney Toronto

The McGraw·Hill Companies

Cataloging-in-Publication Data is on file with the Library of Congress

Copyright © 2005 by The McGraw-Hill Companies, Inc. All rights reserved. Printed in the United States of America. Except as permitted under the United States Copyright Act of 1976, no part of this publication may be reproduced or distributed in any form or by any means, or stored in a data base or retrieval system, without the prior written permission of the publisher.

1 2 3 4 5 6 7 8 9 0 DOC/DOC 0 1 0 9 8 7 6 5

ISBN 0-07-144555-2

The sponsoring editor for this book was Kenneth P. McCombs and the production supervisor was Pamela A. Pelton. It was set in Times by International Typesetting and Composition. The art director for the cover was Handel Low.

Printed and bound by RR Donnelley.

This book is printed on recycled, acid-free paper containing a minimum of 50% recycled, de-inked fiber.

McGraw-Hill books are available at special quantity discounts to use as premiums and sales promotions, or for use in corporate training programs. For more information, please write to the Director of Special Sales, McGraw-Hill Professional, Two Penn Plaza, New York, NY 10121-2298. Or contact your local bookstore.

Information contained in this work has been obtained by The McGraw-Hill Companies, Inc. ("McGraw-Hill") from sources believed to be reliable. However, neither McGraw-Hill nor its authors guarantee the accuracy or completeness of any information published herein, and neither McGraw-Hill nor its authors shall be responsible for any errors, omissions, or damages arising out of use of this information. This work is published with the understanding that McGraw-Hill and its authors are supplying information but are not attempting to render engineering or other professional services. If such services are required, the assistance of an appropriate professional should be sought.

CONTENTS

Preface — vii
Acknowledgments — xi

Chapter 1. Six Sigma in Service Organizations — 1

    1.1 Introduction to the Service Industry — 1
    1.2 Success Factors for Service Organizations — 6
    1.3 Overview of Six Sigma — 12
    1.4 Six Sigma for Service — 21

Chapter 2. Design for Six Sigma Road Map for Service — 25

    2.1 Introduction — 25
    2.2 Why Use Design for Six Sigma in the Service Industry? — 27
    2.3 Design for Six Sigma Phases for Service Product — 30
    2.4 Design for Six Sigma Phases for Service Process — 42

Chapter 3. Value Creation for Service Product — 47

    3.1 Introduction — 47
    3.2 Value and Its Elements — 48
    3.3 Maximize Customer Value by Service Product Design — 55

Chapter 4. Customer Survey Design, Administration, and Analysis — 57

    4.1 Introduction — 57
    4.2 Survey Instrument Design — 65
    4.3 Administering the Survey — 72
    4.4 Survey Sampling Method and Sample Size — 73

Chapter 5. Customer Value Management — 83

    5.1 Introduction — 83
    5.2 Market-Perceived Quality Profile — 84
    5.3 Market-Perceived Price Profile — 88
    5.4 Customer Value Map — 89
    5.5 Competitive Customer Value Analysis — 94
    5.6 Customer Value Deployment — 94

Chapter 6. Quality Function Deployment — 101

    6.1 Introduction — 101
    6.2 History of QFD — 103
    6.3 QFD Benefits, Assumptions, and Realities — 103
    6.4 QFD Methodology Overview — 104
    6.5 Kano Model of Quality — 111
    6.6 QFD Analysis — 112
    6.7 Example — 113
    6.8 QFD Case Study: Yaesu Book Center — 122
    6.9 Summary — 126

Chapter 7. Value Engineering — 129

- 7.1 Introduction — 129
- 7.2 Information Phase — 133
- 7.3 Creative Phase — 164
- 7.4 Evaluation Phase — 166
- 7.5 Planning Phase — 172
- 7.6 Reporting Phase — 174
- 7.7 Implementation Phase — 176
- 7.8 Value-Engineering Case Studies — 180

Chapter 8. Brand Development and Brand Strategy — 187

- 8.1 Introduction — 187
- 8.2 The Anatomy of Brands — 190
- 8.3 Brand Development — 206

Chapter 9. Theory of Inventive Problem Solving (TRIZ) — 227

- 9.1 Introduction — 227
- 9.2 TRIZ Fundamentals — 231
- 9.3 TRIZ Problem-Solving Process — 243
- 9.4 Technical Contradiction Elimination and Inventive Principles — 246
- 9.5 TRIZ Applications in the Service Industry — 254
- 9.6 Business Inventive Principles — 255
- Appendix A: Contradiction Table of Inventive Principles — 284
- Appendix B: Business Contradiction Matrix — 290

Chapter 10. Design and Improvement of Service Processes—Process Management — 299

- 10.1 Introduction — 299
- 10.2 Process Basics — 303
- 10.3 Process Types and Process Performance Metrics — 311
- 10.4 Process Mapping — 355
- 10.5 Lean Operation Principles — 362
- 10.6 Process Management Procedures — 377
- 10.7 A Process Management Case Study — 380

Chapter 11. Statistical Basics and Six Sigma Metrics — 393

- 11.1 Introduction — 393
- 11.2 Descriptive Statistics — 394
- 11.3 Random Variables and Probability Distributions — 399
- 11.4 Quality Measures and Six Sigma Metrics — 405

Chapter 12. Theory of Constraints — 413

- 12.1 Introduction — 413
- 12.2 Basic Concepts in the Theory of Constraints — 416
- 12.3 Theory of Constraints Implementation Process — 421
- 12.4 Change Management — 427

References — 431
Index — 437

PREFACE

Six Sigma is the fastest-growing business management system in industry today. It has been credited with saving billions of dollars for companies over the past 10 years. Developed by Motorola in the mid 1980s, the methodology became well known only after Jack Welch from GE made it a central focus of his business strategy in 1995. In the last few years, the Six Sigma movement started spreading from manufacturing industries into various service sectors, such as banking, insurance, hospitals, schools, and many other service organizations. Many service organizations that implemented Six Sigma reported huge successes.

One of the new developments in Six Sigma is Design for Six Sigma (DFSS). DFSS is a systematic methodology that uses tools, training, and project management discipline to optimize the design process of products, services, and processes in order to achieve superior designs to maximize customer value at Six Sigma quality levels. In contrast to regular Six Sigma, characterized by DMAIC (define, measure, analyze, improve, and control), which emphasizes process improvement without fundamental design change, DFSS emphasizes the importance of the design. DFSS contends that only superior design can create products or services with high customer value, low design vulnerability, and high quality. In recent years, DFSS is getting more attention because of its perceived benefits.

Can DFSS be applied to the service industry? Based on the author's extensive research, the answer is a resounding yes.

The service industry exhibits some distinct features that are not found in the manufacturing industry. Based on the work of Sasser, Olsen, and Wyckoff (Sasser et al. 1978), these distinct features include:

1. Many services are intangible; they are not things like hardware.
2. Many services are perishable; they cannot be inventoried.
3. Services often produce heterogeneous output.
4. Services often involve simultaneous production and consumption.

However, no matter what type of service organizations they are, there are three aspects of services that are detrimental to service quality and customer satisfaction (Ramaswamy 1996):

Service Product Service product refers to the service output's attributes or the service items provided to the customers. For example, in restaurant service, the service product includes meals, use of dining utensils, tables, and chairs, music played if needed, and so on. In healthcare service, the service product includes diagnosis, treatment, and care items.

Service Delivery Process Service delivery process refers to the process that delivers or maintains the service products for customers. For example, in a car rental center, the service process includes all steps needed to rent a car to renters. These steps include collecting the driver's license and credit card, checking car availability, filling and printing the contract, obtaining customer signature, delivering the car key and contract to the customer, locating the car, and so on.

Customer-Provider Interaction In a service process, there is also a human interaction aspect, that is, the interaction between customers and service providers. The quality of this interaction will greatly influence customer satisfaction. For example, in the car rental business, the representative should greet customers politely, ask customers their preference of cars, and patiently explain all the options.

Clearly, a customer-value-based superior design and planning in a service product will make services more attractive to customers, and therefore attract more customers and create more revenue for service organizations. Superior designs in service delivery processes will increase the efficiencies of service processes and reduce cost, and therefore increase the profit for service organizations. Excellent designing, planning, and managing of customer-provider interactions will also certainly improve customer satisfaction level and will help to retain customers.

The primary objective of this book is to provide a systematic framework for implementing DFSS in various service industries. From the above discussion, it is clear that DFSS in the service industry should support the following two key activities:

1. Design and planning of service products
2. Design and management of service delivery processes

Chapter 1 of this book begins with the discussion of several key features of service industries and key success factors for service organizations. Chapter 1 also introduces the concept of Six Sigma and how it should be implemented in the service industry.

Chapter 2 introduces DFSS and discusses how itshould be implemented in service industries. In this book, DFSS for service includes two distinct aspects—DFSS for service products and DFSS for service delivery process. The DFSS roadmap for service products and DFSS roadmap for service delivery process are discussed separately.

Chapters 3 through 9 are methodology chapters on DFSS for service products, which discuss important methods that are useful in DFSS for service products. Specifically, Chapter 3 discusses the concept of customer value and how to create value by service product design. Value creation is a key component for business success.

In order to design the services that are attractive to customers, we first need to know "what customers want." Chapter 4 discusses customer survey design, administration, and analysis. The customer survey is an important activity for obtaining the voice of customers.

Chapter 5 discusses customer value management, which is an important technique to design the survey and obtain key information to develop service designs that are attractive to customers and are competitive in the market place.

Chapter 6 presents the quality deployment method (QFD), a powerful method to guide and plan activities to achieve customer desires. QFD was originally developed in Japan and is now widely used all over the world. Two examples of applying QFD in service industries are presented.

Chapter 7 presents the method of value engineering. Value engineering is an effective method of designing products or services that can satisfy design objectives, yet minimize cost.

Chapter 8 discusses brand development and brand strategy. The success of a service organization is largely dependent on its brand image and customer opinion. Designing of a service product should be consistent with its desired brand image.

Chapter 9 presents the theory of inventive problem solving (TRIZ), which was developed in the former Soviet Union. TRIZ is a very powerful method that makes innovation a routine activity. TRIZ was first developed for technical innovation. Recently, there have been some good extensions of TRIZ into business innovation practices. This chapter will start with general discussions of TRIZ, followed by TRIZ practices in service industries.

Chapters 10 and 12 are methodology chapters on DFSS for service processes. Chapter 10 gives a very comprehensive discussion of service process design and improvement. All service delivery processes can be classified into the following 10 categories as follows:

- Office processes
- Service factory
- Pure service shop
- Retail service store
- Professional service
- Telephone service
- Project shop
- Transportation service
- Logistics and distribution
- Purchasing and supply chain

In this chapter, we discuss each of these processes in detail, and present many effective process diagnosis, design, and improvement methods, such as value stream mapping and lean operation principles. A detailed service process redesign case study is presented at the end of this chapter.

Chapter 12 discusses the theory of constraint, which is an excellent method to analyze and improve service processes in an efficient manner.

Chapter 11 is a reference chapter, which provides the necessary statistical background for service DFSS practitioners.

This book presents DFSS for the service environment in a very clear way and provides practical guidance for Six Sigma practitioners in service industries.

Kai Yang

ACKNOWLEDGMENTS

In preparing this book I received advice, help, and encouragement from several people. I would like to thank Dr. O. Mejabi, who helped me in writing on service process management and preparing a good case study. I would also like to thank David Reeve for his help in the value engineering area and Dr. Rajesh Jugulum and Dr. Jayant Trewn for their valuable input.

Readers' comments and suggestions would be greatly appreciated. I will give serious consideration to your suggestions for future editions. Please contact me at: ac4505@wayne.edu; kyang@simplexsystems.com; http://simplexsystems.com/LeanPD.htm. Also, I am conducting public and in-house Six Sigma and DFSS workshops and provide consulting services.

ABOUT THE AUTHOR

Kai Yang, Ph.D., has wide experience in quality and reliability engineering. The Executive Director of Enterprise Excellence Institute, a renowned quality engineering organization based in West Bloomfield, Michigan, he is co-author of the influential *Design for Six Sigma: A Roadmap for Product Development*. He is also Professor of Industrial and Manufacturing Engineering at Wayne State University, Detroit.

Chapter 1

Six Sigma in Service Organizations

1.1 Introduction to the Service Industry

Entities in the service industry are called service organizations. Many service organizations are profit-earning business enterprises, such as restaurants, hotels, and retail stores; some service organizations are nonprofit organizations, such as universities and post services. In any service organization, however, one or more kinds of services are provided to customers. The service industry exhibits some distinct features that are not found in the manufacturing industry. Based on the work of Sasser, Olsen, and Wyckoff (1978), these distinct features include

1. Many services are intangible; they are not things like hardware.
2. Many services are perishable; they cannot be inventoried.
3. Services often produce heterogeneous output.
4. Services often involve simultaneous production and consumption.

However, behind these apparent differences, there are also many similarities between the manufacturing and service industries. Figure 1.1 shows a generic business operation model for manufacturing-oriented companies.

A manufacturing-oriented company will provide one or many kinds of products to its customers. In any manufacturing-oriented company, there will always be a core operation, which is usually the product development and manufacturing process. Besides the core operation, there are also many other business processes, such as business management, financial operation, marketing, personnel, and supplier management.

Figure 1.2 illustrates a business operation model for many service organizations. In this model, the service organization has a headquarter and many

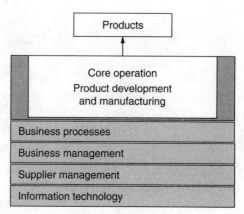

Figure 1.1 Business Operation Model for Manufacturing-Oriented Companies

branches. Each branch is a service delivery process. The service process delivers services to customers. The services provided to customers, no matter how intangible, can be treated as *service products*. For example, in the restaurant business, the meal provided to customers and the music played could be treated as service products; in the insurance business, the insurance policy and processed claims can also be treated as service products. In service organizations, usually the service delivery processes and services are closely related; many of them are in the same place. For example, in the restaurant business, the service delivery process includes the kitchen operation, order taking, the hostess, and the cashier; in the insurance business, the service delivery process includes the insurance agent, insurance policy paperwork processing, insurance claim processing, and information systems.

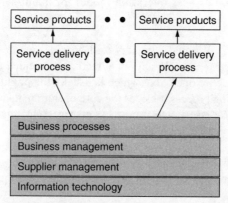

Figure 1.2 Business Operation Model for Many Service Organizations

Of course, there are many other types of service organizations. Some service organizations are one-shop organizations. Others, such as Amazon.com, interact with customers mostly via the Internet. For these organizations, the service process is centralized and the customers are everywhere in the world. It is a "one shop for the whole world" organization. Based on the classification by Schmenner (1994), and Harrell and Tumay (1995), there are 10 kinds of service processes as follows:

1. Office
2. Service factory
3. Pure service shop
4. Retail service store
5. Professional service
6. Telephone service
7. Project shop
8. Transportation service
9. Logistics and distribution
10. Purchasing and supply chain

However, no matter what type of service organizations you look at, there are three aspects of services that are detrimental to service quality and customer satisfaction (Ramaswamy 1996). These are

Service Product Service product refers to the service output attributes to the customers, or the service items provided to the customers. For example, in restaurant service, the service product includes the meals; use of dining utensils, tables, and chairs; and music played if needed. In health-care service, the service product includes diagnosis, treatment, and care items.

Service Delivery Process Service delivery process refers to the process that delivers service products to customers or maintains the service products. For example, in a car rental center, the service process includes all steps needed to rent a car to a customer. These steps include collect driver's license and credit card, check car availability, fill and print contract, obtain customer signature, deliver car key and contract to customer, and locate the car.

Customer-Provider Interaction In service process, there is also a human interaction aspect, that is, the interaction between customers and service providers. The quality of this interaction will greatly influence customer satisfaction. For example, in the car rental business, the representative should greet customers politely, ask customers their car preference, and patiently explain all the options.

Table 1.1 gives a comprehensive summary of the features of service products, service delivery process, and customer-provider interaction for

Table 1.1 Service Types and Service Features

Service Type	Features	Examples	Service Products	Service Delivery Process	Customer-Provider Interaction
Office processes (transaction process)	Sequence of paperwork, data entries, and decisions	Insurance, mortgage and loan.	Finished paperwork	Paperwork processing	Usually happens at the beginning and the end of a transaction
Service factory	Front room and back room, high equipment requirement	Restaurant, copy centers	Meals, copies, binders	Back room is similar to factory.	Important. Usually happens during the whole process
Pure service shop	Front room and back room, highly customized service	Hospitals, repair shop, court	Diagnosis and treatments, repairs, rulings	Multiple steps; service process varies from customer to customer.	Important. Usually happens during the whole process
Retail service store	Large facility, many choices of goods, customer self-service	Supermarket, hardware stores	Selection of goods, nice layouts and labeling	Purchasing and shipping, inventory management, checkout	Usually happens at checkout and during in-store help

Professional service	Usually small number of experts, high cost on labor	Tax service, consulting, architectural service	Knowledge-based output (documents, reports, designs)	Each service is a project, but there is a lot of customer input.	Important. Usually happens during the whole service process
Telephone service	Telephone interactions, no face-to-face	Call centers, mail order center, technical support	Advice, reservations, orders	Call routing, phone system	Very important
Project shop	One big project at a time	Software development, R&D projects	Software, R&D discoveries, patents	Project, long duration possible	Usually happens at the beginning and the end
Logistic and distribution	Ship goods from place to place	FedEx, moving company	Deliver goods to correct destinations without damage	Routing, scheduling, sorting, bookkeeping, all types of transportation	Usually happens at the time of customer order and customer receiving
Transportation	Ship people from place to place	Airlines, buses	Move people to right place with right time and take care of them during travel	Routing, scheduling, transportation	Important. Usually happens during the whole process
Purchasing and supply	Purchasing and inventory keeping	Purchasing and supplier management department and firms	Purchase and supply parts and goods at right time, right price and deliver to right places	Vendor identification evaluation, and selection; contracting, ordering and purchasing; payment	Important. Usually happens during the whole process

10 types of service processes. More detailed discussions on service processes are given in Chap. 10.

1.2 Success Factors for Service Organizations

For profit-earning service organizations, profitability is one of the most important factors for success. High profitability is determined by strong sales and overall low cost in the whole enterprise operation. It is common sense that

$$\text{Business profit} = \text{revenue} - \text{cost} \tag{1.1}$$

In addition,

$$\text{Revenue} = \text{sales volume} \times \text{price} \tag{1.2}$$

Here *price* means the sustainable price, that is, the price level that customers are willing to pay with satisfaction.

Many researchers (Sheridan 1994, Gale 1994) have found that both sales volume and sustainable price are mostly determined by customer value. As a matter of fact, it is customers' opinions that will determine a product's fate. Customers' opinions will decide the price level, the size of the market, and the future trend of this product family. When a product has a high customer value, it often is accompanied by an increasing market share, increasing customer enthusiasm toward the product, word-of-mouth praises, a reasonable price, a healthy profit margin for the company that produces it, and increasing name recognition.

Sherden (1994) and Gale (1994) provided a good definition for customer value. They define the customer value as perceived benefit (benefits) minus perceived cost (liabilities), or specifically as

$$\text{Customer value} = \text{benefits} - \text{liabilities} \tag{1.3}$$

The benefits include the following categories:
1. Functional benefits
   a. Product functions, functional performance levels
   b. Economic benefits, revenues (for investment services)
   c. Quality and reliability
2. Psychological benefits
   a. Prestige and emotional factors, such as brand-name reputation
   b. Perceived dependability (for example, people prefer a known brand product over an unknown brand product)

c. Social and ethical reasons (for example, environmentally friendly brands)
   d. Psychological awe (Many first-in-market products or services not only may provide unique functions but also give customers a tremendous psychological thrill; for example, the first copy machine really impressed customers)
   e. Psychological effects of competition
3. Service and convenience benefit
   a. Availability (ease with which the product or service can be accessed)
   b. Ease of obtaining correctional service in case of product problem or failure

The liabilities include the following categories:

1. Economical liabilities
   a. Price
   b. Acquisition cost (such as transportation cost, shipping cost, time and effort spent to obtain the service)
   c. Usage cost (additional cost to use the product or service in addition to the purchasing price, such as installation)
   d. Maintenance costs
   e. Ownership costs
   f. Disposal costs
2. Psychological liabilities
   a. Uncertainty about product or service dependability
   b. Self-esteem liability of using unknown brand product
   c. Psychological liability of low-performance product or service
3. Service and convenience liability
   a. Liability due to lack of service
   b. Liability due to poor service
   c. Liability due to poor availability (such as delivery time, distance to shop)

Even for nonprofit service organizations, it is not desirable to lose money. In addition, it is a natural goal for all service organizations to maximize customer value. Who doesn't want their customers to be satisfied?

Clearly, higher customer value means higher revenue, and profitability is key for business success. Therefore, business success has to be achieved by the two following success factors:

- Maximize customer value
- Minimize cost

Then the next important question is, in a service organization, how can these two success factors (that is, maximizing customer value and minimizing cost) be achieved? This question can be answered by studying the relationship between these two success factors and the three important aspects of service (service product, service delivery process, and customer-provider interaction).

Table 1.2 summarizes how service product, service delivery process, and customer-provider interaction will affect customer value and total cost. The table shows that excellent service product, excellent service delivery process, and excellent customer-provider interaction are really the keys to delivering high customer value with low cost.

According to Rohit Ramaswamy (1996), excellent service product, service delivery process, and customer-provider interaction can be accomplished by superior service design and superior service delivery. Figure 1.3 shows the contents of service design and service delivery activities and their relationships.

Service Product Design Service product design refers to the design of service output attributes to the customers, or the service items provided to the customers. For example, in restaurant service, the service product design includes the menu design, decisions on what kind of dining utensils to use, and service protocols.

Service Facility Design Service facility design refers to the design of the physical layout of the facility where the service is delivered. For example, in restaurant service, the facility design includes the design of the restaurant's kitchen, dining hall interior, decoration, layout, and lighting. The quality of the facility design will directly affect the service process delivery; for example, in the restaurant business, the design and layout of the kitchen directly affects the quality and speed of the meal service. The quality of the facility's front room (the portion of the facility that is visible to customers) design will directly affect customers' consumption process of the service, as well as customers' perception of the quality of the service.

Service Process Design Service process design refers to the designing of the service process that is needed to deliver service products to customers or maintain the service products. For example, in a car rental center, the service process includes all steps needed to rent a car to a customer. These steps include collect driver's license and credit card, check car availability, fill and print contract, obtain customer signature, deliver car key and contract to customer, and locate the car.

Table 1.2 How Service Quality Affects Customer Value and Total Cost

		Customer Value						
		Benefits			Liabilities			
		Convenience	Psychological	Functional	Convenience	Psychological	Price	Total Cost
Service Product		Important Having a good service location and service plan makes a difference.	Very Important Good service product design coordinated with brand development will create a good psychological impact on customers. Quality and reliability make customers feel secure.	Very Important The design and offering of the right service product is the most important key to providing functional benefits to customers.	Important Same reason as for convenience benefits	Very Important Same reasons as for psychological benefits	Very Important What goes into the service product design will absolutely affect the cost and price.	Very Important What goes into the service product will absolutely affect the cost.
Service Delivery Process		Very Important Good and efficient service delivery saves customers much trouble.	Important Efficient, and timely service delivery makes customers satisfied and happy.	Important Service delivery process has to deliver the service product reliably and consistently.	Very Important Same reason as for convenience benefits	Important Same reason as for convenience benefits	Very Important Efficient process will lower the cost, so it may lower the price.	Very Important Efficient process will lower the cost.

(*Continued*)

Table 1.2 How Service Quality Affects Customer Value and Total Cost (*Continued*)

	Customer Value						
	Benefits			Liabilities			
	Convenience	Psychological	Functional	Convenience	Psychological	Price	Total Cost
Customer-Provider Interaction	Related Better customer-provider communication will help customers.	Very Important Person-to-person interaction directly affects what customers think.	Related Adequate interaction will make customers feel good about services, and communication helps in delivering services that fulfill customer needs.	Related Same reason as for convenience benefits	Very Important Same reason as for psychological benefits	Related somewhat	Related somewhat

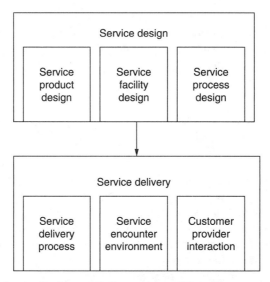

Figure 1.3 Service Design and Delivery (*Adapted from Ramaswamy 1996*)

Service Delivery Process The service delivery process is the execution of the designed service process steps in order to deliver the desirable service products to customers. The ideal service delivery process can deliver the routine service smoothly, effortlessly, and predictably; it also has the flexibility to deal with abnormal service requirements, personalized services, and difficult situations.

Service Encounter Environment The service encounter environment is the environment in which the service is delivered and the customer-provider interaction takes place. For example, in a hospital, the service environment includes all the places where patients will possibly stay, such as offices, emergency rooms, and hospital beds. In many service operations, service encounters happens in front rooms. A clean, well-lit, and comfortable service encounter environment will make customers feel good.

Customer-Provider Interaction In service process, there is also the human interaction aspect; for example, in the car rental business, the representative should greet customers politely and ask customers their car preference. The whole service process includes both the execution of process steps and human interaction.

From the preceding discussion, we can see that the excellent service product, service delivery process, and customer-provider interaction can be accomplished by superior service design as well as flawless and efficient service delivery. It is the author's belief that Six Sigma can be a tremendous help in service design and service delivery. The fundamental aspects of Six Sigma are discussed in Sec. 1.3.

1.3 Overview of Six Sigma

1.3.1 What Is Six Sigma?

Six Sigma is a business strategy that provides businesses with the tools to improve the capability of their business processes. In Six Sigma, a process is the basic unit for improvement. A process could be a product or a service that a company provides to outside customers, or it could also be an internal process within the company, such as a billing process or a production process. In Six Sigma, the purpose of process improvement is to increase a process' performance and decrease its performance variation. This increase in performance and decrease in process variation will lead to a reduction in defects and an improvement in profits, employee morale, quality of product, and eventually to business excellence.

Six Sigma is the fastest growing business management system in industry today. It has been credited with saving billions of dollars for companies over the past 10 years. Developed by Motorola in the mid-1980s, the methodology only became well known after Jack Welch, from GE, made it the central focus of his business strategy in 1995.

Compared with other quality initiatives, the key difference of Six Sigma is that it applies not only to product quality, but also to all aspects of business operation by improving key processes. For example, Six Sigma can be used to help create well-designed, highly reliable, and consistent customer billing systems, cost control systems, and project management systems.

1.3.2 Six Sigma in a Nutshell

Six Sigma is not just statistical jargon; it is a comprehensive business strategy with multiple aspects. Figure 1.4 illustrates the whole picture of Six Sigma. There are five aspects of Six Sigma: fundamental beliefs, organizational infrastructure, training, project execution, and methods and tools. We present a general overview of all these aspects.

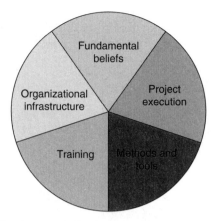

Figure 1.4 The Five Aspects of Six Sigma

Fundamental Beliefs

Product, Process, and People

No matter what kind of business enterprise we look at, whether it is a manufacturing-oriented company, for which the business operation model illustrated by Fig. 1.1 will apply, or a service-oriented company, for which the business operation model illustrated by Fig. 1.2 will apply, we can see clearly that excellent products and excellent processes are the key to business success. What is a process? Caulkin (1989) defines it as being a "continuous and regular action or succession of actions, taking place or carried on in a definite manner, and leading to the accomplishment of some result; a continuous operation or series of operations." Keller et al. (1999) defines the process as "a combination of inputs, actions and outputs." Figure 1.5 gives a general description for all kinds of processes.

Clearly, the process model shown in Fig. 1.5 can be used to characterize almost all kinds of business operations, such as service processes, product development, financial transactions, and customer billing. Of course, all processes are designed and operated by people. Therefore, in order to achieve business excellence, the only factors that really matter are excellent products, processes, and people (3 Ps).

Do the Right Things, and Do Things Right

The sentence "Do the right things, and do things right" best captures the essence of Six Sigma. Do the right things means that whether it is a product or a process, it has to do the right thing for the customer. For a product, it

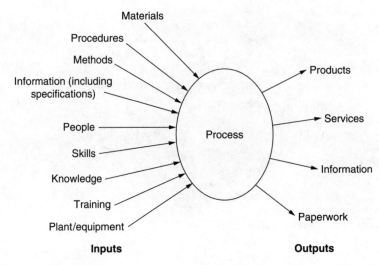

Figure 1.5 A Diagram of a Process (*Oakland, 1994*)

means that the product has to be designed perfectly so it captures maximum customer value; the product should deliver superior performance to its customers. Do things right means that a product or a process should be able to perform consistently and defect-free.

Actually, the name Six Sigma came from statistical terminology. Sigma, or σ, means "standard deviation." For a normal distribution, the probability of falling within a ±6 sigma range around the mean is 0.9999966. In a production process, the *Six Sigma standard* means that the process will produce defectives at the rate of 3.4 defects per million units. Clearly Six Sigma indicates a degree of extremely high consistency and extremely low variability. In statistical terms, the purpose of Six Sigma is to reduce variation to achieve very small standard deviations.

A perfect product or process is one that will do the right things, and do things right. A perfect example is an Olympic gymnast. If an athlete wants to win a gold medal, he or she must first do right things; that is, he or she must be able to design and execute absolutely world-class routines (10.0-point performance). The routine has to beat those of all competitors, and impress the judges and audience. A 9.0-point routine, no matter how flawless and consistent, will not do the job. Secondly, the athlete has to do things right every time. If the athlete can sometimes do an excellent job, but sometimes does a poor job, he or she will not be able to win the gold medal.

Organizational Infrastructure

In order to achieve perfect products and processes in a business enterprise, an organization needs to build an infrastructure to manage and execute Six Sigma improvement activities. The hierarchy of members in a typical Six Sigma organizational infrastructure is illustrated in Fig. 1.6.

Champion The Champion is responsible for coordinating the business road map to achieve a Six Sigma quality goal. He or she will select Six Sigma projects, execute control, and alleviate roadblocks for the Six Sigma projects in his or her area of responsibility.

Master Black Belt The Master Black Belt is a mentor, trainer, and coach of Black Belts and others in the organization. He or she will bring the broad organization up to the required Six Sigma professional competency level.

Black Belt A Black Belt is a team leader implementing the Six Sigma methodology on projects. He or she introduces the methodology and tools to team members and the broader organization.

Green Belt A Green Belt is either an important team member who helps the Black Belt, or a leader of successful, small, focused departmental projects.

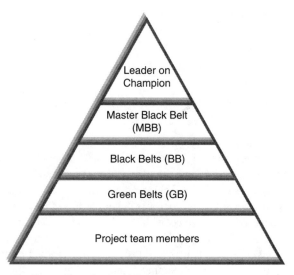

Figure 1.6 Six Sigma Organizational Infrastucture

Project Team Members Project team members participate on the project teams and support the goals of the project, typically in the context of their existing responsibilities. Team members are expected to continue to utilize learned Six Sigma methodology and tools as part of their normal job.

Training

All Six Sigma professionals undergo various trainings depending on their duties. For example, Black Belt training usually takes 4 weeks. Black Belt candidates learn some basics of business processes, project management, team leadership skills, process maps, and many statistical methods. In order to get a Black Belt certificate, the Black Belt candidate needs to complete one or several projects, and these projects should show verifiable financial benefits to the company.

Project Execution

Six Sigma activities are featured by doing projects, a lot of projects. The goal for each project is usually about improving one process at a time.

In Six Sigma there are sophisticated operation procedures for the following project aspects:

- Projection selection
- Projection flowchart
- Project management
- Project evaluation

Project Selection

The Six Sigma leadership team will select projects based on the following characteristics:

1. The project should have a strong tie to business bottom lines. The projects should have a substantially positive impact on profits, have strategic importance, and enhance customer satisfaction and loyalty.
2. Project results should be visible and measurable. For a regular Six Sigma process improvement project it is desirable to see meaningful results within 6 months. The project goals can be quantified and measured.
3. Project should be feasible. The project scope should not be too big, the project tasks should not be too difficult to be handled by the current Six Sigma project team members, and the potential solutions should be relatively low cost.

Project Flowchart

A typical Six Sigma process improvement project usually follows a DMAIC project flowchart. The DMAIC means the following five project steps: define, measure, analyze, improve, control. Specifically, they are

1. Define the problem and customer requirements.
2. Measure the defects and process operation.
3. Analyze the data and discover causes of the problem.
4. Improve the process to remove causes of defects.
5. Control the process to make sure defects do not reoccur.

Project Management

During each project execution, periodical reviews are conducted to find out project progress, identify the project bottlenecks, and resolve problems. Many details of the project, such as time, resource, work force, and task completion are recorded and monitored carefully.

Project Evaluation

On completion of each Six Sigma project, the real financial benefits will be tracked and verified by financial accounting personnel within the company.

Methods and Tools

Many methods and tools are used in the regular Six Sigma process improvement (DMAIC) activities. Six Sigma Green Belts and Black Belts will go through rigorous trainings to learn these methods and tools and apply them in the projects. The commonly used methods and tools include

Business process map: Process flowcharts.
Project management: Critical path method (CPM), project evaluation, and review techniques (PERT), Microsoft project management.
Team and leadership: Team works, team communication, and facilitation.
Probability and statistics: Probability distributions, mean, variance, hypothesis testing, confidence intervals, and so on.
Simple graphic tools: Histogram, scatter plot, Pareto charts, and so on.
Advanced statistical tools: Linear regression, design of experiments, multivari charts, control charts, process capabilities, measurement system analysis, and so on.
Basic Lean Manufacturing: Seven wastes, Kaizen, and so on.

Summary of Six Sigma

1. Six Sigma is a comprehensive strategy that provides businesses with organizational infrastructure, discipline, training, and tools to improve the capabilities of their business processes. Compared with other quality initiatives, the key difference of Six Sigma is that it not only applies to product quality, but also to all aspects of business operation. Six Sigma is a method for business excellence.
2. Six Sigma tries to achieve business excellence by improving products, processes, and people. By excellence we mean do the right things, and do things right.
3. The Six Sigma organization infrastructure consists of Six Sigma professionals, such as the Champion, Master Black Belts, Black Belts, and Green Belts. All Six Sigma professionals are trained with Six Sigma tools and methods.
4. Six Sigma activities are project-based activities. Six Sigma tries to achieve business excellence by doing one project at a time. It has sophisticated project selection, project flowchart, project management, and project evaluation procedures. Each project should provide visible benefits to a company's bottom line with verifiable financial results. The project activities are led and executed by members of the Six Sigma organizational infrastructure.

1.3.3 Design for Six Sigma and Lean Six Sigma

The Six Sigma movement started with regular Six Sigma process improvement activities featured by DMAIC. This DMAIC strategy does not involve any changing or redesigning of the fundamental structure of the underlying process. It involves finding solutions to eliminate the root causes of performance problems in the process and of performance variation, while leaving the basic process intact. The goals for DMAIC projects are usually related to reducing defects, variations, and costs from poor quality.

However, after several years of application, this Six Sigma process improvement strategy (DMAIC) encountered a few problems:

1. In many cases, the Six Sigma DMAIC strategy is applied to many processes. It takes a lot of effort to make some processes become more capable, less variational, and defect-free. However, it is often found later that many of these processes are redundant and wasteful. From a lean manufacturing point of view, these processes should have been eliminated or simplified in the first place.
2. Some processes are fundamentally flawed in their design, and DMAIC-based projects will not yield sufficient improvement for these processes.

3. DMAIC-based strategy mostly focuses on variation and defect reduction. So it mostly accomplishes "do things right." It does not address the problems of higher performance levels and higher customer value nor does it adequately address the issue of "do the right thing."

In order to overcome these three deficiencies, two other Six Sigma strategies were proposed in recent years. One is Design for Six Sigma (DFSS) (Chawdhury 2002, Yang and El-Haik 2004); the other is Lean Six Sigma (George 2003).

Design for Six Sigma

Design for Six Sigma was first proposed to be applied to the process of new product development for the manufacturing industry. It is a systematic methodology that uses tools, training, project management, and discipline to optimize the design process of products, in order to achieve superior designs to maximize customer value at Six Sigma quality levels.

In the manufacturing industry, DFSS is needed because

- The design decisions made during the early stages of the design life cycle have the largest impact on total cost and quality of the system. It is often claimed that up to 80 percent of the total cost is incurred in the concept development phase (Fredrikson, 1994).
- Poor design concepts adopted in the early design stage are easy to correct at the early stage of the product development cycle, but are very costly to correct at later stages.
- Superior customer value, creative concept, and robust performances of the products are intrinsically determined in the early design stage.

DFSS can

- Design products with maximum customer value.
- Do design right upfront, to avoid costly design-build-test-fix cycles.
- Bring creativity in design.
- Reduce design vulnerabilities.
- Make design robust.
- Shorten lead times, cut development and manufacturing costs, and lower total life cycle cost to improve the quality of the design.

Besides statistical methods, DFSS uses many other system design methods such as quality function deployment (QFD), theory of inventive problem solving (TRIZ), axiomatic design, value engineering, and the Taguchi method. DFSS is also a project-based activity. The DFSS projects usually take a longer time to finish, but they also have greater impacts. The most popular DFSS project procedures are

1. *IDOV (Identify, Design, Optimize and Verify):* The IDOV project procedure is usually used for new designs.
2. *DMADV (Define, Measure, Analyze, Design, and Verify):* The DMADV procedure is usually used for redesign projects.

The DFSS strategy can also be applied to processes, because in many cases the original design of the process is fundamentally flawed; merely patching the holes in the process will not yield satisfactory performances. DFSS for Process is a Six Sigma approach that will involve changing or redesigning the fundamental structure of the underlying process. The goal of DFSS for Process is to design or restructure the process in order to intrinsically achieve maximum performance.

DFSS for Process is needed

- When a business chooses to replace, rather than repair, one or more core processes.
- When a leadership or Six Sigma team discovers that simply improving an existing process will never deliver the level of quality customers are demanding.
- When the business identifies an opportunity to offer an entirely new product or service.

Technical approaches for DFSS for Process are thoroughly discussed in Chap. 10.

Lean Six Sigma

Lean Six Sigma (George 2003) is a combination of lean manufacturing (Womack 2000, Liker 2004) and regular Six Sigma process improvement strategy featured by DMAIC. Lean manufacturing practice is based on the Toyota production system, which features the following principles.

1. There are many wastes (muda) in production processes. The wastes are the activities that do not add value to the products. These are
   - Overproduction
   - Unnecessary inventory
   - Unnecessary transportation
   - Unnecessary movement
   - Waiting
   - Defective product
   - Overprocessing

   The muda should be eliminated from process steps. A value stream map is an effective way to find muda and guide process improvement

2. The ideal production process should be a continuous one-piece flow.
3. A "pull" system should be used to avoid overproduction.
4. "Quick setup time" techniques should be used to ensure a smooth process flow and to handle multiple product line 5.
5. Cellular manufacturing should be used to reduce travel distance and ensure quick flow.

These five principles will be thoroughly discussed in Chap. 10.

The Lean Six Sigma approach is purely a process-based approach; it does not involve product design activities. In Lean Six Sigma implementation practice, usually lean manufacturing principles are used first to eliminate unnecessary process steps and even unnecessary processes. Clearly, this includes reducing process costs (by eliminating non-value-added steps) and improving process efficiency (by reducing process lead time and increasing throughput). Then DMAIC activities will follow to reduce the variations in the process and improve process capability.

1.3.4 The Roles of DMAIC, DFSS, and Lean in Business Excellence

Based on the discussion in Sec. 1.2, the two key factors for business success are

1. Maximize customer value
2. Minimize cost

Customer value is defined as the total benefits minus the total liabilities from the products to customers. There are functional, psychological, and convenience benefits. Most, if not all, of these benefits have to be designed into the products. Therefore, DFSS is the most important activity in increasing the total benefits portion of the customer value. DMAIC relates to the variation reduction and defect reduction and to the functional benefit the product provides for customers. It also relates to reducing the cost of poor quality. Lean manufacturing practice is mostly concerned with cost reduction and efficiency improvement practices; it does not relate to product design. In summary, Table 1.3 lists the relative importance of various Six Sigma activities in improving customer value and reducing cost.

1.4 Six Sigma for Service

As discussed in Sec. 1.3, Six Sigma is not just statistical jargon. It is a comprehensive business strategy that has fundamental beliefs, organizational infrastructure, training, projects, methods, and tools. To a larger

Table 1.3 Relative Importance of Six Sigma Activities in Improving Customer Value and Reducing Cost

Maximize customer value	DFSS, DMAIC, Lean
Minimize cost	Lean, DFSS, DMAIC

Note: Six Sigma activities are listed in order of importance, the first item being of greatest importance.

extent, it involves extensive cultural changes both within organizations and in relationships with suppliers. Under the Six Sigma framework, it is very easy to add such new twists as Lean Six Sigma and DFSS into companywide Six Sigma activities. You can simply let a few existing Six Sigma professionals take some additional lean manufacturing and DFSS training, and initiate Lean Six Sigma and DFSS projects on top of existing Six Sigma projects. Therefore, Six Sigma is very flexible in adding and adapting emerging methods and business approaches to existing Six Sigma activities. In the actual implementation, most companies customize Six Sigma approaches to their own special circumstances.

Six Sigma started in the manufacturing industry and has been spreading steadily into the service industry, such as health care, banking, and insurance. Many Lean Six Sigma and DMAIC type projects are reported in various service industries relating such diverse objectives as error reduction, cost cutting, cycle time, and lead time reduction (George 2003).

As we discussed earlier, there are three important aspects of service quality:

1. Service product
2. Service delivery process
3. Customer-provider interaction

Currently, most reported Six Sigma activities in the service industry involve using either Lean Six Sigma or DMAIC in improving the service delivery process. Not much has been done in designing and improving the service product and customer-provider interaction. On the other hand, service delivery processes contain many varieties of process types, as introduced in Table 1.1 and will be thoroughly discussed in Chap. 10. Besides lean manufacturing approaches, many other methods, such as process management and discrete event simulation can also be used to help in redesigning and improving the service delivery process.

Recall from Fig. 1.3 that the two most important activities in service are

1. Service design
   a. Service product design
   b. Service facility design
   c. Service process design
2. Service delivery
   a. Service delivery process
   b. Service encounter environment
   c. Customer-provider interaction

It is clear that almost all current Six Sigma activities in service industries only involve the service delivery process portion of service delivery activity. Not many Six Sigma activities have been reported in service design. However, if the quality of the service product design and service process design is poor, this can be very detrimental to the success of service organizations.

Just as many approaches of Six Sigma that were originally developed in the manufacturing industry can be adapted into the service industry and achieve great success, DFSS can also be adapted into the service industry and achieve amazing results. Because the manufacturing industry is global and is under hypercompetition, many service industries, such as health care, are so inefficient that there are many "low-hanging fruits" ready to be picked, yet since the service sector accounts for a big portion of the gross domestic product (GDP) in the developed world, the potential for Six Sigma to gain in the service industry is very big.

This book is fully devoted to discussing DFSS for service industries. The DFSS approach in this book has two aspects: DFSS for service product and DFSS for service process. Chapter 2 will discuss the DFSS project procedures for both service product and service process. Chapters 3 through 9 will cover the technical aspects of DFSS for service product. Chapter 10 discusses the technical aspects of DFSS for service process. Chapter 11 is a reference chapter on necessary statistical techniques used in this book, and Chap. 12 discusses an important technique in service process management, the theory of constraints.

Chapter 2

Design for Six Sigma Road Map for Service

2.1 Introduction

This chapter provides an introduction to Design for Six Sigma (DFSS) theory, implementation processes, and applications. The material presented herein is intended to give the reader an understanding of what DFSS is and of its uses and benefits. After reading this chapter, readers should have a sufficient knowledge of DFSS to assess how it could be used in relation to their jobs and to identify their needs for further learning.

The theoretical foundation for DFSS comes from several sources:

- Customer-oriented design
- System design and creative design
- Taguchi method and "fire-prevention" philosophy

2.1.1 Customer-Oriented Design

The customer-oriented design is a development process of transforming customers' wants into design solutions that are useful to the customer. In simple words, customer-oriented design translates the *voice of the customer* (VOC) into products and services. Firms listening to the VOC have discovered that the financial results are tremendous, with returns occurring roughly one to two years after deployment. Companies as varied as Cummins, 3M, Samsung, and Bank of America are publicized examples. Bank of America alone noted that it has captured over a billion dollars due to added revenue generation.

Several important methods that relate to customer-oriented design originated from Japan, including quality function deployment (QFD)

(Cohen, 1988, 1995), the Kano model, and the KJ method. Quality function deployment is a work template that can be used to derive product design parameters and specifications from the voices of customers. The voices of customers are usually obtained from customer surveys, customer focus groups, or other means.

However, the mere knowledge of "what customers want our product or service to do for them" is not enough to create successful products or services that generate high sale volumes, premium prices, and an excellent reputation, because customers' opinions on products or services not only are influenced by products and services' functionalities, but also by many psychological factors, as well as convenience factors. The customer value concept that we discussed in Chap. 1 (Gale 1994, Sheridan 1994) is an excellent model that relates customer opinions to product and service values. Brand identity and brand reputation (Chap. 8) is an important part of the product and service values. Therefore, the design of the product or service should be coordinated with brand development.

For service industries, the relationship between customers and service providers is even closer than that of manufacturing industries; customer-oriented design should be even more important for the success of service organizations.

2.1.2 System Design and Creative Design

After we fully understand the voices of customers, we need to create the designs that can make the voices of customers into realities, that is, products and services. We also need to be able to produce these products and services efficiently and at low cost. This design generation task could be very challenging, especially if we deal with high technologies. The important theories in this area include axiomatic design (Suh 1990) and the theory of inventive problem solving (TRIZ) (Altshuller 1984). Axiomatic design theory tries to develop several generic design principles that can be used to evaluate designs and discover design vulnerabilities. TRIZ is based on thorough studies of patents and discoveries and has developed a rich pool of principles, methods, and knowledge base to generate creative designs. Again, axiomatic design and TRIZ were developed in manufacturing-related areas; however, many researchers (Mann 2004) are extending TRIZ into nontechnical areas.

For service industries, creativity is also very important for business success. The creative business operation models, such as those of Amazon.com and

Domino's Pizza, have created tremendous successes and revenues for these companies. Chapter 9 of this book will discuss the TRIZ method and its extensions in business management.

2.1.3 Taguchi Method and Fire Prevention Philosophy

The method of robust design as suggested by G. Taguchi (1985, 1990, 1996) is also called the Taguchi method. In this method, a good design is one where products from the design are robust to usage conditions, environmental conditions, and manufacturing variation. The robust design is also based on the idea of "design it right the first time." The Taguchi method's emphasis is on doing a good job in the early design stage so that you can prevent problems in the manufacturing stage and after the product goes to market. We can also call this approach *fire prevention*, instead of "fire fighting" where fire fighting is the practice of correcting errors and mistakes; caused by improper designs.

The Taguchi method is an integration of some sound design principles with experimental design methods (DOE). Ideal function, orthogonal array experiment, and signal-to-noise ratio (S/N) are among several important techniques in the Taguchi method (Yang 2003).

The techniques used in the Taguchi method may not be easily used in service industry circumstances. However, Taguchi's emphasis on doing a better job in design and on fire prevention makes a lot of sense in the service industry as well.

2.2 Why Use Design for Six Sigma in the Service Industry?

Nobody in the service industry would dispute the importance of service quality, customer orientation, and efficiency. There are vast amounts of literature and case studies on service quality (Ramaswamy 1996). However, instead of focusing on the millions of customers served satisfactorily and without incident, the literature often concentrates on the heroic efforts made by some individual employees to provide extraordinary services during some special and unique service encounter. The importance of service product design and service process design is rarely mentioned.

In the service industry, service managers have not paid much attention to planned and systematic service design. Most often, services are put together haphazardly, relying on a mixture of judgment and past experience

(Ramaswamy, 1996). In a study on the approaches used to develop new services in the financial sector, de Brentani (1993) said, "... companies tend to use a hit-and-miss approach when planning new services where ideas are generated and defined in a haphazard fashion, limited customer research is carried out prior to planning the design, service designs often lack creativity and precision and do not incorporate the appropriate technology, testing for possible fail points is rarely done, and market launch is often characterized by trial and error." Service managers compensate for the lack of sound service design by trying to please the customers through ad hoc activities; the process improvement is mostly hit-and-run. These are clearly expensive methods and are not competitive in the long run.

Many service sectors, such as health care, suffer from low efficiency and extremely high cost. In many developed countries, especially the United States, health-care costs have become a heavy burden to the whole national economy and global competitiveness.

Fortunately, many service sectors, especially, health care and banking, are adopting Six Sigma approaches in their businesses. Many of them are starting to reap the benefits from Six Sigma. Clearly, Do the right things and Do things right are all it takes to make successful service products and service processes. However, merely using DMAIC-based process improvements will not be sufficient to raise the service quality and efficiency to the best possible level.

As we discussed in Chap. 1, there are two important aspects in service operation: service design and service delivery. Service design refers to the elements that are planned into service, including the features offered by service, the nature of facilities where service is provided, and the processes through which the service is delivered. The quality of the service design determines the ability of the service to effectively and efficiently supply the performance level expected by customers. The service design will determine what is offered in the service and service performance level (Do the right things), as well as the stability and reproducibility of the service performance (Do things right).

Service delivery refers to the manner in which the service is offered during customer encounters. It is the system operation aspect of the service process. DMAIC-based process improvement focuses on the service delivery aspect of the service operation. Similar to the limitations of the DMAIC-based approach in the manufacturing industry, if the service design is basically flawed, the DMAIC-based approach will not be able to make a fundamental difference in guiding design or design changes.

However, as we discussed in Chap. 1, DFSS is a Six Sigma–based approach in guiding design activities. Specifically, DFSS is a systematic methodology that uses tools, training, project management, and discipline, to optimize the design process during the designing of products, services, and processes, in order to achieve superior designs to maximize customer value at Six Sigma quality levels. DFSS is also a problem-preventing methodology that guides the new design and redesign processes by using a DFSS road map that focuses on defect prevention and value creation. This is accomplished through Six Sigma (DFSS) projects.

Since good service design has a big impact on service quality, it is natural for a Six Sigma service company to extend Six Sigma activities in the service design area in order to raise its service performance level, service quality, and service efficiency to a whole new level. This is the essence of DFSS in the service industry.

However, when implementing DFSS in the service industry, we have to realize that unlike the manufacturing industry, the service industry has many special features as follows:

1. Some manufactured products have a long product development lead time [from a few months (consumer goods) to tens of years (defense industry)] and heavy research and development (R&D) expenditure. The design and build time in the service industry can be much shorter, and R&D expenditure is usually much less. Of course, there could be exceptions; for example, software development can be a costly and lengthy process.
2. There are more direct customer interactions with service providers than with providers in the manufacturing industry; the intangible psychological factors, such as personal image, word of mouth, and brand image usually play a larger role.
3. Service industries usually work in a more volatile marketplace, where the market situation changes very quickly. There are usually frequent service redesigns.
4. When service redesign happens, usually when the service product (service offerings) changes, the corresponding service delivery process will also change. However, there are plenty of cases where only the service process changes.

Therefore, the implementation of DFSS in the service industry has to take into account these special features. The DFSS road map introduced in this book is tailored to fit the special needs of the service industry. It has the following features:

1. Since redesign occurs frequently in the service industry, the project road map that we selected is DMADV (define, measure, analyze, design, verify) instead of IDOV (identify, design, optimize, verify), because IDOV is more appropriate for completely new design projects in the manufacturing industry.
2. Since service product design and service process design usually go hand in hand, we are providing DFSS road maps for both the service product design and service process design.

2.3 Design for Six Sigma Phases for Service Product

DFSS for service product has the following five phases:

1. Define (D) the project goals.
2. Measure (M) and determine customer needs and specifications.
3. Analyze (A) the design options to meet customer needs.
4. Design (D) (in detail) the service product to meet customer needs.
5. Verify (V) the design performance and ability to meet customer needs.

2.3.1 Phase 1: Define

The objective of this phase is to define the DFSS project goals and scope. This phase has the following steps.

Step 1: Draft Project Charter

- Business case
- Goals and objectives of the project
- Milestones
- Project scope, constraints, and assumptions
- Team memberships
- Roles and responsibilities
- Preliminary project plan

Step 2: Identify the Key Customers of Services

In this step, customers are fully identified and their needs collected and analyzed, with the help of customer surveys.

Step 3: Identify Customer and Business Requirements

In this step, customers are fully identified and their needs collected and analyzed, with the help of quality function deployment (QFD) and Kano analysis.

Then the most appropriate set of critical-to-satisfaction (CTS) metrics are determined in order to measure and evaluate the design.

The detailed subtasks in this step include

- Identify methods of obtaining customer needs and wants.
- Obtain customer needs and wants and create a VOC list.
- Translate the items on the VOC list into functional and measurable requirements.
- Finalize requirements.
- Identify CTSs as critical-to-quality, critical-to-delivery, critical-to-cost, etc.

DFSS tools used in this phase include

- Market and customer surveys
- Quality function deployment
- Kano analysis

Example 2.1: Customer Needs for Restaurant Service
This example (from Ramaswamy 1996) shows how VOC data can be used to derive key CTS metrics in the restaurant business. The VOC data are often disorganized, nonspecific, and nonquantitative. Table 2.1 is the list of customer needs for restaurant service.

An affinity diagram or the KJ method (Shigeru 1988) can be used to analyze and organize the voices of customers into a CTS tree. A CTS tree is a refined multilevel attribute that characterizes some critical characteristics for customer satisfaction. An affinity diagram is a well-known method that was developed in Japan, and it is a standard technique used in Six Sigma. By using the affinity diagram, the following three-level preliminary CTS tree is derived from the VOC data in Table 2.1 and is listed in Table 2.2.

Therefore, the first-level CTS factors are

Satisfying food
Clean and attractive surroundings
Good service

The second-level CTSs explain the first-level CTSs, and they are the aggregated categories of original customer statements. For a complete set of CTSs, we still need quantitative measures. For example, "short wait for table" is one third-level CTS, but how short is short? 5 minutes? or 10 minutes? Similarly, the CTS "food tastes good" does not indicate whether our food tastes good enough now. Also, we need to know the relative importance of each CTS item; for example, for the average customer which is more important, the taste or the nutrition of the food? All this needed information can be found by many means, such as by a specially designed customer survey,

Table 2.1 List of Customer Needs for Restaurant Service

1. Food tastes good	20. Can order quickly
2. Unusual items on menu	21. Know how long a wait for table
3. Hot soup, cold ice cream	22. Food is healthy
4. Feel full after the meal	23. Menu items easy to understand
5. Don't feel overfull after meal	24. Prompt delivery after ordering
6. Food looks appetizing	25. Get what was ordered
7. Food courses arrive on table at right time	26. Get the correct bill
8. Don't feel hungry one hour after meal	27. Billed as soon as meal is over
9. Clean restrooms	28. Shouldn't feel rushed out of restaurant
10. Clean tables	29. Make me feel at home
11. Clean plates and silverware	30. Order additional items quickly
12. Clean, well-dressed employees	31. Errors and problems quickly resolved
13. Light not too bright	32. Errors and problems satisfactorily resolved
14. Light not too dim	33. Staff willing to answer questions
15. Shouldn't feel too crowded in space	34. Greeted immediately on being seated
16. Don't want noisy atmosphere	35. Waiter should be patient while ordering
17. Want smoke-free atmosphere	36. Fill water glass promptly without asking
18. Wide choice of food	37. Polite, friendly staff
19. Enough time to read menu	38. Short wait for table

Source: Ramaswamy (1996).

Table 2.2 Three-Level Preliminary CTS Tree for Restaurant Service

First Level	Second Level	Third Level
Satisfying food	Tasty food	Food tastes good Balance of flavors Hot soup, cold ice cream Food looks appetizing Food is healthy
	Enough food	Feel full after meal Don't feel overfull after meal Don't feel hungry one hour after meal
	A lot of variety	Wide choice of food Unusual items on menu
Clean and attractive surroundings	Clean facility	Clean restrooms Clean tables Clean plates and silverware Clean, well-dressed employees
	Comfortable atmosphere	Light not too bright Light not too dim Shouldn't feel crowded in space Don't want noisy atmosphere Smoke-free atmosphere
Good service	Friendly and knowledgeable staff	Make me feel at home Staff willing to answer questions Polite, friendly staff Waiter should be patient while ordering Menu items easy to understand Shouldn't feel rushed out of restaurant Fill water glass promptly without asking Enough time to read menu
	Quick and correct service	Short wait for table Know how long a wait for table Can order quickly Greeted immediately on being seated Prompt delivery after ordering Get what was ordered Order additional items quickly Food courses arrive on table at right time

(*Continued*)

Table 2.2 Three-Level Preliminary CTS Tree for Restaurant Service (*Continued*)

First Level	Second Level	Third Level
	Accurate billing	Get the correct bill Billed as soon as meal is over
	Problems and complaints addressed effectively	Errors and problems quickly resolved Errors and problems satisfactorily resolved

Source: Ramaswamy (1996).

competitor benchmarking, or the "mysterious customer" approach. The mysterious customer approach is to hire either employees or temporary helpers as mysterious customers. These mysterious customers are paid to visit competitors' facilities, as well as your own facility, and fill out a number of specially designed questionnaires after each visit. Some stopwatch activities may also be included to time such measures as the service waiting time and the time needed to deliver the meal. With the help of mysterious customers, clear, quantitative measures can be developed. For example, if in the best competitor's place, the waiting time is no longer than 5 minutes, and if waiting time is really important in customers' eyes, then you need to set the goal to have the waiting time be less than 5 minutes. If mysterious customers found that your competitor offers better tasting food, then you need to work on how to make your food taste better. Many times, these kinds of activities are carried out in the second phase of DFSS, the measure phase.

2.3.2 Phase 2: Measure

The goal of this phase is to measure and determine customer needs and specifications. This phase has the following steps.

Step 1: Identify, Quantify, and Prioritize Customer Needs in Order of Importance

Specifically, we need to establish the measurable metric for each CTS and establish the ranking or priority score of each CTS.

Step 2: Identify the Performance Metrics Required by the Service That Meet These Customer Needs

If a customer need is only a fuzzy description, for example, in the restaurant service case, "short wait for table" is a customer need, but what is it, precisely? And how short is short? We may want to develop some detailed, quantifiable performance metrics such as

1. Time between arrival and seating
2. Time between seating and menu delivery
3. Time between menu delivery and order taking
4. Time between order taking and meal delivery

Of course, there are some customer needs that are difficult to measure and quantify, for example, "taste of food." However, we can always develop evaluation standards such as tasting scores, just like wine tasting scores. Together with competitive benchmarking, we can develop workable performance metrics for all important customer needs.

Step 3: Measure Our Current Performances and Measure and Analyze Competitors' Performances by Using the Performance Metrics Developed in Step 2

After the performance metrics are determined, it is time to measure our current performance levels. If possible, we should measure and compare our performances with our major competitors' performances.

Step 4: Determine the Relationship between Performance Metrics and Customer Satisfaction and Determine the Specifications for Performance Metrics

Not every performance metric is equally important; some performance metrics are much more important than others in terms of their contribution to customer values. Chapters 3 to 5 of this book discuss in detail how to design customer surveys to determine the relationship between performance metrics and customer value and develop priorities among performance metrics. Chapter 5 also discusses how to develop the specifications for performance metrics. For example, if "time between order taking and meal delivery" is a key performance metric, and we find that most of our competitors' average time between order taking to meal delivery is less than 15 minutes, then naturally we should set our specification on the time between order taking and meal delivery to be less than 15 minutes on average.

Example 2.2: Measure the Performances of Restaurants

This example (adapted from Ramaswamy 1996) is a continuation of Example 2.1, assuming that the following performance metrics are developed to measure the performances of restaurants:

- Degree of waiter patience
- Degree of waiter responsiveness

- Degree of waiter knowledge
- Degree of waiter friendliness
- Time between seating and menu delivery
- Time between menu delivery and order taking
- Time between ordering and meal delivery
- Percent of bills produced without errors

Mysterious customer studies and competitive benchmarking on several competitors yield the results shown in Table 2.3.

Clearly this example illustrates how we can develop two different kinds of measurable performance metrics. One is an evaluation score type, such as the degree of waiter patience; the other is a measurable performance metric, such as the time between ordering to meal delivery. Competitor benchmarking can be used to design our performance specifications, as illustrated in Our Desired Performance row in Table 2.3. The performance gaps on performance metrics can be used to guide our redesign practice.

The DFSS tools used in this phase are

- Customer survey
- Customer value management
- Basic statistical analysis, such as mean, standard deviation, and confidence interval

2.3.3 Phase 3: Analyze

The objective of this phase is to generate and analyze the design options to meet customer needs. This phase has the following steps.

Step 1: Translate Customer Requirements and Performance Metrics to Service Product Functions and Functional Requirements

Customer requirements (CTSs) give us ideas about what will make customers satisfied, but they can't be used directly as the requirements for our service product design. We need to identify the relevant service product functions that deliver these customer needs and translate customer requirements to service functional requirements. The QFD technique (described in Chap. 6) and the value engineering technique described in Chap. 7 can be used to identify service product functions and identify functional requirements.

Example 2.3: Restaurant Service Again
For example, in the restaurant service example, customers would like shorter waiting times, such as the waiting time between menu delivery and order taking, and the time from order taking to meal delivery. But these customer requirements are not service product functions. From a functional analysis

Table 2.3 Results of Mysterious Customer Studies and Competitive Benchmarking

	Degree of Waiter Patience	Degree of Responsiveness	Degree of Knowledge	Degree of Friendliness	Time Between Seating	Time Between Menu Delivery and Order	Time Between Ordering and Meal Delivery	Percent of Bill Produced Without Errors
Performance gap	One grade short	One grade short	Two grades short	One grade short	Achieved	3-min gap	2-min gap	5% gap
Our desired performance	Exceptional	Exceptional	Exceptional	Exceptional	<5 min	<5 min	10 min	95%
Our restaurant	Excellent	Excellent	Good	Excellent	<5 min	<8 min	12 min	90%
Vive la France	Good	Exceptional	Excellent	Good	<10 min	<10 min	20 min	>90%
Downtown steakhouse	Excellent	Excellent	Excellent	Exceptional	<5 min	<8 min	15 min	>90%
Sarah's Seafood House	Good	Good	Good	Good	<5 min	<5 min	10.5 min	>92%

point of view (value engineering), the service functions performed from menu delivery to order taking include the following:

1. (Customers) obtain information on available food items
2. (Customers) select food items
3. Record selected food items

The corresponding functional requirements are

1. Customers can obtain information on available food items easily, quickly, clearly and with full explanation.
2. Customer can select food items easily, quickly, and with explanation.
3. Selected food items can be recorded easily, quickly, and mistake-free.

The service functions performed from order taking to meal delivery include the following:

1. Transmit (ordered food items) information to kitchen.
2. Prepare food items (in kitchen).
3. Cook food items.
4. Put food items in container.
5. Deliver cooked food items to the right customer.

The corresponding functional requirements are

1. Ordered food items' information can be transmitted to kitchen quickly and mistake-free.
2. Food items can be prepared quickly, and correctly.
3. Food items can be cooked correctly and quickly.
4. Food items will be put in correct containers and in the right way.
5. Cooked food items can be delivered to customers quickly, in the right condition and mistake-free.

Step 2: Generate Design Alternatives for the Services

After the determination of service product functions and functional requirements, we need to generate design ideas to satisfy these functional requirements. There are two possibilities; one is that there are existing ideas that can be used to develop this design. The other possibility is that brand-new creative design ideas can be generated. The value engineering technique (Chap. 7) and the theory of inventive problem solving (TRIZ) (Chap. 9) are good techniques to develop creative designs. The type of design alternatives generated and the design requirements should also take into consideration the company's brand development strategy and the company's desirable brand image. Brand development is discussed in Chap 8.

Example 2.4: Restaurant Service

This example (adapted from Ramaswamy 1996) is a continuation of Example 2.3. Of course, all the service functions and functional requirements can be accomplished by "good old ways." However, there are several nonconventional design alternatives. For the functions

1. (Customers) obtain information on available food items
2. (Customers) select food items
3. Record selected food items

the design alternatives could be

1. *Customer-operated terminal system:* A hand-touch screen is available at the vicinity of each dinner table, the menu appears on the screen, and explanations and pictures of each item can be shown. Customers can make their selections on the terminal; the selections can be confirmed at the terminal and sent to the kitchen. The screen will indicate when the meal is ready. Of course, the customer can choose to have a waiter, but if customers choose to use the touch screen, the tip can be waived. The potential benefits of this design include shortened ordering time and reduced ordering errors.
2. *Menu designed with check boxes:* Customers can check the items they want and give the menu to the waiter. The potential benefits of this design include an easy and quick customer ordering process.
3. *Waiter food expertise training:* Waiters receive food expertise training including the history of menu items, ingredients used, and cooking techniques. The potential benefit of this design alternative is the ability to provide better information about food items to customers.
4. *Dim sum–type ordering and delivery:* The foods are cooked and put in containers, the containers are put on carts, and the waiters move the carts around the dining area allowing customers to see immediately what they will get and to order on the spot.

Step 3: Evaluate Design Alternatives

Several design alternatives might be generated in this last step. We need to evaluate them and make a final determination on which concept will be used. Many methods can be used for design evaluation, including the Pugh concept selection technique, and design reviews. After design evaluation, a winning concept will be selected. During the evaluation, many weaknesses of the initial set of design concepts will be exposed and the concepts will be revised and improved.

DFSS tools used in this phase are

- TRIZ
- QFD
- Value engineering
- Brand development
- Design review
- Simulation

2.3.4 Phase 4: Design

The goal of this phase is to develop a detailed service product design to meet customer needs. The result of this phase is an optimized design with all functional requirements released at the Six Sigma performance level. As the concept design is finalized, there are still a lot of design parameters that can be adjusted and changed. With the help of computer simulation, design review, failure modes and effects analysis (FMEA), and some simple statistical analysis, the detailed design will be determined.

It is common that the detailed design on service products cannot be completed without completion of the corresponding service process design or redesign. Service process design (or redesign) can follow the DFSS phases for service process that will be discussed in Sec. 2.4.

A good design should have the following properties:
- Superior performance
- High performance capability
- Robustness

Superior performance means the new design can achieve high performance metrics. For example, in the restaurant service case, if we redesign the ordering procedure, a good design would be one that has shorter ordering times and fewer mistakes. If we redesign the kitchen workflow process, a good design would be one that has shorter cooking times, better-tasting food, and less waste in the kitchen.

High performance capability means the newly designed service can perform consistently and has very few mistakes. For example, McDonald's is famous for its high consistency in worldwide operations; no matter where you go in the world, you can expect that the same food item in one McDonald's will be cooked in the same way and will have a similar taste to that in another McDonald's and the kitchen productivity will be the same in both.

Robustness means that a newly designed service will perform consistently for various operating conditions and various types of customers.

The following are some general design principles that can help in developing service designs that have high capability and robustness (Ramaswamy 1996):
- Selection of technologies with large capacities so that resources are not stressed

- Use of modular designs where it is easy to add incremental capacity units
- Multiple and backup service centers that provide a better ability to manage volumes
- Accurate distribution of input volumes by using automated telephone systems
- Transferring some functional responsibilities to customers through better partnerships
- Designing processes to address special events or natural disasters
- Automation of routine and repetitive steps
- Process simplification and reduction of steps
- Reduction of paperwork flow through information technology
- Convenient electronic access to documentation
- Procedures for efficient cross-organizational and cross-process handoffs

The DFSS tools used in this phase are

- Design and simulation tools
- Value engineering
- Process management (Chap. 10)
- Statistics, capability analysis (Chap. 11)
- Design review
- Robustness assessment

2.3.5 Phase 5: Verify

The goal of this phase is to verify the design performance and ability to meet customer needs. This phase has the following activities.

Step 1: Pilot Test and Refining

No service should go directly to market without first piloting and refining. Here we can use design failure mode effect analysis (DFMEA) as well as pilot and small-scale implementations to test and evaluate real-life performance. The detailed activities in this step include the following:

- Select the members of the testing team.
- Select the customers to be involved in the test (by invitation).
- Select the locations for the test.
- Specify the design characteristics to be tested.
- Define the performance measures.
- Conduct test, collect data, and analyze performance.
- Determine causes of performance problems.
- Correct deficiencies in design, implementation, and testing.

Step 2: Validation and Process Control

In this step we will validate the service process capability of the new service and make sure that the actual process capability is acceptable. The process control procedures will be established.

Step 3: Full Commercial Rollout and Handover to New Process Owner

As the new service is validated and process control is established, we will launch a full-scale commercial rollout, which together with the supporting processes can be handed over to design and process owners, complete with requirements settings and control and monitoring systems. The DFSS tools used in this phase are

- Process capability analysis
- Statistical data collection and analysis
- Poka-Yoke, mistake-proofing
- Process control plan

2.4 Design for Six Sigma Phases for Service Process

Many service process design tasks are process redesign. So a DMADV road map is used. Chapter 10 of this book has a very comprehensive and in-depth discussion on service process designs. DFSS for service process also has five phases:

1. Define (D) the project goals.
2. Measure (M) process performance metrics and determine performance requirements.
3. Analyze (A) existing design and generate alternative process design options to meet the performance needs.
4. Design (D) (in detail) the process and evaluate design alternatives.
5. Verify (V) the process performance and ability to meet performance needs.

2.4.1 Phase 1: Define

The objective of this step is to define the DFSS project goals and scope. This phase has the following steps.

Step 1: Draft Project Charter

- Business case
- Goals and objectives of the project

- Milestones
- Project scope, constraints, and assumptions
- Team memberships
- Roles and responsibilities
- Preliminary project plan

Step 2: Identify the Key Customers of the Process

In this step, customers are fully identified and their needs collected and analyzed. In process design, the customers could be internal and/or external. For example, the customers of a restaurant kitchen include both internal customers, such as waiters, and external customers, such as the restaurant patrons. The identification of key customers will help to develop appropriate process performance metrics.

Step 3: Develop a Process Map

In this step, the detailed process map(s) should be developed by using the process mapping techniques discussed in Chap. 10. There are several techniques, such as process flow chart, IDEF0 process map, value stream map, and simulation model. Several different process maps can be established simultaneously. For example, for the same process, you can draw a regular process flowchart and a value stream map as well as a discrete event simulation model.

Step 4: Define Process Performance Metrics

In this step, the key process performance metrics should be defined by the DFSS team with the help of customers and management. Chapter 10 provides a very thorough discussion on process types and corresponding process performance metrics. It can be used to help identify and define appropriate performance metrics. The DFSS tools used in this phase include

- Process flowchart
- IDEF0 map
- Value stream map
- Simulation model
- Process management

All these tools are discussed in Chap. 10.

2.4.2 Phase 2: Measure

The goal of this phase is to measure process performance metrics and determine performance requirements.

Step 1: Measure the Current Process Performance Levels Based on Process Performance Metrics

For example, in the restaurant service case, if the performance metrics for the kitchen workflow process are

- Time from ordering to meal delivery
- Taste of food

Then the time from ordering to meal delivery can be measured by a stopwatch; a sample (such as 50) of such time measurements can be taken and recorded. The mean, standard deviation, and range can be used as the basic statistical estimates for performance evaluation. The histogram of this data set can be used as the basis for simulation model parameters. The taste of food can be measured by a scaled taste evaluation.

Step 2: Determine Performance Requirements That Meet These Customer Needs

In this step, the required target values for all process performance metrics should be established. These target values could be based on competitive benchmarking. The DFSS tool used in this phase is basic statistical analysis.

2.4.3 Phase 3: Analyze

The objective of this phase is to analyze the existing design and generate alternative process design options to meet the performance needs. This phase has the following steps.

Step 1: Perform a Process Diagnosis for the Current Process

The goal of process diagnosis is to identify the key weaknesses of the process and provide the guidelines for process redesign and improvements. The following approaches are often used in process diagnosis:

1. *Value stream map analysis:* A value stream map can expose non-value-added activities and process efficiency problems. By using lean operating principles to analyze the current state value map, possible improvement ideas can be generated.
2. *Process map analysis:* A real detailed process map may expose "hidden factories," that is, unnecessary loops and steps. This process map analysis may help to generate process improvement ideas.
3. *Process analysis based on process types:* The knowledge outlined in Sec. 10.3 can also be used to analyze the possible weaknesses of the process. For example, if we find that our process is an office process but we use a job shop type of layout, then we can immediately know

that this process type is inefficient for our office process and probably we should change to using several lean work cells.
4. *Cause-and-effect diagram:* Also called fishbone diagram analysis.
5. *Data collection:* Collecting such data as the waiting time, process time, and equipment and operator utilization for each process step may help to identify the weak links and bottlenecks of the process.
6. *Process simulation:* For many service processes, discrete event simulation can be a very useful tool to evaluate the current process and identify weak links and bottlenecks.

Step 2: Generate New Design Alternatives

After the process diagnosis step, the weaknesses and bottlenecks of the process are found. Now is the time to propose the process change and generate the new designs. The new design can be generated based on

1. *Applying lean operation principles:* Such as the future state value stream map derived by applying lean operation principles on current state value map.
2. *Brainstorming:* The DFSS team can use brainstorming to generate new designs.
3. *Process knowledge:* The process knowledge described in Sec. 10.3 can be used to generate design ideas. For example, if we identify our process as a project shop–type process, then the redesign solution should be based on project management techniques, such as redividing the work breakdown structure, generating a different project network, and redistributing the resource allocation.

The DFSS tools used in this phase include

- Process management
- Lean operation principles
- Value stream map
- Simulation

2.4.4 Phase 4: Design

The goal of this phase is to design (in detail) the process and evaluate design alternatives. A discrete event simulation experiment can be used as a valuable tool to try out each design alternative. The evaluation of the simulation results will help us to select the best design alternative. The DFSS tools used in this phase include

- Process management
- Simulation

2.4.5 Phase 5: Verify

The goal of this phase is to verify the process performance and ability to meet performance needs. The following steps are needed in this phase.

Step 1: Pilot Test and Refining

A pilot process should be implemented. Its performances will be measured and evaluated in detail. Refinements may be found in different areas, from the sequence of steps in the process, to the configuration of tools and resources selected, and even the documentation of the process manual. The validation plan and all the necessary follow-through should be overseen by the DFSS team in order to determine that all the process needs will be met.

Step 2: Validation and Process Control

In this step we will validate the service process capability of the new service and make sure that the actual process capability is acceptable. The process control procedures will be established.

Step 3: Full-Scale Rollout and Handover to New Process Owner

As the new service is validated and process control is established, we will launch a full-scale rollout. The new process can be handed over to process owners, complete with requirement settings and control and monitoring systems.

Chapter 3

Value Creation for Service Product

3.1 Introduction

The bottom line for every company is not its short-term profitability but the value of its products in the eyes of customers, often called customer values. Short-term profitability reflects a company's recent history and past strengths, but without everlasting enthusiasm from its customers, a company may not last.

As a matter of fact, it is customers' opinions that determine a product's fate. Customers' opinions decide the price level, the size of the market, and the future trend of a product family. A product with high customer value often has increasing market share, increasing customer enthusiasm toward the product, word-of-mouth praises, a reasonable price, a healthy profit margin for the company that produces it, and increasing name recognition. Clearly, the ability to design and deliver service products that have high customer value is the key to success for service organizations.

There are plenty of books and articles that discuss issues related to customer value. The famous book *Market Ownership* by William Sherden (1994) has an excellent chapter on customer value. Bradley Gale's book *Managing Customer Value* (1994) also presents workable methods to survey and deploy customer value into product and service design. Many value engineering books and articles (Park 1999) provide detailed value definition and quantitative methods for value analysis and cost reduction. Quality function development (QFD) (Cohen 1988, 1995) is an excellent method developed in Japan that can be used to deploy customer wants into product design, including service product design. Using well-designed surveys is a basic way to obtain customers' opinions. There are some excellent articles in the literature regarding customer survey design (Rea and Parker 1992). Customer value is also highly related to brand recognition. Usually, customers are willing to pay more to buy a product with a well-established brand name than a similar product with no name recognition, so brand building should be an important strategic consideration in service product design and customer value enhancement. There are books and articles on brand strategy and its relationship with customer value. Innovation and uniqueness is another huge factor for customer value. Customers may be willing to pay a premium price for a unique product or a first-of-its-kind product.

In this book, we will develop a comprehensive strategy that integrates all these wonderful methods in order to create superior customer values for service products. This chapter outlines this customer value creation strategy and provides overviews for all these methods. Section 3.2 will formally define customer value and its components. Section 3.3 will give an overview to all relevant methods that can be used to design customer values into service products. Section 3.4 will give a value creation road map to service product design.

3.2 Value and Its Elements

Value is one of the most frequently used words, yet the concept of value is one of the most confusing. The nature of value has been extensively studied by many researchers. One school of such researchers are value engineering or value analysis professionals.

Based on Park (1999), one of the leaders in the field of value engineering, "Cost is a fact, it is a measure of the amount of money, time, labor, and any other expenses necessary to obtain a requirement. Value, on the other hand, is a matter of opinion of the buyer or customer as to what the product is worth, based on what it does to him/her. In addition, a person's measure of value is constantly changing to meet a specific situation."

The Merriam-Webster Online dictionary (www.m-w.com) lists the following definitions for value:

1. A fair return or equivalent in goods, services, or money for something exchanged
2. The monetary worth of something : marketable price
3. Relative worth, utility, or importance

The *American Heritage Dictionary* defines value as follows:

1. An amount, as of goods, services, or money, considered to be a fair and suitable equivalent for something else; a fair price or return.

2. Monetary or material worth.
3. Worth in usefulness or importance to the possessor; utility or merit.

Value is also related to worth. *Worth* is defined as the quality that renders something desirable or valuable or useful.

Many value engineering researchers discuss the nature of value extensively. Fallon (1980) states that worth is a simple concept; it becomes value when it is related to cost. He further states that cost is a necessary component of value. O'Brien (1986) further defines value as the ratio of worth and cost. In this definition, worth is defined as an appraisal of the properties of product, so it is essentially an appraisal of the function of the product. In other words, value is the ratio of function to the cost. In this case, the function is defined as what the product does for the customer.

Many value engineering researchers and practitioners have developed several precise definitions for value. Several of them are listed here.

(Bryant 1986)

$$\text{Value } V = \frac{\text{wants} + \text{needs}}{\text{resources}}$$

$$= \frac{\text{sell functions} + \text{use functions}}{\text{dollar} + \text{people}}$$

(Harris 1968)

$$V = \frac{\text{worth}}{\text{effort}}$$

(Kaufman 1981)

$$V = \frac{\text{functions}}{\text{cost}}$$

(Wasserman 1977)

$$V = \frac{\text{function}}{\text{cost}} = \frac{\text{utility}}{\text{cost}} = \frac{\text{performance}}{\text{cost}}$$

(Fallon 1980)

$$V = \frac{\text{objectives}}{\text{cost}}$$

In each of these definitions, the denominator is a unit that can be measured by dollars, effort, resources, work force, and so on. Eventually they can all be converted to dollars. We can say that all these definitions converge to the denominator that is a measure of cost. The numerators converge to a measure of functions, or performances. Therefore in value engineering, value is measured primarily as a function-to-cost ratio. A product with better functionality and lower cost gives a higher value.

In value engineering, if a product or method can accomplish a given function with the lowest cost compared with all competitors, then this product or method is called the *best value*. Clearly, a higher function-to-cost ratio is important for increasing value in the eyes of customers. However, in this author's opinion, the function-to-cost ratio alone is not sufficient to provide an adequate measure for customer value for the following two reasons:

1. It is a convention that people also like to measure value in dollars. In value engineering, sometimes the best value of a function is also defined as the lowest cost to accomplish that function, that is, in dollars. But this definition is not consistent with the definition that value is function divided by cost. (Does this imply that function has the unit of dollars squared?)

2. There are many cases where two products have exactly the same function, but people are willing to pay different prices for them. For example, the Toyota Corolla is actually exactly the same car as the GEO Prizm, but people are willing to pay $300 more to buy a Toyota Corolla. As another example, the same item in a neighborhood convenience store will usually sell at a significantly higher price than that in a large discount chain store. So the function-to-cost ratio alone cannot explain the value in the eyes of customers adequately.

Sherden (1994) and Gale (1994) provided a broader definition for customer value. In their view, the customer value is defined as perceived benefit (benefits) minus perceived cost (liabilities), specifically,

$$\text{Value} = \text{benefits} - \text{liabilities}$$

The benefits include

1. Functional benefits
   a. Product functions, functional performance levels,
   b. Economic benefits, revenues (for investment services)
   c. Reliability and durability
2. Psychological benefits
   a. Prestige and emotional factors, such as brand name reputation
   b. Perceived dependability (for example, people prefer a known-brand product to an unknown-brand product)

c. Social and ethical reasons (for example, environmentally friendly brands)
   d. Psychological awe (many first-in-market products not only may provide unique functions but also give customers tremendous psychological thrill, for example, the first copy machine really impressed customers)
   e. Psychological effects of competition
3. Service and convenience benefits
   a. Availability (ease with which the product or service can be accessed)
   b. Ease with which correctional service in case of product problem or failure can be obtained

The liabilities include

1. Economical liabilities
   a. Price
   b. Acquisition cost (such as transportation cost, shipping cost, time and effort spent to obtain the service)
   c. Usage cost (additional cost to use the product or service in addition to the purchase price, such as installation)
   d. Maintenance costs
   e. Ownership costs
   f. Disposal costs
2. Psychological liabilities
   a. Uncertainty about product or service dependability
   b. Self-esteem liability of using unknown brand product
   c. Psychological liability of low-performance product or service
3. Service and convenience liabilities
   a. Liability due to lack of service
   b. Liability due to poor service
   c. Liability due to poor availability (such as delivery time, distance to shop)

Clearly, this definition of customer value contains much more information than simply function and cost, and it is also in the unit of dollars, or monetary worth.

3.2.1 Value and Other Commonly Used Metrics

There are other product metrics that can very easily be mixed in with the concept of value. These metrics include price, performance, cost, and quality. Let us discuss the similarities and differences of value and these metrics.

Value and Price

Some economists define *value* as price; however, as we discussed earlier, price is only one factor that affects value. Specifically, price is one important element in customer costs (economic liabilities). In general, customers' acceptance of a high selling price of a product indicates that the product has superior benefits in the eyes of the customers. A higher price may provide a higher profit margin to the company that sells the product; however, the sales volume of this product can be sustained only if customers think that this product will provide more benefits than costs, that is, high customer value. Higher customer value can come from providing more benefits for lower prices. The gap between customers, perceived benefits versus costs determines the magnitude of the overall attractiveness to customers and, thus, determines the size of the market, sales volume, and market share.

Value and Performance

Performance is also called "function." The *function* of a product is what the product is supposed to do for customers. As we discussed earlier, function is only one component of benefits; superior customer value is the difference between all benefits minus all costs. Therefore, a product with more and better functions may not always be a better value. Also, a function can create value only if that function is what customers definitely need. In 1979 the American Can Company thought that a product with more and better functions would always sell, so it designed a stronger paper towel called BOLT. It looked and performed like cloth, and it was sold at a higher price than regular paper towels. However, this product turned out to be a total failure, because the customers did not perceive a benefit in a paper towel that could be washed. Actually, in many cases, customers usually enjoy a product change that adds more functions without increasing price, and very often they would be delighted to see a product change that reduces some functions together with a much lower price. Adding and improving functions with a higher price is usually a risky strategy.

Value and Cost

The company's costs in producing a product usually do not relate to customer value. The company could provide a product with a lot of features and high cost that customers do not appreciate. However, reducing costs will provide either a better profit margin or more room for price reduction.

Value, Quality, and Perceived Quality

Like value, *quality* is also a very tricky concept to define. Different people, even different quality gurus, define quality differently. The American Society for

Quality (ASQ) defines quality as: "A subjective term for which each person has his or her own definition. In technical usage, quality can have two meanings: 1. the characteristics of a product or service that bear on its ability to satisfy stated or implied needs. 2. a product or service free of deficiencies."

The first meaning in this definition, "the characteristics of a product or service that bear on its ability to satisfy stated or implied needs" refers to customer preferred performance and function. The second definition, "a product or service free of deficiencies," is definitely related to dependability and reliability. In comparison to customer value, we can see that value is a much broader concept. Only some aspects of value are related to quality.

Quality is also determined by personal opinion; it is largely subjective and psychologically related. This is the meaning of "A subjective term for which each person has his or her own definition" in the ASQ's quality definition. For example, a drug with a brand name may have exactly the same functionality and the same manufacturing quality as a generic drug; however, a substantial portion of consumers will still consider the drug with a brand name to have higher quality. The overall customer opinion on the quality level of a particular product or service is also called *perceived quality*. The perceived quality level is a better indicator of the customer value. There are primarily two components in perceived quality. One is the technical component that relates to such factors as performance, functionality, dependability, and defective level; the other is the psychological component of quality, such as brand image.

3.2.2 The Versatility and Dynamics of Value

Since value is a matter of opinion, it may mean different things to different people. We call this the versatility of value. Value may also change over time with the change of people's minds and lifestyles, and we call it the dynamics of value.

The versatility of value reflects the fact that the marketplace consists of different people; it is difficult to find even two people who have exactly the same opinion. In marketing science, people can be divided into market segments. For a particular kind of product or service, the customers in each segment display similar behaviors and opinions. Some products or services can only find customers in a niche market, that is, a particular market segment. For example, a good state-of-the-art computer engineering book can only be sold at university campuses and to computer engineers; this book is useless to an animal trainer. Some products or services can find customers in mass markets; for example, the products and services such as vegetables, fruits, pencils, and personal banking services can find customers all over the social spectrum.

In order to be successful in the presence of versatility of value and vastly different customer opinions, it is important to know the types of market that your product or service is in and understand what values from your product or service can be brought to customers.

The value of even a single product or service can be broken down into several categories:

- *Use value.* Properties that make something work (also called functional value)
- *Esteem value.* Properties that make something desirable to own (also called emotional or psychological value)
- *Exchange value.* Properties that make it possible to exchange one thing for another

For example, the use value of an airline ticket is the ability to let a customer take an airplane from point A to point B; even a coach class ticket is able to provide that use value. A business class ticket provides a little bit more functional value, such as better seating and better food, but it provides substantial esteem value, that is, the feeling of "I am special, I have special status." A coach class ticket deals with a mass market; the business class ticket deals with a niche market.

If we want to enhance the value of our product or service so we can be more successful in the marketplace, we need to determine what will make it more valuable. But value is a matter of customer opinion. We need to understand what makes people motivated and excited.

Abraham Maslow developed a simple scale to define the psychological needs of people. He called this scale the *hierarchy of needs*, as illustrated in Fig. 3.1. Dr. Maslow said that people are motivated to do different things at different levels of psychological development or different levels of society. He divided these motivational factors into five different basic needs. As each need is satisfied, other higher needs arise. Although the lower-level needs may never disappear, they become weaker or less important. A person may have several needs at the same time, but one need is dominant.

Dr. Maslow's theory provides a lot of insights about customers' buying motivation. For example, customers in developing countries usually prefer products that address basic needs, robust in harsh user conditions, low cost,

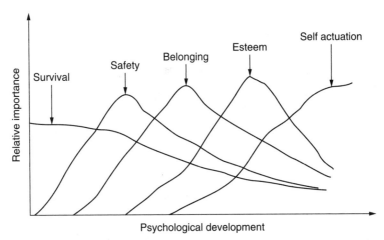

Figure 3.1 Value and Psychological Development

and without fancy features. This fact shows that survival and safety needs are predominant for these consumers. In more affluent countries, printing some sports star's figure on cereal boxes might be a very effective way to increase sales because it addresses the esteem and belonging needs for many children.

The customer value is also very dynamic; that is, it changes over time. Many factors can change customers' preferences and tradeoffs, such as economic conditions, new technologies, and the evolution of customer psychological development.

The oil price change will affect people's attitude toward the types of cars they like. Bad economic times, fierce competition, and a tough job market can make discount chain stores the favorite shopping place. A strong competitor's emergence in a market segment will greatly change the expectations for a particular product or service.

3.3 Maximize Customer Value by Service Product Design

One of the most important keys for a service organization to succeed in the marketplace is to design customer values into service product design. In order to accomplish this, the service organization needs to do the following:

1. Find out what customers really need by marketing research.
2. Analyze customer values.
3. Deploy customer values into service product design.

There are many methods that can be used to accomplish these tasks. They are summarized here.

3.3.1 Methods to Identify Customer Needs

The customer survey is an important tool to get information about what customers really want. Chapter 4 gives a detailed description of all important customer survey issues and techniques, such as selecting survey samples, designing and conducting survey questionnaires, and analyzing survey data.

3.3.2 Methods to Analyze Customer Values

Customer value management (Gale 1994) contains several powerful techniques to quantitatively identify the relative importance of various factors, such as functions, availability, and psychological aspects, that contribute to customer values. Customer value management will be discussed in Chap. 5.

3.3.3 Methods to Deploy Customer Values into Service Product Design

Quality function deployment (QFD) is a Japanese-initiated technique that deploys quality into design. QFD uses a matrix-like template, called a *house of quality*, to facilitate the design process. Chapter 6 discusses QFD in a service industry framework.

Value engineering is a well-established, step-by-step technique that has been used to design products with superior functions and lower cost. Chapter 7 discusses the value engineering technique in a service industry framework.

Brand name recognition is an important part of value. The product with good brand recognition will usually enhance the product value greatly. In service product design, brand name building should be an important consideration. Chapter 8, 'Brand Development and Brand Strategy' gives a road map to brand name building.

The theory of inventive problem solving (TRIZ) is an established method used to guide innovation in product design. In recent years, there are many articles in the literature that discuss applying TRIZ in the service industry. Chapter 9 provides a brief discussion of TRIZ in a service industry environment.

Chapter 4

Customer Survey Design, Administration, and Analysis

4.1 Introduction

One of the most important success factors in designing superior service products is to know your customers. For a Six Sigma company, the design decision should be driven by the voices of customers. For the service industry, obtaining detailed customer data is even more important because maintaining customers' loyalty is extremely vital for the survival of a service institution. The customer survey is an essential tool that can yield information about customer expectations and customer satisfactions and assist in creating strategies for improvement.

Commercial enterprises use customer survey findings to formulate marketing strategies and design their products. Television and radio programs are evaluated and scheduled largely based on the results of consumer surveys. Libraries, restaurants, financial institutions, and recreational facilities use customer surveys to gather information about customer composition and customer desires.

Customer surveys of only a sample of customers are taken to learn about the whole population of a customer base. The sample may contain as few as 30 (a small sample) to as many as 30,000 or more (a very large sample) people. The data obtained on the survey are then analyzed. The survey method is based on sound statistical principles, and over 70 years of modern survey practices show that the predictions based on information from a survey of a relatively small sample are usually quite accurate.

4.1.1 Customer Survey Types

There are three types of customer survey methods: mail out, telephone, and in-person.

Mail-Out Surveys

The mail-out survey involves mailing printed questionnaires to a sample of predesignated potential respondents. Respondents are asked to complete the questionnaire and mail it back to the survey researcher.

The advantages of the mail-out survey include the following:

1. *Low cost:* Other survey techniques require trained interviewers, and in-person interviews may incur high travel costs.
2. *Convenience:* The questionnaire can be completed at the respondent's convenience.
3. *Privacy:* Because there is no personal contact, the respondents may feel more comfortable and their privacy is preserved.
4. *Lack of time pressure:* The respondents can take their time to complete the questionnaire and consult their personal records if necessary.

The disadvantages of the mail-out survey include the following:

1. *Lower response rate than with other methods:* The mail-out survey response rate usually ranges from 20 to 30 percent in the worst case to 85 to 90 percent in the best case. Therefore, an adequate number of mail-out questionnaires and many follow-ups are needed to achieve the desired sample size.
2. *Comparatively long time period:* Questionnaires from the mail-out survey usually take a few weeks to be returned.
3. *Self-selection:* The survey researcher only receives the returned questionnaires. The people who choose to fill out and return the questionnaires may not adequately represent the population. For example, people who have low reading and writing proficiency may never return their questionnaires; thus they will not be represented in the sample.

In-Person Interviews

In-person interviews are conducted by an interviewer who talks directly to respondents to get the information.

The advantages of the in-person interview include the following:

1. *High response rate:* The response rate of the in-person interview is much higher than that of mail-out surveys.
2. *Ability to contact hard-to-reach population:* Certain groups are difficult to reach by mail or telephone; the in-person interview is the only way to reach them.

3. *Greater complexity:* Because of the direct interaction between the interviewer and respondents, complex questions can be asked and further explained if necessary.
4. *Shorter time to finish the survey*

The disadvantages of the in-person interview include following:

1. *High cost:* In-person interviews can be very costly due to work force requirements, travel expenses, interviewer training, and so on.
2. *Interviewer bias:* The interviewer may subconsciously introduce personal bias and affect respondents' choices in answering questions.
3. *Greater stress:* The interviewing process is usually very stressful for both the interviewer and respondent.

Telephone Surveys

The telephone survey is a method of collecting information through telephone interviews between a trained interviewer and respondents.

The advantages of the telephone survey include the following:

1. *Rapid response:* The telephone interview usually gets results more quickly than that of in-person interviews and mail-out surveys.
2. *Lower cost:* The cost of a telephone survey is usually much less than that of an in-person interview, and it can be less costly than a mail-out survey.
3. *Privacy:* The respondents' privacy is better preserved than with in-person interviews.

The disadvantages of the telephone survey include the following:

1. *Less control:* The interviewer has less control in the interview process than the interviewer of an in-person interview since the respondent can hang up the phone easily.
2. *Stress:* Like telemarketing, receiving a stranger's call may be very annoying for many respondents.
3. *Lack of visual material:* A telephone interview cannot use visual aids, such as maps, charts, and pictures to explain the survey questions.

Other Methods Used to Gather Information

Besides customer surveys, there are several other methods that can be used to gather customer information. These methods include secondary research, direct measurement, and direct involvement.

Secondary Research

The customer information may already exist somewhere, such as in libraries, government agencies, and, more recently, on the Internet. In secondary research, the researcher tries to retrieve this information. Data mining techniques (Berry and Linoff 2000, Edelstein 1999) have been developed to dig important information from these huge data sources and gain valuable clues to guide sales and promotion efforts. Data mining is a process of analyzing data and summarizing it into a useful, informational format. It is primarily used by companies with a strong customer focus, such as retail, financial, communication, and marketing organizations. It enables these companies to determine relationships among internal factors such as price, product positioning, and staff skills, and external factors such as economic indicators, competitions, and customer demographics. Data mining enables companies to determine the impact of these factors on sales, customer satisfaction, and corporate profitability, and develop marketing and sales strategies to enhance corporate performance and cut down on losses.

Direct Measurement

This technique involves direct counting, testing, or measuring of data. Typical examples of direct measurement are testing cholesterol levels, monitoring customer arrival times and duration in a service institution, and recording and counting the type and number of errors in insurance claims.

Direct Involvement

This technique is practiced by Toyota. The key idea is to ask the product design leaders to actually play the role of a consumer and practice the product usage process. One story (Liker 2004) stated that a Japanese design leader, who had never been to the United States, was assigned to design a car for the North American market. To overcome his lack of knowledge of this market, he actually traveled to the United States, rented a car, and drove through all 50 U.S. states and 13 Canadian provinces to experience the actual driving and car usage conditions in North America. Because of this first-hand experience, he made a few very good changes in the car design.

4.1.2 Stages of the Customer Survey

Customer survey research is a well-established area. A step-by-step procedure is available to guide the whole customer survey process. In this section, the stages of a typical customer survey are outlined and discussed. We first give a brief overview of each of the following stages:

Stage 1: Establishment of goals and objectives of the survey
Stage 2: The survey schedule and budget
Stage 3: Establishment of an information base
Stage 4: Determination of population and sampling frame
Stage 5: Determination of sample size and sample selection procedure
Stage 6: Design of the survey instrument
Stage 7: Pretest of survey instrument
Stage 8: Selection and training of survey interviewers
Stage 9: Implementation of the survey
Stage 10: Data analysis and report

Stage 1: Establishment of Goals and Objectives of the Survey

The most important, and often the most challenging, part of survey design is clarifying survey intent and scope. This is where knowledge about your own business is essential. Survey specialists cannot do this step for you because they are not the experts of your own business. Only you know what is most important to your business or program.

The following aspects are essential in establishing the goals and objectives of your survey.

Determine Survey Purpose. The most critical part of survey design is a clearly defined statement of purpose and a well-structured view of what you will do with your newly acquired information. Surveys are decision-making tools. They have little value if you are not clear on the decisions your survey will support. It might be helpful to complete the following phrases in order to figure your real intent in this survey study.

- I want to do a survey because _____.
- I intend to use the information I am seeking by _____.
- The information to be gathered will enable me to decide _____.
- I am prepared to implement change as a result of this survey because _____.

Identify Who Will Use the Results and How the Results Will be Communicated. Surveys are communication tools. It is important to understand who will use the survey results and what type of information they respond to. Avoid the common pitfall of assuming others share your tastes in data. Although marketing staff may be comfortable with focus group results, a technical audience may want statistics. Leaders often want both numbers and human-interest perspectives when making decisions. Thus, a quantitative survey paired with selected case studies can be the most compelling for this group. Completing the following statements may help you to define the survey audience.

- The users of survey results include _____.
- The group(s) I will provide with information include _____.
- They want the information in order to _____.
- The way(s) they like to receive information include _____.

Identify What Specific Information Is Needed and When. This next level of refinement identifies and double-checks the information needed from the survey process. Completing the following statements may help you prioritize your information needs.

- The information I really need is _____.
- I need this information because _____.
- My top-priority information needs are _____.

Develop specific research questions that need to be answered. For example, if you are looking for information on a proposed product, identify the new product's elements and the concerns you have about each of them. Also, consider how your choice of survey information will protect the participants' privacy.

Once you have prioritized your information needs, determine when surveying should begin and how often it should occur. Answering the following questions will help you decide.

- How soon will I need information from this survey?
- How often does significant change occur in my process, in my customers or their lives, or in the environment in which this process operates?
- How often do I need updated information to manage my process or program?
- How often do survey users need updated information?

Determine how often to survey by weighing the potential benefit derived from resurveying versus the investment in resources—both yours and your customers. As a rule, the time period between surveys should be short enough to give you reliable information, yet long enough that your customers will not feel bothered. Several factors drive survey frequency, including survey length and complexity, changes in your customer base, changes in your service or product delivery process, and seasonality of services or products.

Stage 2: The Survey Schedule and Budget

After the goals and objectives of the customer survey are determined, the survey researcher should establish a budget and timetable for the duration of this survey project. Ideally, the money and time resources devoted to your survey will be driven by the importance of the decisions you will make based on the data. Consider the following questions when scoping your resource requirements.

- What is the value of the information I am seeking?
- What are the potential consequences of the decisions I will make based on this information?
- What is the cost of not having survey data?
- What staff and other resources are currently available?
- What staff and other resources do I need?

Then determine the resource requirements by identifying the staff and financial resources that can be devoted to the project. It is important to agree about resource issues to ensure you have the capability to deliver a survey that meets your overall expectations. You may have internal resources, such as trained interviewers or data analysts. If not, consider whether you are willing to contract out for these services. Limitations on resource requirements will shape the entire survey design process. The timetable should be flexible enough to accommodate unforeseen delays.

Stage 3: Establishment of an Information Base

Before developing a survey instrument (questionnaire), it is necessary to gather information about the subject matter under investigation from interested parties and individuals. The purpose of this stage is to develop the information base for the questionnaire. A focus group meeting is usually the key activity in this stage. The focus group consists of carefully selected customer representatives, company marketing representatives, product development people, and so on. A typical size of a focus group ranges from 5 to 12. In this focus group meeting, based on the goals and objectives of the survey established in stage 1, an exhaustive list of raw questions will be developed and they will serve as a basis to develop the survey instrument.

Stage 4: Determination of Population and Sampling Frame

The population is the entire set of people, organizations, households, etc., that are addressed by your survey research. For example, for a fast-food chain, the relevant population will be fast-food eaters. For a suburban hospital, the population will be residents in the neighboring area. The portion of the population that can be identified to be interviewed is called the *sampling frame*.

For example, the population of fast-food eaters may include all people except homeless and sick people. But if a telephone interview is to be conducted, only the people with known telephone number can be reached, so "people with a telephone number" will be the sampling frame for fast-food eaters. The concepts of population and sampling frame will be discussed further in Sec. 4.4.

Stage 5: Determination of Sample Size and Sample Selection Procedure

The survey researcher will have to select a sample that adequately represents the population under study. In general, larger samples will yield greater accuracy than small samples in terms of analysis results. The sample size is usually determined by balancing between analysis accuracy and the increased cost and time due to larger sample size. Once the sample size is determined, the method of sampling will be determined. The commonly used sampling methods include random sampling, stratified random sampling, and cluster sampling. The sample size determination and sampling methods will be discussed in Sec. 4.4.

Stage 6: Design of the Survey Instrument

The development of a survey instrument or questionnaire is a key component of the customer survey. At this stage, a survey researcher must design a series of unbiased, well-structured questions that can systematically obtain the information based on the survey goals and objectives developed in stage 1. The input from the focus group is a major source for survey questions. The development of a questionnaire can be an extremely detailed and time-consuming process. The best-designed questionnaire should be short and concise with well-worded questions. Long and wordy questionnaires will often result in a lower response rate and a higher survey cost. The design of the survey instrument is discussed in Sec. 4.2.

Stage 7: Pretest of the Survey Instrument

After a draft questionnaire is designed, it is important to pretest the questionnaire with a small group of respondents. During this pretest, poorly worded questions will be identified and refined. The refined questionnaire will have a better quality.

Stage 8: Selection and Training of Survey Interviewers

For telephone and in-person interviews, trained interviewers are required. The source of interviewers can be college students, part-time workers, and so on. Interviewers should be familiar with the questionnaire and know how to handle uncooperative respondents.

Stage 9: Implementation of the Survey

The administering of the survey instrument is a very crucial stage of the customer survey. It is very important to abide by the sampling procedure to ensure the validity of the survey. It is also very important to stick to the time schedule of the survey process. Care must be taken to ensure the privacy of the respondents and minimize their inconvenience. The implementation of the survey is further discussed in Sec. 4.3.

Stage 10: Data Analysis and Report

The data from returned questionnaires will be summarized and analyzed by statistical methods, and the final findings of the analysis will be reported. Section 4.5 will discuss details of data analysis and report.

4.2 Survey Instrument Design

One of the key tasks in the customer survey is the questionnaire development process. The main issues in questionnaire development include the type of questions, the wording of questions, the sequence of questions, and the length of the questionnaire.

4.2.1 Types of Questions

There are two types of questions in the customer survey: closed-ended questions and open-ended questions.

Closed-Ended Questions

Closed-ended questions provide a fixed list of alternative responses and ask respondents to select one or more alternatives as the best answer(s) for the question. Closed-ended questions have many different formats.

Multiple-Choice Questions

A multiple-choice question has a list of answers and usually asks for a fact. Only one answer is supposed to be applicable. Here is an example of a multiple choice question:

> *What department do you work in?*
> a. *Sales*
> b. *Marketing*
> c. *Manufacturing*
> d. *Research*

When you construct this kind of question, you must be careful to make sure the list of answers that you offer to your respondents is exhaustive and mutually exclusive.

In the example question, if someone works in the personnel department, he or she will not be able to find an appropriate answer. In such a case, this question has an inadequate number of answers. It is also recommended that the total number of answers in a multiple-choice question be no more than 10 to 12.

Checklist or Inventory Questions

A checklist or inventory question asks respondents what subset of items on a list would apply to them. Here is an example of a checklist question:

> Please indicate from what sources you obtain information about new music and movies? Check all that apply.
>
> _____ Radio   _____ Television   _____ Internet
> _____ Newspaper   _____ Friend   _____ Magazine
> _____ Other (please specify)

Clearly in this question, more than one answer may apply.

Rating Questions

A rating question asks the respondent to use a given scale to judge something. Here is an example of a rating question:

> Please rate your instructor's teaching ability in the following categories on a 1 to 5 scale, where 1 is very poor, 3 is average, and 5 is excellent.
>
> | **Rating** | **Category** |
> |---|---|
> | _____ | Course contents |
> | _____ | Instruction |
> | _____ | Office hours |

The Pros and Cons of Closed-Ended Questions

The advantages of closed-ended questions include

1. The set of answers in a closed-ended question is uniform, so it is easy to compare the differences among respondents.
2. The uniformity in the set of answers for each question will make computer data entry easier.
3. The fixed list of answers tends to make the question clearer to the respondent.

The disadvantages of closed-ended questions include

1. Closed-ended questions compel respondents to choose the closest representation of their actual response on the list of prespecified answers, which may deviate from their true opinions.
2. When respondents are unsure of what is the best answer, they may choose a random answer, which will lead to errors.

Level of Measurements in Closed-Ended Questions

Before being analyzed, the survey data should be organized as variables. A variable is a specific characteristic of the population, such as age, sex, preference, and rating. Depending on the design of the questions and answers, the variables used in the survey have different measurement properties, referred to as levels of measurement or measurement scales. The commonly used measurement scales include nominal, ordinal, and interval. In survey design, there is a specially designed interval scale called the Likert scale. We are going to discuss these in detail.

Nominal Scale

The nominal level of measurement simply places the survey answers into categories. For example, a variable such as political party preference in the United States can be categorized as three classes, Democrat, Republican, or Independent. In nominal scale, survey data can be placed into categories and their frequency of occurrences counted. There is no ranking or ordering of the categories.

Ordinal Scale

The ordinal level of measurement goes one step beyond the nominal scale; it ranks the categories by a certain criterion. For example, the education levels of people can be classified into the following categories: high school graduate or lower, two-year college degree, bachelor's degree, master's degree, and Ph.D. degree. Clearly, we can rank these education levels; a Ph.D. is certainly higher than a master's, for example, but it is difficult to define a numerical difference between these educational achievements.

The Interval Scale

The interval level of measurement gives the greatest amount of information about the variables. It labels, orders, and uses numerical units of measure to indicate the exact value of each category. For example, variables such as income, age, and weight are in interval scales.

Likert Scale

The Likert scale (after Rensis Likert) is used for the measurement of attitudes and opinions. A Likert scale may contain several items such as strongly agree, agree, neutral, disagree, or strongly disagree. Here is an example:

Netscape is easier to use than Microsoft Internet Explorer.

1. Strongly disagree
2. Disagree
3. Neutral
4. Agree
5. Strongly agree

Sometimes, a numerical scale is explicitly displayed in the questionnaire:

What is your general impression of how the Port city government affects your business?

Highly negative				***Highly positive***
1	*2*	*3*	*4*	*5*
___	___	___	___	___

Likert items are ordinally scaled. It is not assumed that the difference between the choices of strongly agree and agree is the same size as the difference between the choices of agree and neutral. However, in survey data analysis, it is general practice to treat the Likert scale as an interval scale. For example, in college course evaluations, there are many Likert scale questions about a professor's course teaching; the scores from all students for each question are averaged as the evaluation score. Clearly this treatment assumes that the Likert scale is an interval scale.

Open-Ended Questions

Open-ended questions deal with situations where the list of possible answers is very long or it is very difficult to construct an exhaustive list. Here are some examples:

What is your favorite place to go for summer vacation? _____

How long have you and your family lived in your current place? _____

What is the first foreign language you learned? _____

Open-ended questions allow respondents to provide longer, more complex answers than for closed-ended questions. There are several disadvantages to open-ended questions. First, open-ended questions will elicit some irrelevant answers. Second, it takes a lot more effort to analyze open-ended questions, since the number of distinct answers from all respondents may be very high and messy, Thirdly, open-ended questions are difficult to analyze by statistical methods, because statistical methods require some degree of data standardization.

Overall, it is highly recommended that most survey questions be in the form of closed-ended questions to ensure a higher response rate, shorter questionnaire completion time, and easier data analysis.

4.2.2 The Wording of Survey Questions

The wording of survey questions is very important for a successful customer survey. Good survey questions should be

1. Clear, easily understandable, and stated in a direct and straightforward way
2. Specific and precisely stated so the respondent knows exactly what is being asked
3. Unambiguous and unequivocal so there is only one way to understand or interpret what the question is asking about
4. Simple and brief rather than complicated, cluttered, and long-winded
5. Stated in terms that your respondents are likely to be familiar and comfortable with, without using complex technical terminology, jargon, or overly sophisticated wording

Good survey questions should *not* be

1. *Leading*. They should not draw the respondent toward a specific answer or make some answers clearly unattractive or undesirable.
2. *Multipurpose*. They should not ask about two or more things together in the same question.
3. *Threatening*. They should not make the respondent uncomfortable or put the respondent in a difficult or compromising position.

Example 4.1: Here Is an Example of a Multiple Purpose Question

Do you believe the development of the I-696 freeway entrance will affect the image and property value for our whole subdivision?

Yes____   No____

This question is difficult to answer, because a yes or no answer indicates that the respondent feels that both the image and property value will be affected in the same way. However, the respondent might feel that the image of the community will increase, because of the better freeway access, but the property value will go down due to increased traffic, noise, and commercial development.

Example 4.2
If a survey question is worded as "What is your income?", it will generate all kinds of answers, such as annual income, hourly pay, monthly income, or total household income. This is an example of an ambiguous question. A better wording of the question would be

What is your total annual income before taxes?

a. *Below $20,000*
b. *$20,001 to $40,000*
c. *$40,001 to $60,000*
d. *$60,001 to $80,000*
e. *$80,001 to $100,000*
f. *Over $100,000*

4.2.3 Order of Questions in Surveys

The order in which questions are presented can affect the overall customer survey significantly. A poorly organized questionnaire can confuse the respondents, bias their responses, and jeopardize the quality of the survey.

It is a good idea to start the survey with some easy introductory questions. Here are some examples:

Do you own a car?

Yes_____   No_____

Do you have an e-mail address?

Yes_____   No_____

How long have you lived in the current property?

_____ (years)

You should save more complicated questions that may require some careful thought for later after you warm up the respondents with introductory questions.

There are several typical organizational patterns for survey questions:

1. *Chronological order:* The questions are in sequential or temporal order, for instance, from most recent to least recent.
2. *Funnel pattern:* From general to specific.

3. *Inverted funnel:* From specific to general.
4. *Tree pattern:* The questions branch out in different directions, depending on the respondent's answers to early questions.

The question organizational pattern should be chosen based on the goals and objectives of the customer survey. Usually, topically related questions should be grouped together.

Example 4.3: The Following Is an Example of a Good Grouping of Questions

1. *How would you describe the current relationship between labor and management?*

   Good _____   Fair _____   Poor _____

2. *During the past five years, do you think this labor-management relationship has*

   Improved _____   Remained the same _____   Worsened _____

3. *In what way do you think the labor-management relationship can be improved?*

These three questions are related and in logical order. However, let's we put the following three questions together.

1. *Do you or your coworker participate in a company-sponsored suggestion program?*

   Yes _____   No _____

2. *During the last five years, do you think this labor-management relationship has*

   Improved _____   Remained the same _____   Worsened _____

3. *Would you be interested in a job training program?*

   Yes _____   No _____

The respondents will be able to answer these questions, but they may get disoriented after answering 10 or more of these kinds of misplaced questions.

4.2.4 Questionnaire Length

The questionnaire should be as concise as possible while covering the necessary subjects based on the goals and objectives of the survey. Practice has shown that when the survey becomes too long, the response rate and the quality of the answers will go down significantly.

As a general guideline, telephone interviews should be less than 20 minutes; mail-out questionnaires should not take more than 30 minutes to answer;

the in-person interview should be limited to less than an hour. Ideally, the telephone survey should take between 10 to 15 minutes, mail-out surveys should take about 15 minutes, and in-person interviews should take less than 30 minutes.

4.3 Administering the Survey

Once the survey instrument is designed, pretested, and revised, it is time to administer the survey. For different survey methods, that is, mail-out, telephone, and in-person, the way of administering the survey will be different. In this section, we are going to discuss how to administer the survey for each of these cases.

4.3.1 Administering a Mail-Out Survey

In a mail-out survey, the questionnaire should be designed in the form of a booklet in order to ensure a professional appearance. Any resemblance to an advertisement brochure should be strictly avoided. The professional appearance of the questionnaire is very important to ensure a satisfactory response rate. There should be adequate spacing between questions. A good cover letter is very important to explain the purpose of the survey. The questionnaire should be designed such that it is very convenient for respondents to mail it back.

There are two ways to present the questionnaires to respondents. Questionnaires can be personally delivered to respondents. This method is more costly in terms of time and effort, but it is likely to result in a higher response rate, a more rapid response, a higher percentage of completed questions in the questionnaire, and perhaps more valid and accurate responses. The other method is the direct mailing of questionnaires to respondents; this method will usually result in a lower response rate. Some remedies for this include follow-up mailings, or follow-up phone calls. Usually these follow-ups should be done three to four weeks after the questionnaires are sent by mail. Direct mailing plus the follow-ups usually will achieve a 50 to 60 percent response rate. Additional follow-ups may raise the response rate to over 70 percent.

4.3.2 Administering a Telephone Survey

The telephone survey is less complex to implement than the mail-out survey. The most important aspect of the telephone survey is the selection and training of telephone interviewers. A good source of possible interviewers is university students, especially graduate students. The interviewers usually first study the questionnaire by themselves. Then they are trained in pretest results, potential tough issues of the questionnaire, and many general ethical issues.

Interviewers should not introduce any bias in the interview process and should not express any opinions in response to the answers from the respondents.

Many companies that conduct telephone interviews use computer-assisted telephone systems where the interviewer sits at a computer that dials the telephone number and puts the questions to be asked on the screen so the interviewer can read them to the respondents. The software that manages this can take care of the data entry and coding, as the interviewer uses the computer's keyboard or mouse to indicate the respondent's answers to the questions.

4.3.3 Administering an In-Person Survey

In-person interviews are the most expensive to conduct, in terms of both time and money, and the most intrusive method. A major strength of the in-person interview is the ability to deal with complex topics. Because you can see how respondents react to the questions as you ask them, you will have a better idea how well they understand the questions and what confuses them. You will also have opportunities to resolve any glitches in the interview process.

The selection and training of in-person interviewers is even more important than that of telephone interviewers. The selection methods and training of interviewers are almost the same as that of telephone interviewers.

4.4 Survey Sampling Method and Sample Size

The main goal of a customer survey is to produce an accurate picture about the population based on the information drawn from a scientifically selected subset from that population. Sampling is necessary because it is impractical to seek information from every member of the population. In this sampling process, first we need to determine our population. Then we need to define a sampling frame that is a list of elements in the population that may be selected in the sample. Thirdly, we need to identify a sampling method, that is, a method to select a subset from the sampling frame. Before we do the sampling, we need to identify what is the adequate sample size for this survey in order to ensure the credibility of the survey data analysis. In this section, we are going to discuss these issues.

4.4.1 Population and Sampling Frame

The first consideration in survey sampling is the specification of the unit of analysis. The *unit of analysis* is the individual, object, institution, or group that bears relevance to the survey study. For example, for a fast-food chain owner, the individual consumer could be the unit of analysis; for a mortgage lending operation, each household could be the unit of analysis; for a medical equipment supplier, each hospital or clinic could be the unit of analysis. The *population* is defined to be the collection of units of analysis that findings of the survey will apply to. For example, the population of fast-food chain customers is the collection of all potential individual customers that the chain can reach; the population of the customers for a mortgage lending operation is the collection of all the households that the lending operation could do business with; the population of the customers for a medical equipment supplier is the collection of all potential hospitals and clinics that could do business with this supplier.

However, in any population, usually not all the units of analysis can be identified and reached. For example, if a population is to be the people living within a metropolitan area, then the unit of analysis will be each single resident. From a practical point of view, it is unlikely that all the residents of this metropolitan area can be identified and reached. People are born and people die; people move in and out. There are people who do not have telephones or stable living places. Usually only a portion of the population is identifiable and reachable; this portion of the population is often called a *working population*. From the working population, it is possible to develop a list of units of analysis that can be readily reached in our customer survey. This list is called the *sampling frame*. For example, if the population is all the residents in a metropolitan area, then the working population could be all residents that can be reached by phone, and the sampling frame could be the residents listed in the local telephone directory. Some other possible sources for the sampling frame include voter lists; utility (gas, electric, water, and so on) customer lists; motor vehicle registrants; magazine and newspaper subscriber lists.

With most sampling frames you will have to deal with some of the following problems:

1. *Missing elements:* Legitimate members of the population not included in the sampling frame. For example, in some polls of U.S. elections in 2004, only traditional phone users were polled; people with cell phones only were not selected in the poll list. Therefore, a sizeable portion of young professionals was left out.
2. *Foreign elements:* Some people's names are listed in the sample frame, but they are actually no longer in the population. For example, people could have moved out a while ago but their names are still in the phone directory.
3. *Duplicated elements:* Population members listed more than once in the sample frame.

In all these situations you need to determine how many missing, foreign, and duplicated elements are in the sampling frame and how big a proportion these wrong elements are as a percentage of the whole group of sampling frame elements. If this proportion is large and it will affect the accuracy of the poll, you should consider the possibility of using a different sampling frame. For example, as stated before, in some of the opinion polls of the U.S. 2004 election, people with cell phones only were excluded in the opinion poll sampling frame. If the portion of people excluded was a sizeable portion of voters and their opinions were significantly different than those of traditional phone users, then this opinion poll might be rather unreliable.

4.4.2 Sampling Methods

Probability Sampling versus Nonprobability Sampling

Sampling methods can be classified into probability sampling and nonprobability sampling. Probability sampling is used when you would like to draw conclusions on the whole population based on the data you collected in the sample. If your goal is just to learn something about the sample and you do not intend to draw conclusions on the whole population, then probability sampling is not necessary.

There are two characteristics of probability sampling:

1. The probability of selection is equal for all elements of the sampling frame at all stages of the sampling process.
2. The selection of one element from the sampling frame is independent of the selection of any other element.

For example, consider a sampling frame of 1000 people whose names are written on equal-sized pieces of paper and where these paper pieces are thoroughly mixed and selected one by one without the names on the paper pieces being seen. If we assume that 100 people will be selected for this sample, then the probability of selecting any person in the first draw is 1/1000; the selection of any person in the second draw is 1/999; ... ; and selection of any person in the 100th draw is 1/901. Though the probability of selecting a particular person is slightly different in each draw, within each draw, the probability of selection for all the available people is the same; this is consistent with the first rule of probability sampling: The probability of selection is equal for all elements of the sampling frame at all stages of the sampling process. Also, the probability of selecting a particular person is clearly independent of previous drawings of other people, so this sampling practice is an example of probability sampling.

Chapter Four

There are several methods of probability sampling: random sampling, systematic sampling, stratified random sampling, and cluster sampling. We discuss each of these probability sampling methods and nonprobability sampling in detail.

Random Sampling

The best-known probability sampling method is random sampling. In the random sampling method, each unit in the sampling frame is assigned a distinct number. Then the units are chosen at random by a process that does not favor certain numbers or certain patterns of numbers. The chosen units will become the sample. A commonly used method to randomly choose units from the sampling frame is the use of the table of random numbers. Table 4.1 shows a portion of a table of random numbers.

Suppose there are 1000 people in the sampling frame and we want to select a random sample of 30 people. Each person will be assigned a number ranging from 000 to 999. Using Table 4.1, we can then arbitrarily select three digits from the five digits given. For example, we can choose the last three digits. In this case we will select the people with numbers 073, 849, 761, 622, 905, 276, 837, ... ,033.

For large samples the use of a random number table will become tedious and time-consuming, so computer-generated random numbers can be used to select a random sample.

Systematic Sampling

Systematic sampling is an adaptation of the random sampling method. It is used when the sampling frame is quite large and the sampling units cannot be easily numbered. For example, if a sampling frame has 3,000,000 people

Table 4.1 A Portion of a Table of Random Numbers

	1	2	3	4	5	6
1	77073	51849	15761	85622	38905	72276
2	20837	95047	50724	16922	04405	30858
3	37504	15645	36630	28216	10056	97628
4	40392	58557	60446	11553	60013	38037
5	53408	14205	33152	70651	17314	93033

on the list and a sample of 1500 people is required, a random sampling approach might be unrealistic, simply because numbering 3,000,000 people is already a big task. If the original list of these 3,000,000 is randomly distributed, we can select sample units by selecting them from the list at fixed intervals (every nth entry, for example, every 20th car on the highway, every 50th customer in a store). In this example, if we want to select 1500 people, because 3,000,000/1500 = 2000, we can select 1 out of every 2000 people in the sampling frame. In this case, if we start with a random starting point and then select a person after we count every 2000th sampling unit, this procedure will create a random sample of 1500 people.

Stratified Random Sampling

Stratified sampling assumes that the sampling frame consists of several mutually exclusive groups, called *strata*. In stratified random sampling, the total number of samples is divided among strata by a predetermined proportion. Then, random samples are taken from each stratum. For example, in a community, assume that 60 percent of the population is white, 15 percent is black, 15 percent is Hispanic, and 10 percent is Asian. If a sample of 1000 people is needed, the stratified sampling method will divide these 1000 people into four ethnic groups based on the proportion in the population. So 600 samples will be allocated to whites, 150 samples to blacks, 150 samples to Hispanics, and 100 samples to Asians. Then these 600 people in the "white" strata will be randomly selected from the white sampling frame, 150 samples of blacks will be randomly selected from the black sampling frame, and so on.

Cluster Sampling

Cluster sampling deals with the situation in which there is a hierarchy of sampling units. The primary sampling unit is a group (or cluster), such as counties, cities, schools, or subdivisions. The secondary sampling units are the individual elements within these clusters from which the information is to be collected. For example, if we want to study the needs of first and second graders, it is difficult to directly locate the sampling frame from a raw population list, such as a telephone directory. It is easy to identify the clusters, such as public and private schools, in which there are first grade and second grade classrooms. After we select a subset of classrooms, we can randomly select sample units from these classrooms.

Nonprobability Sampling

In nonprobability sampling, the probability that a particular unit will be selected is unknown. In this case, we cannot generalize the finding within the sample to the population because we cannot assume any valid statistical relationship between the sample and population or use such a useful probability distribution model as the normal distribution. However, nonprobability sampling can still be helpful. It is much easier to select a sample and get a feel of what a portion of customers may think. For example, nonprobability sampling can be used to quickly select a small sample of respondents (say 30) to pretest a survey instrument. Although the conclusion from these 30 people cannot be generalized to the general population, a lot of shortcomings of the survey instrument can be identified.

There are a few commonly used nonprobability sampling methods. The most commonly used sampling method is the *sidewalk survey*. The interviewer, for instance, may interview passersby near a shopping center, assuming the general population is the shoppers. In this approach, the sampling frame is not explicitly identified and numbered; the probability of selecting any particular passerby is unknown. The advantage of this approach is the ability to get a lot of information quickly. The other commonly used nonprobability sampling technique is *snowball sampling*. Snowball sampling is particularly beneficial in instances where it is difficult to identify potential respondents. Once a few respondents are identified and interviewed, they are asked to identify others who might qualify as respondents. Soon the list of respondents will be increased.

4.4.3 Sample Size Determination

One critical question in a survey project is how many units in a sample are needed so that the analysis result derived by this sample can be generalized to the whole population. The answer to this question depends on two key factors. One key factor is what level of accuracy is required in this study; the greater level of accuracy required in the study, the larger the sample size needed. The other factor is the cost and time that we would like to spend in this survey study; a larger sample size will certainly mean higher cost and longer time. Therefore, the sample size is mostly determined by the tradeoff between desired level of accuracy and cost and time.

Determination of Sample Size for Variables Expressed in Proportions

In survey data analysis, many variables are expressed in terms of proportions. For example, we could ask customers:

Do you like the service of ABC Bank?

Yes _____ *No* _____

The proportion of people in the survey sample answering yes, which is often called the sample proportion $\hat{P}$, is frequently used as the statistical estimate of population proportion p, where p is the real proportion of customers who like ABC Bank's service. Of course, we would like $\hat{p}$ as close to p as possible. From the properties of the normal distribution, and if the random sampling method is used, the probability distribution of $\hat{p}$ is

$$\hat{p} \sim N\left(p, \frac{p(1-p)}{n}\right) \tag{4.1}$$

The $100(1 - \alpha)\%$ confidence interval for p is

$$\hat{p} \pm Z_{\alpha/2}\sqrt{\frac{p(1-p)}{n}} \approx \hat{p} \pm Z_{\alpha/2}\sqrt{\frac{\hat{p}(1-\hat{p})}{n}} \tag{4.2}$$

We can use $\Delta_p = Z_{\alpha/2}\sqrt{\frac{p(1-p)}{n}} \approx Z_{\alpha/2}\sqrt{\frac{\hat{p}(1-\hat{p})}{n}}$ to represent the half width of the confidence interval for p. The magnitude of Δ_p represents the accuracy of $\hat{p}$ as an estimator of p, because

$$P(\hat{p} - \Delta_p \leq p \leq \hat{p} + \Delta_p) = (1-\alpha)100\% \tag{4.3}$$

Δ_p is also called the margin of error.

Example 4.4
In a customer satisfaction survey, the preliminary results indicate that the proportion of unsatisfied customers is very close to the proportion of satisfied customers. What sample size is needed if we want the accuracy of the survey to be within ±3 percent of the true proportion, with 95% confidence?

In this case, clearly $p \approx 50\%$, from the problem, statement, we want

$$\Delta_p = Z_{\alpha/2}\sqrt{\frac{p(1-p)}{n}} = 3\%$$

Therefore

$$n = \left(\frac{Z_{\alpha/2}\sqrt{p(1-p)}}{\Delta_p}\right)^2 \tag{4.4}$$

is the sample size formula for this case. Specifically,

$$n = \left(\frac{1.96 \times \sqrt{0.5(1-0.5)}}{0.03}\right)^2 = 1067$$

for this example, where $Z_{0.025} = 1.96$.

So a sample of 1067 or more people is needed to ensure the accuracy of ±3 percent.

Table 4.2 Minimum Sample Sizes for Proportions

Confidence Interval (Margin of Error, %)	Sample Size	
	95% Confidence	99% Confidence
±1	9,604	16,590
±2	2,401	4,148
±3	1,067	1,844
±4	601	1,037
±5	385	664
±6	267	461
±7	196	339
±8	151	260
±9	119	205
±10	97	166

Table 4.2 lists the relationship between sample size, margin of error, and confidence level.

Determination of Sample Size for Variables Expressed in Proportions When the Population Is Small

The sample size rule specified by Eq. (4.4) is based on the assumption that the population size is infinite or very large. In some survey studies, however, the population size is rather limited. For example, the customer base for a medical equipment supplier will consist of a number of hospitals and clinics; the population size will be in hundreds in the best circumstance. If the population size, say N, is known, then according to Rea and Parker (1992), the sample size n can be calculated by

$$n = \frac{Z_{\alpha/2}^2[p(1-p)]N}{Z_{\alpha/2}^2[p(1-p)]+(N-1)\Delta_p^2} \quad (4.5)$$

Example 4.5
In a customer satisfaction survey, the preliminary results indicate that the proportion of unsatisfied customers is very close to the proportion of satisfied customers, and the population size is $N = 2500$. What sample size is needed if we want the accuracy of the survey to be within ±3 percent of the true proportion, with 95% confidence?

Using Eq. (4.5)

$$n = \frac{Z_{\alpha/2}^2[p(1-p)]N}{Z_{\alpha/2}^2[p(1-p)] + (N-1)\Delta_p^2} = \frac{1.96^2 \times (0.5 \times 0.5)(2500)}{1.96^2 \times (0.5 \times 0.5) + 2499 \times (0.03)^2} = 749$$

This sample size is smaller than that of Example 4.4.

Determination of Sample Size for Interval-Scale Variables

In survey analysis, some variables are interval-scale variables. For example, personal income, age, and evaluation scores based on the Likert scale are all interval-scale variables. The population means of these interval-scale variables μ are usually of interest. The sample mean of the interval-scale variable $\bar{x}$ is often used as the statistical estimate of population mean μ. Similarly, we would like μ as close to μ as possible. From the properties of the normal distribution, and if the random sampling method is used, the probability distribution of $\bar{x}$ is

$$\bar{x} \sim N\left(\mu, \frac{\sigma^2}{n}\right) \tag{4.6}$$

The $100(1-\alpha)\%$ confidence interval for μ is

$$\bar{x} \pm Z_{\alpha/2} \frac{\sigma}{\sqrt{n}} = \bar{x} \pm \Delta_\mu \tag{4.7}$$

where Δ_μ is the margin of error for μ.

By using the relationship

$$\Delta_\mu = Z_{\alpha/2} \frac{\sigma}{\sqrt{n}} \tag{4.8}$$

we can derive the sample size rule:

$$n = \frac{Z_\alpha^2 \sigma^2}{\Delta_\mu^2} \tag{4.9}$$

Example 4.6
In a survey study of household incomes for county Y, the preliminary estimate of average household income is $40,000 and the standard deviation is estimated to be $6000. If we would like to determine a survey sample size so that the margin of error for the average household income is no more than $1000, what is the minimum sample size, if a confidence level of 95% is desired?

Using Eq. (4.9),

$$n = \frac{Z_\alpha^2 \sigma^2}{\Delta_\mu^2} = \frac{1.96^2 \times 6000^2}{1000^2} = 139$$

Therefore, a minimum sample size of 139 households is required.

Determination of Sample Size for Interval-Scale Variables When the Population Is Small

The sample size rule specified by Eq. (4.9) is based on the assumption that the population size is infinite or very large. In some survey studies, however, the population size is rather limited. If the population size, say N, is known, then according to Rea and Parker (1992), the sample size n can be calculated by

$$n = \frac{Z_{\alpha/2}^2 \sigma^2}{\Delta_\mu^2 + Z_{\alpha/2}^2 \sigma^2/(N-1)} \tag{4.10}$$

Example 4.7
In a survey study of household incomes for county Y, the preliminary estimate of average household income is $40,000 and the standard deviation is estimated to be $6000. If we would like to determine a survey sample size so that the margin of error for the average household income is no more than $1000, and it is known that the total number of households in county Y is 5000, what is the minimum sample size, if a confidence level of 95% is desired?

By using Eq. (4.10)

$$n = \frac{Z_{\alpha/2}^2 \sigma^2}{\Delta_\mu^2 + Z_{\alpha/2}^2 \sigma^2/(N-1)} = \frac{1.96^2 \times 6000^2}{1000^2 + (1.96^2 \times 6000^2)/4999} = 135$$

Chapter 5
Customer Value Management

5.1 Introduction

In Chap. 3 we illustrated that customer value is the difference between benefits and liabilities, that is,

$$\text{Value} = \text{benefits} - \text{liabilities}$$

where the benefits consist of

1. Functional benefits
2. Psychological benefits
3. Service and convenience benefits

and the liabilities consist of

1. Economical liabilities (customer costs)
2. Psychological liabilities
3. Service and convenience liabilities

For any product or service, the company that provides more value to its customers than its competitors will eventually gain in sales and profitability. However, for each particular product or service, the profile of benefits and liabilities will be very different, and customers view each item's benefits and liabilities with different relative importance. When a product or service has several competitors, it is necessary to do better in the areas that customers view as very important.

B. Gale (1994) developed a systematic approach that tries to maximize customer value in providing products and services. This approach consists of the following steps:

1. Conduct customer surveys in order to get the information about relative importance ratings of the aspects of a particular product or service that matter to customers, and customers' ratings of the company's product

or service together with those of competitors; Gale called this step of compiling the *market-perceived quality profile*.
2. Collect information about the price of the company's product or service or combined customer costs, together with competitors' prices or combined customer costs. Gale calls this step of compiling the *market-perceived price profile*.
3. Complete a comprehensive customer value evaluation of the company's product or service compared with competitors, Gale calls this step of compiling the *customer value map*.
4. Complete an area-to-area competitive analysis, so the critical areas can be identified in order to gain a competitive advantage.
5. Deploy the critical improvement into the product or service design.

In this chapter, we discuss this customer value management in detail. Section 5.2 discusses all aspects of establishing a market-perceived quality profile. Section 5.3 gives details about how to establish a market-perceived price profile. Section 5.4 discusses the customer value map, Sec. 5.5 discusses area-to-area competitive analysis, and Sec. 5.6 covers the QFD-like approach to improve product service design to maximize market-perceived customer value.

5.2 Market-Perceived Quality Profile

In the customer value management approach, all non-cost-related attributes, such as functional benefits, psychological benefits, and service and convenience benefits are considered to be components of market-perceived quality. If a product or service can offer a higher market-perceived quality than all competitors yet have its cost under control, then clearly this product will have a competitive edge. The market-perceived quality profile is a detailed quality scorecard that provides quantitative quality ratings of a company's own product versus competitors' products on all important non-cost-related attributes.

According to Gale (1994), creating the market-perceived quality profile involves the following steps:

1. In forums such as focus groups, ask customers in the targeted market, including both your customers and competitors' customers, to list the factors, other than price, that are important in their purchase decisions.
2. Establish how the various nonprice factors are weighted in the customer's decision, usually by simply asking customers to tell you how they weigh the various factors, distributing 100 points among them.

3. Ask well-informed customers, including both yours and those of your competitors, how you and your competitors perform on the various quality attributes. Then for each attribute, divide the score of the product or service you are studying by the scores of competitors' products. This gives you the performance ratio on that attribute. Multiply each ratio by the weight of that attribute. And add the results to get an overall market-perceived quality score.

Table 5.1 is an example of a market-perceived quality profile for frozen chicken (Gale 1994). In this table, the first column lists all the important non-cost-related quality attributes in the chicken business: yellow bird, meat-to-bone, and so on. The second column of this table lists the relative importance rating for the attributes listed in column 1; the relative importance ratings add up to 100 percent. For example, the yellow bird attribute accounts for 10 percent of relative importance; the fresh attribute accounts for 15 percent of relative importance. These importance ratings are obtained from a specially designed customer survey. The third column of this table lists the average customer rating of our business's product for each quality attribute. The highest possible score is 10.0; the lowest possible score is 1.0. These scores are also computed based on a specially designed customer survey. Column 4 of this table lists the average customer ratings of competitors' products for each customer attribute. The fifth column lists the ratio of column 3 to column 4, that is, the average customer rating of our product versus the average customer rating of competitors' products. Clearly, if this ratio is less than one, then it means that in this quality attribute category, our product performs worse than that of our average competitor; if this ratio is greater than one, then it means that in this quality category, our product performs better than that of our average competitor. The values in the last column, column 6, are obtained by multiplying the value from column 5 (the ratio) times that in column 2 (the relative importance score). Clearly, if all the ratios are equal to one, then it means that our product is an average product in comparison with those of competitors. Then the total score in column 6 will be equal to 100, and our market-perceived quality score will be 100. The product with a market-perceived quality score of larger than 100 is considered to be a competitive product; the higher the score, the more competitive the product is.

The information used to compile the market-perceived quality profile, such as that of Table 5.1, can also be obtained by conducting a special kind of customer survey. A design of this kind of customer survey form is illustrated in Table 5.2.

The survey population should include the customers of our company, as well as all consumers who purchase this kind of product or service, including the customers of competitors. For example, if our company is

Table 5.1 Market-Perceived Quality Profile in Chicken Business

Quality Attributes	Customer's Weight of Attributes (Total = 100)	Industry Comparison		Quality Scores	
		Perdue (Our Business) (1 = Lowest, 10 = Highest)	Average Competitor	Ratio (Ours/Competitor) (ratio > 1.0 means "better than competitor")	Customer Weight × Ratio
Yellow bird	10	8.1	7.2	8.1/7.2 = 1.13	11.3 = 1.13 × 10
Meat-to-bone	20	9.0	7.3	1.23	24.6 = 1.23 × 20
No pinfeathers	20	9.2	6.5	1.42	28.4
Fresh	15	8.0	8.0	1.00	15.0
Availability	10	8.0	8.0	1.00	10.0
Brand image	25	9.4	6.4	1.47	36.8
Total	100				126.1 = market-perceived quality score

Table 5.2 Survey Form for Customer Values

Quality (Nonprice) Attributes	Importance Weights (Add Up to 100)	Performance Scores 1–10 (1 = Lowest, 10 = Highest)			
		Company A (Our Company)	Company B (Competitor 1)	Company C (Competitor 2)	Company D (Competitor 3)
1					
2					
3					
4					
5					
6					
7					
8					
9					
10					
Sum of importance weights =	100				
Price (perceived transaction price)					
Price competitiveness					

Table 5.3 Quality Profile Studies of Gallbladder Operations: Endo-Surgery versus Traditional Surgery

Quality Attributes	Customer's Weight of Attributes (Total = 100)	Industry Comparison		Quality Scores	
		Endo	Traditional	Ratio	Customer Weight × Ratio
At home recovery period	40	1–2 weeks	6–8 weeks	3.0	120
Hospital stay	30	1–2 days	3–7 days	2.0	60
Complications rate	10	0–5%	1–10%	1.5	15
Postoperative scar	5	0.5–1 inch	3–5 inch	1.4	7
Operation time	15	0.5–1 hour	1–2 hours	2.0	30
Total	100				232 = market-perceived quality score

McDonald's, then the survey population should be the customers for all fast-food chains and should include the customer population of Burger King, Wendy's, and so on.

Clearly, if we collect enough finished survey forms from customers, we will be able to fill in all the information needed in a market-perceived quality profile study. Besides customer survey data, functional data can also be used in a market-perceived quality profile study. Table 5.3 gives such an example.

5.3 Market-Perceived Price Profile

For some industries, such as retailing, the price of a particular item is one-shot, so it is very clearly understood. In this case, the price comparison is simply one dollar amount versus another dollar amount. For many other businesses, the overall customer cost structure is rather complicated. For example, the cost-related factors in purchasing a car might involve trade-in allowance, rebate, and finance rate, besides the purchase price of the car. In such a case, the construction of the market-perceived price profile is necessary, because it will integrate all the cost factors and compile a combined price score.

The construction of the market-perceived price profile is very similar to that of the market-perceived quality profile. Customers are asked to list the factors that affect their perception of a product's cost. Table 5.4 is an example of a market-perceived price profile in the luxury car market.

However, it is people's convention that the lower the price, the better. Having a higher customer satisfaction score value in price level is counter-intuitive. In the example of Table 5.4, Acura's market-perceived price score is 118.7. Since it is more than 100, this means that the Acura's overall price level is more attractive (lower) than other competitors. However, using the inverse score 84.2 is more intuitively appealing.

If we just compare a one-shot purchasing price, a simple price ratio can be used. For example, if Acura's price is \$35,200, and the average competitor's price is \$40,000, then the price ratio is \$35,200 \$40,000 = 0.88. If a percentage score is used, then the relative price ratio is $0.88 \times 100 = 88$. If a relative price ratio of a product is less than 100, then its price level is lower than that of its competitors' products.

5.4 Customer Value Map

The customer value map is a very useful tool to identify the competitive position of a particular product in comparison with other competitors' products. A product is competitive if it has high customer benefit and low customer cost. The customer benefit can be well represented by the market-perceived quality score that we discussed in Sec. 5.2, and the customer cost position can be well represented by the relative price ratio that we discussed in Sec. 5.3.

The customer value map is a two-dimensional plot of the market-perceived quality score on the horizontal axis versus the relative price ratio on the vertical axis. Figure 5.1 illustrates what a customer value map looks like. In this customer value map, each dot represents a particular product. The position of the dot depends on the values of its market-perceived quality

Table 5.4 Market-Perceived Price Profile: Luxury Cars

Price Satisfaction Attributes	Customer's Weight of Attributes (Total = 100)	Industry Comparison		Quality Scores	
		Acura (Our Business) (1 = Lowest, 10 = Highest)	Average Competitor	Ratio (Ours/Competitor) (ratio > 1.0 means "better than competitor")	Customer Weight × Ratio
Purchase price	60	9	7	1.29 = 9/7	77.4 = 1.29 × 60
Trade-in allowance	20	6	6	1.0	20 = 1.0 × 20
Resale price	10	9	8	1.13	11.3
Finance rates	10	7	7	1.00	10.0
Total	100				118.7 = market-perceived price score
Relative price ratio					84.2 = (1/118.7) × 100

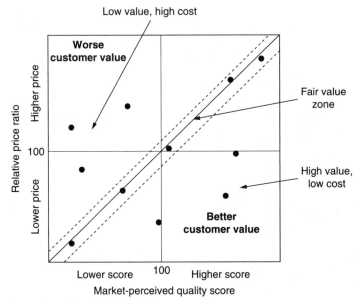

Figure 5.1 Customer Value Map

score and relative price ratio. The diagonal line in the customer value map represents where the market-perceived quality score is equal to the relative price ratio. For example, if a product has a market-perceived quality score equal to 80, and its relative price ratio is equal to 80, then the dot representing this product will be on the diagonal line. This is a low-value, low-price product. Similarly, if a product has a market-perceived quality score of 120, and the relative price ratio is also 120, then this product will also be on the diagonal line and it is a high-price, high-value product. Overall, the region around the diagonal line can be called the fair-value zone. The products in the fair-value zone can be considered to be average products. The products in the lower right-hand corner of the customer value map are featured by a lower relative price ratio and a higher market-perceived quality score. We can call these products high-value, low-price products. They have a superior competitive position and are poised to gain market share. Products in the upper left-hand corner of the customer value map are featured by a high relative price ratio and a low market-perceived quality score. We can call these low-value, high-price products. They have an inferior competitive position in the marketplace and are vulnerable to losing market share.

Example 5.1
Here is an example of a toaster customer value analysis. Table 5.5 lists the market-perceived quality scores and prices for 15 brands of toasters. By using the data from Table 4.5, we can draw the customer value map shown in Fig. 5.2.

From the customer value map, the products above the diagonal line are the ones that have a low customer value. Products 1 and 3 (Cuisinart CPT-60 and KitchenAid) have a good market-perceived quality score but are very high in price. Product 9 (Oster) has a below-average market-perceived quality score but has a high price. Products 5, 6, and 11 (Cuisinart CPT-30,

Table 5.5 Market-Perceived Quality Profile and Price Profile for Toasters

Name of Toaster	Market-Perceived Quality Score	Toaster Price	Relative Price
1. Cuisinart CPT-60	128	$70	215
2. Sunbeam	119	28	85
3. KitchenAid	117	77	237
4. Black & Decker	112	25	77
5. Cuisinart CPT-30	109	40	123
6. Breadman	107	35	108
7. Proctor-Silex 22425	104	15	46
8. Krups	101	32	98
9. Oster	94	45	138
10. Toastmaster B1021	91	16	49
11. Proctor-Silex 22415	87	35	108
12. Toastmaster B 1035	84	21	65
13. Betty Crocker	84	25	77
14. Proctor-Silex 22205	83	11	34
15. Rival	80	13	40
Average	100	$33	100

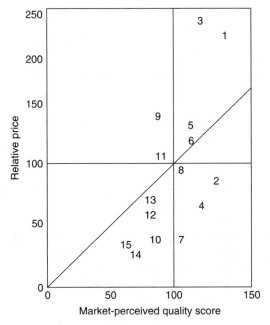

Figure 5.2 Customer Value Map of Toasters

Breadman, and Proctor-Silex 22415) are also relatively high in price and low in performance, but they are close to fair-value zone. Products 2, 4, and 7 (Sunbeam, Black & Decker, and Proctor-Silex 22425) are low-price, high-performance products. Products 10, 12, 14, and 15 (Toastmaster B1021, Toastmaster B 1035, Proctor-Silex 22205, and Rival) are low-price, reasonable-performance products. Products 8 and 13 (Krups and Betty Crocker) have better than average customer value but they are in the fair-value zone.

Overall, the products that are located in the lower-right portion of the chart have a better customer value; the further products deviate from the fair value line, the better is their customer value. In Fig. 5.2, products 2, 4, and 7 are in the lower-right portion and have the largest distances from the fair-value line, so they have the best customer value. Product 2 has a higher price, so it is a best-value product at a higher price level. Product 7 has a lower price, so it is a best-value product at a low price level. Similarly, the products located in the upper-left corner of the chart have a worse customer value; the further products deviate from the fair-value line, the worse their customer value.

5.5 Competitive Customer Value Analysis

Competitive customer value analysis is a graphical display chart that can be used to compare your product versus competitors' products in important aspects of customer value. This analysis will show you which areas are the most important to focus on to improve your product most effectively.

Here we can use the following example to show how competitive customer value analysis works. Table 5.6 shows a market-perceived quality profile of two printers, printer A (our printer) and printer B (competitor's printer). Table 5.7 shows the market-perceived price profile of these two printers. Figure 5.3 shows a head-to-head customer value area chart that compares printers A and B. Each bar represents a market-perceived quality characteristic. The horizontal dimension of the bar shows how much our product is better or worse than our competitor's product. The thickness of each bar is proportional to the relative importance of each characteristic. So the total area in white represents our advantage; the total shaded area represents our disadvantage. Our goal will be to maximize the white area and minimize the shaded area in the most effective way.

Figure 5.4 shows a head-to-head market-perceived price ratio chart for printers. Again, each bar represents a customer cost component, its horizontal dimension represents how much better or worse our product's price compares with our competitor's price, and the bar's thickness represents the relative importance of that cost component in the eyes of customers. So the total area in white minus the total shaded area represents our product's cost advantage; the larger the cost advantage, the more competitive is our product in price.

5.6 Customer Value Deployment

After our competitive customer value analysis and relative price ratio analysis, we need to find an effective way to overcome our disadvantages and strengthen our existing advantages in order to improve our customer values and win over the competition. To do that, we need to identify the critical areas of the company that are related to our key market-perceived quality factors and market-perceived customer cost areas.

A quality function deployment (QFD) like template can be very useful in deploying key customer values into our process improvements. Table 5.8 shows the customer value deployment matrix for the printer case. In this

Table 5.7 Market-Perceived Price Profile: Printers

Price Satisfaction Attributes	Customer's Weight of Attributes (Total = 100)	Industry Comparison		Quality Scores	
		Printer A (Our Printer)	Printer B (Competitor's Printer)	Relative Price Ratio	Customer Weight × Ratio
Purchase price	40	$508	$585	0.87 = 508/585	34.8 = 0.87 × 40
Service and repair	30	$60/year	$65/year	0.92 = 60/65	27.6
Toner	20	$235/year	$235/year	1.0	20
Paper	10	$124/year	$101/year	1.23	12.3
Total	100				94.7 = relative price ratio

Customer Value Management 97

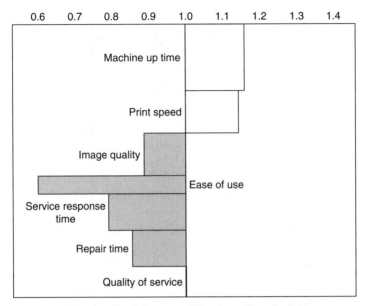

Figure 5.3 Head-to-Head Customer Value Area Chart for Printers

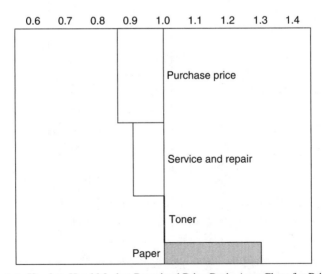

Figure 5.4 Head-to-Head Market-Perceived Price Ratio Area Chart for Printers

Table 5.8 Customer Value Deployment Matrix for Printer Example

Quality Attributes	Design	Manufacturing	Quality Control	Sales and Service	Distribution	Marketing
Machine uptime	3	9	9			
Print speed	9	3	3			
Image quality	9	9	9			9
Ease of use	9			9		3
Service response time				9	9	
Repair time				9	9	
Quality of service				9		

matrix, correlation scores of 9, 3, 1, and 0 are used. A score of 9 means "very much related," a score of 3 means "related," a score of 1 means "slightly related," and a score of zero means "not related."

For example, in the "machine uptime" category, quality control and manufacturing are very critical in ensuring printer dependability; product design is also related to the dependability of the printer. In the "ease of use" category, of course, design is very important in creating a printer that is easy to use. However, sometimes there is a gap between a customer perceived quality image and the real quality level. For example, printer A may actually be easy to use, but because of poorly written customer instructions, poor service support, and poor marketing, a significant portion of customers may have developed a stereotype that printer A is hard to use. The right way to overcome this problem may not be to redesign the printer. Instead, a comprehensive strategy that includes improving customer service, rewriting customer instructions, and developing the right marketing message might be the right way.

Chapter 6

Quality Function Deployment

6.1 Introduction

In the context of Design for Six Sigma (DFSS), quality function deployment (QFD) is best viewed as a planning tool that relates a list of delights, wants, and needs of customers to design technical functional requirements. With the application of QFD, possible relationships are explored between quality characteristics expressed by customers and *substitute quality requirements* expressed in engineering terms (Cohen 1988, 1995; Clausing 1994). In the context of DFSS, we will call these requirements *critical-to* characteristics, which include subsets like *critical-to-quality*, and *critical-to-delivery* characteristics. In the QFD methodology customers define the product using their own expressions which usually do not carry any significant technical terminology. The *voice of the customer* (VOC) can be discounted into a list of needs used later as input to a relationship diagram, which is called QFD's *house of quality*.

Knowledge of customer needs is a must requirement in order for a company to maintain and increase its position in the market. Correct market predictions are of little value if the requirements cannot be incorporated into the design at the right time. Critical-to-innovation and critical-to-market characteristics are critical because companies that are first to introduce new concepts at six-sigma (6σ) levels usually capture the largest share of the market. Wrestling market share away from a viable competitor is more difficult than wrestling market share away from a first producer into a market. One major advantage of a QFD is attainment of the shortest development cycle, which is gained by companies with the ability and desire to satisfy customer expectations. Another significant advantage is the improvement gained in the design family of the company resulting in increased customer satisfaction.

The team should take the time required to understand customer wants and to plan the project more thoughtfully. Using QFD, the DFSS team will be able to anticipate failures and avoid major downstream changes. Quality function deployment prevents downstream changes by an extensive planning effort at the beginning of the DFSS design or redesign project. The team will employ marketing and product planning inputs to deploy customer expectations through design, process, and production planning and all across functional departments. This will assure resolution of issues, lean design, and focus on those potential innovations (delighters) that are important to the customer.

Figure 6.1 shows that a company using QFD places more emphasis on responding to problems early in the design cycle. Intuitively, it incurs more time, cost, and energy to implement a design change at production launch than at the concept phase because more resources are required to resolve problems than to preclude their occurrence in the first place.

Quality function deployment translates customer needs and expectations into appropriate design requirements. The intent of QFD is to incorporate the voice of the customer into all phases of the product development cycle, through production and into the marketplace. With QFD, quality is defined by the customer. Customers want products, processes, and services that throughout their lives meet customers' needs and expectations at a cost that represents value. The results of the process being customer driven are total quality excellence, greater customer satisfaction, increased market share, and potential growth.

The real value of QFD is its ability to direct the application of other DFSS tools like statistical process control (SPC) and robustness to those entities

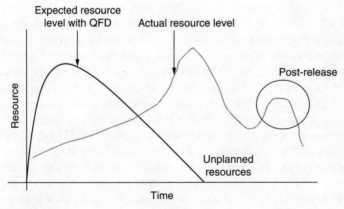

Figure 6.1 QFD Effect on Project Resources

that will have the greatest impact on the ability to design and satisfy the needs of the customers, both internal and external. Quality function deployment is a *zooming* tool that identifies the significant design elements on which to focus design and improvement efforts and other resources. In the context of QFD, planning is key and is enhanced by reliable information in benchmarking and testing.

The objectives of this chapter are to

1. Provide Black Belts, Green Belts, and other readers with the knowledge and skills they need to define quality function deployment
2. Identify the four key elements of any QFD chart
3. Provide a basic understanding of the overall four phases of QFD methodology
4. Define the three quality features of the Kano model

6.2 History of QFD

Quality function development was created by Mitsubishi Heavy Industry at Kobe Shipyards in the early 1970s. Stringent government regulations for military vessels coupled with the large capital outlay per ship forced Kobe Shipyard's management to commit to upstream quality assurance. The Kobe engineers drafted a matrix, which related all the government regulations, critical design requirements, and customer requirements to company technical controlled characteristics of how the company would achieve them. In addition, the matrix also depicted the relative importance of each entry, making it possible for important items to be identified and prioritized to receive a greater share of the available company resources.

Winning is contagious. Other companies adopted QFD in the mid-1970s. For example, the automotive industry applied the first QFD to the rust problem. Since then, QFD usage has grown as a well-rooted methodology into many American businesses. It has become so familiar because of its adopted commandment: Design it right the first time.

6.3 QFD Benefits, Assumptions, and Realities

The major benefit of QFD is customer satisfaction. QFD gives customers what they want, such as a shorter development cycle. Failures and redesign peaks (Fig. 6.1) are avoided during prelaunch, and know-how knowledge as it relates to customer demand is preserved and transferred to the next design team.

Before QFD can be implemented, a multidisciplinary DFSS team needs to be in place and more time should be spent upstream understanding customer needs and expectations and defining the product or service in greater detail.

There are many initial realistic concerns that must be addressed in order to implement QFD successfully. For example, departments represented in the team do not tend to talk to one another. In addition, market research information is not technically or design focused, and QFD is more easily applied to incremental design than to brand creative design. The traditional reality that problem prevention is not rewarded as well as problem solving will be faced initially be the DFSS team. This "reality" will fade away as the team embarks on its project using the rigor of DFSS.

6.4 QFD Methodology Overview

Quality function deployment is accomplished by multidisciplinary DFSS teams using a series of charts to deploy critical customer attributes throughout the phases of design development. QFD is usually deployed in multiple phases. Figure 6.2 shows the typical four-phase deployment in a typical manufacturing setting. The four phases are

Phase 1: CTS planning
Phase 2: Functional requirements
Phase 3: Design parameters planning
Phase 4: Process variables planning

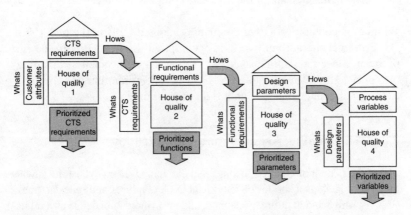

Figure 6.2 Four Phases of QFD in Manufacturing Application

Quality Function Deployment

In a typical service industry setting, QFD can be deployed in two phases, as illustrated in Fig. 6.3.

QFD uses many techniques in an attempt to minimize and ease the task of handling large numbers of functional requirements that might be encountered. Applications in the range of 130 (functions) × 100 (customer features) were recorded (Hauser and Clausing 1988). One typical grouping technique that may be used initially in a QFD study is the affinity diagram. The *affinity diagram* is a hierarchical grouping technique, which is used to consolidate multiple unstructured ideas generated by the voice of the customer. It operates based on intuitive similarities that may be detected from low-level stand-alone ideas (bottom) to arrangements of classes of ideas (up). This bundling of customer features is a critical step. It requires a cross-functional team that has multiple capabilities such as the ability to brainstorm, evaluate, and revolutionize existing ideas in pursuit of identifying logical (not necessarily optimum) groupings and, hence, minimizing the overall list of needs into manageable classes.

Another technique is the tree diagram, which is a step beyond the affinity diagram. The *tree diagram* is used mainly to fill the gaps and cavities not detected previously in order to achieve a more completed structure leading to more ideas. Such expansion of ideas will allow the structure to grow but at the same time will provide more vision into the voice of the customer (Cohen 1988).

The *house of quality* (Fig. 6.4) is the relationship foundation of QFD. Employment of the house will result in improved communication, planning, and design activity. This benefit extends beyond the QFD team to

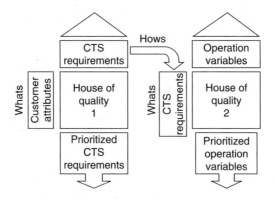

Figure 6.3 Multiple-Phase QFD in Service Application

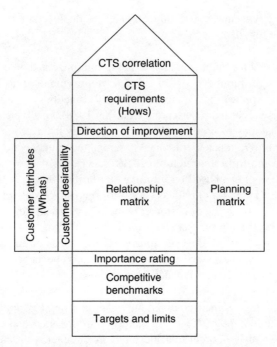

Figure 6.4 House of Quality

the whole organization. Defined customer wants through QFD can be applied to many similar products and form the basis of a corporate memory on the subject of critical-to-satisfaction (CTS) requirements. As a direct result of the use of QFD, *customer intent* will become the driver of the design process as well as the catalyst for modification to design solution entities. In Fig. 6.4, we have the following components that constitute the house of quality (Cohen 1988):

Customer Attributes (Whats)

Customer attributes (*Whats*) are obtained from the voice of the customer through surveys, claim data, warranties, and promotion campaigns. Usually customers use fuzzy expressions in characterizing their needs with many dimensions to be satisfied simultaneously. Affinity and tree diagrams may be used to complete the list of needs. Most of these Whats are very general ideas that require more detailed definition. For example, customers often say they want to purchase a "stylish" or "cool" product. "Being cool" may be a very desirable feature, but since it has different interpretations to different people, it cannot be acted upon directly. Legal and safety requirements or other internal wants are considered extensions to the Whats. The Whats can be characterized using the Kano model (Sec. 6.5).

Hows

Hows are design features derived by the DFSS team to answer the Whats. Each of the initial Whats needs operational definitions. The objective is to determine a set of CTS requirements with which Whats can be materialized. The answering activity translates customer expectations into design criteria such as speed, torque, and time to delivery. For each What, there should be one or more Hows that describe a means of attaining customer satisfaction. For example, a "cool car" can be achieved through a stylish body (different and new), seat design, leg room, lower noise, harshness, and vibration requirements. At this stage only overall requirements that can be measured and controlled need to be determined. These substitute for the customer needs and expectations and are traditionally known as substitute quality characteristics. In this book, we will adopt the critical-to terminology aligning with Six Sigma.

Teams should define the Hows in a solution-neutral environment and not be restricted by listing specific parts and processes. Just itemize the means (the Hows) whereby the list of Whats can be realized. One-to-one relationships do not usually exist in the real world, and many Hows will relate to many customer wants. In addition, each How will have some direction of goodness or improvement as illustrated in the following figure:

Direction of improvement		
Maximize	↑	1.0
Target	●	0.0
Minimize	↓	–1.0

The circle represents the nominal-the-best target case.

Relationship Matrix

The process of relating Whats to Hows often becomes complicated by the absence of one-to-one relationships as some of the Hows affect more than one What. In many cases, they adversely affect one another. Hows that could have an adverse effect on another customer want are important. For example, "cool" and "stylish" are two of the Whats that a customer would want in a vehicle. The Hows that support the "cool" attribute are lower noise, roominess, and seat design requirements among others. These Hows will also have some effect on the "stylish" attribute as well. A relationship is created in the house of quality (HOQ) between the Hows as columns and the Whats in the rows. The relationship in every (What, How) cell can be displayed by placing a symbol representing the cause-and-effect relationship strength in that cell. When employees at the Kobe Shipyards developed this matrix in 1972, they put the local horse racing symbols into their QFD as relationship matrix symbols; for example, double-centered circles mean a strong relationship, one circle means a medium strength relationship, and the triangle indicates a weak relationship. Symbols are used instead of direct numbers because they can be identified and interpreted easily and quickly. Different symbol notations have been floating around, and we found the following to be more common than others:

Standard 9-3-1		
Strong	●	9.0
Moderate	◇	3.0
Weak	▽	1.0

After determining the strength of each (What, How) cell, the DFSS team should take the time to review the relationship matrix. For example, blank rows or columns indicate either gaps in the team's understanding or a deficiency in fulfilling customer attributes. A blank row shows a need to develop a How for the What in that row indicating a potentially unsatisfied customer attribute. When a blank column exists, one of the Hows does not impact any of the Whats. Delivering that How may require a new What that has not been identified, or it might be a waste. The relationship matrix gives the DFSS team the opportunity to revisit its work leading to better planning and therefore better results.

What is needed is a way to determine to what extent the CTS requirement at the head of the column contributes to meeting the customer attribute at the beginning of the row. This is a subjective weighing of the possible cause-and-effect relationships.

To rank order the CTS requirements and customer features, we multiply the numerical value of the symbol representing the relationship by the *customer desirability index*. This product when summed over all the customer features in the Whats array provides a measure of the relative importance of such CTS requirements to the DFSS team and is used as a planning index to allocate resources and efforts, comparing the strength, importance, and interactions of these various relationships. This importance rating is called the *technical importance rating*.

Importance Ratings

Importance ratings are a relative measure indicating the importance of each What or How to the design. In QFD, there are two importance ratings:

1. The customer desirability index is obtained from the voice of the customer activities such as surveys and clinics and is usually rated on a scale from 1 (not important) to 5 (extremely important) as follows:

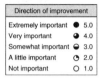

2. The technical Importance Rating is calculated as follows:
   a. By convention, each symbol in the relationship matrix receives a value representing the strength in the (What, How) cell.
   b. These values are then multiplied by the customer desirability index, resulting in a numerical value for the symbol in the matrix.
   c. The technical importance rating for each How can then be found by adding together the values of all the relationship symbols in each column.

   Technical importance ratings have no physical interpretation; their values lie in their ranking relative to one another. They are utilized to determine which Hows take priority and should be allocated the most resources. In doing so, the DFSS team should use the technical importance rating as a main metric coupled with other factors like difficulty, innovation, cost, reliability, timing, and all other measures in their project charter.

Planning Matrix

The planning matrix is used to make comparisons of competitive performance and identification of a benchmark in the context of ability to meet specific customer needs. It is also used as a tool to set goals for improvement using a ratio of performance (goal rating/current rating). Hauser and Clausing (1988) view this matrix as a perceptual map in trying to answer the following question: How can we change the existing product or develop a *new* one to reflect customer intent given that the customer is more biased toward certain features? The product of *customer value*, the *targeted improvement ratio* for the raw (feature), and the *sales point*, which is a measure of how the raw feature affects sales, will provide a weighted measure of the relative importance of this customer feature to be considered by the team.

Hows Correlation (The Roof)

Each cell in the roof is a measure of the possible correlation of two different Hows. The use of this information improves the team's ability to develop a systems perspective for the various Hows under consideration.

The correlation matrix is one of the more commonly used optional extensions to the original QFD developed by Kobe engineers. Traditionally, the major task of the correlation matrix is to make tradeoff decisions by identifying the qualitative correlations between the various Hows. This is a very important function in the QFD because Hows are most often correlated. For example, assume a matrix contains quality and cost objectives. The design engineer is looking to decrease cost, but any improvement in this aspect will have a negative effect on the quality. This is called a negative correlation, and it must be identified so that a tradeoff can be addressed. Tradeoffs are usually accomplished by revising the long-term objectives (How Muchs). These revisions are called *realistic objectives*. Using the negative correlation example just discussed, in order to resolve the conflict between cost and quality, the cost objective would be changed to a realistic objective.

In the correlation matrix, once again, symbols are used for ease of reference to indicate the different levels of correlation as shown in the following figure:

If one How directly supports another How, a positive correlation is produced.

Targets (How Much)

For every How shown on the relationship matrix, a How Much should be determined. The goal here is to quantify the customers' needs and expectations and create a target for the design team. The How Muchs also create a basis for assessing success. For this reason, Hows should be measurable. It is necessary to review the Hows and develop a means of quantification. Target orientation to provide a visual indication of the target type is usually optional. In addition, the tolerance around targets needs to be identified, based on the company marketing strategy and contrasting it with the best-in-its-class competitor. This tolerance will be cascaded down using the axiomatic design method.

Competitive Assessments or Benchmarking

Competitive assessments are used to compare the competition's design with the team design. There are two types of competitive assessments:

1. *Customer competitive assessment:* Found to the right of the relationships matrix in the planning matrix. Voice of the customer (VOC)

activities (e.g., surveys) are used to rate the Whats of the various designs in a particular segment of the market.
2. *Technical competitive assessment:* Found at the bottom of the relationships matrix. It rates Hows for the same competitor from a technical perspective.

The assessments should align; a conflict between them indicates a failure by the team to understand the VOC. In case of a conflict, the team needs to revisit the Hows array and check its understanding of the array. The team should contrast that understanding with VOC data. Further research may be needed. The team may then add new Hows that reflect customer perceptions. Any unexpected items that violate conventional wisdom should be noted for future reference. Situations like this can only be resolved by having the DFSS team involved in the QFD, instead of having only marketing people comparing competitive designs. In this way, the team that is responsible for designing for customer attributes will interpret exactly what those wants are.

6.5 Kano Model of Quality

In QFD, voice of the customer activities such as market research, provide the array of Whats that represent customer attributes. Such Whats are "spoken" by the customer and are called *performance quality* or *one-dimensional*. However, more Whats have to be addressed than just those directly spoken by the customer. As Fig. 6.5 shows, there are also unspoken Whats. Unspoken Whats are the basic features that customers automatically assume the design will have. Such Whats are implied in the functional requirements of the design or assumed from historical experience. For example, customers automatically expect their lawnmower to cut grass to the specified level, but they would not discuss it on a survey unless they had trouble with one in the past. Unspoken wants have a weird property. They do not increase customer satisfaction. However, if they are not delivered, they have a strong negative effect on customer satisfaction.

Another group of unspoken Whats can be categorized as innovations or delighters. These pleasant surprises increase customer satisfaction in a nonlinear fashion. For example, in the automotive industry, van owners were delighted by the second van side door and by baby seat anchor bolts.

Design features may change position on the Kano model over time. In the 1990s, the second side door in a caravan was a pleasant surprise for customers, but now, on most models, the second door is standard and

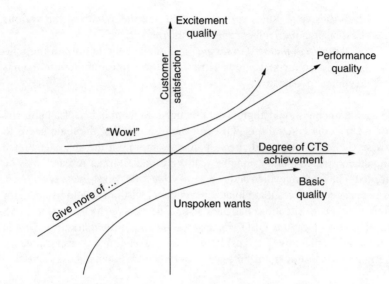

Figure 6.5 Kano Model of Customer Attributes

expected to be installed without a specific request. The ideal DFSS project plan would include all three types of quality features: excitement quality (unspoken latent demands), performance quality (spoken and one-dimensional), and basic quality (unspoken or assumed).

6.6 QFD Analysis

The completion of the first QFD house of quality may give the DFSS team a false impression that its job is completed. In reality, all the team's work to this point has been to create a tool that will guide future efforts toward deploying the VOC into the design. QFD matrix analysis in every phase will lead to the identification of design weaknesses, which must be dealt with as potential strength opportunities to make the product or service best in its class. A relatively simple procedure for analyzing the house of quality phase is to address the following points:

- *Blank or weak columns*: These indicate Hows that do not strongly relate to any customer attribute. The "How" that relates to a blank or weak column could be deleted.
- *Blank or weak rows*: These indicate customer attributes that are not being strongly addressed by a How. In this case, another 'How' or

Hows should be added so that the blank or weak rows are changed to stronger rows.
- *Conflicts*: Determine whether the technical competitive assessment is in conflict with the customer competitive assessment.
- *Significance*: Determine which Hows are significant, that is, those that relate to many customer attributes, safety and regulatory issues, and internal company requirements.
- *Eye-opener opportunities*: If the team's company and competitors are doing poorly, the DFSS team should seize the opportunity to deliver on these sales points, which may be treated as delighters in the Kano model initially.
- *Benchmarking*: The team should take the opportunity to incorporate the competitor's highly rated Hows. It is advisable for the team to modify and incorporate benchmarking and not resort to creation.

6.7 Example

This example is a QFD study conducted by a DFSS team. The following are highlights of the QFD example.

Project Objective

Design a global commercial process with Six Sigma performance.

Project Problem Statement

- Sales cycle time (lead generation to full customer setup) exceeds 182 business days. Internal and external customer specifications range from 1 to 72 business days.
- Only 54 percent of customer service requests are closed by the commitment date. The customers expect 100 percent of their service requests to be completed on time.
- None of the commercial processes is standard or is Six Sigma capable.

Business Case

- There is no consistent, global process for selling to, setting up, and servicing accounts.
- Current sales and customer service information management systems do not enable a measurement of accuracy and timeliness on a global basis.
- Enterprise-wide customer care is a must be requirement; failure to improve the process threatens growth and retention of the portfolio.

Project Goals

- Reduce prospecting cycle time from 16 to 5 business days.
- Reduce discovery cycle time from 34 to 10 business days.
- Reduce the close-the-deal cycle time from 81 to 45 business days (all sales metrics net of customer-wait time).
- Reduce setup cycle time from 51 to 12 business days.
- Increase the percentage of service requests closed by the commitment date from 54 percent (1.6σ) to 99.97 percent (5.0σ).

The following are the QFD steps.

Step 1: Identify the Whats and Hows and their Relationship

The DFSS team identifies customers and establishes customer wants, needs, delights, and usage profiles. Corporate, regulatory, and social requirements should be identified also. The value of this step is to greatly improve the understanding and appreciation DFSS team members have for customer, corporate, regulatory, and social requirements. The DFSS team, at this stage, should be expanded to include market research. A market research professional might help the Black Belt assume leadership during startup activities and perhaps later remain an active participant as the team gains knowledge about customer engagement methods. The Black Belt should put plans in place to collaborate with identified organizations and/or employee relations to define tasks and plans in support of the project and to train team members in customer processes, and forward-thinking methods such as brainstorming, visioning, and conceptualizing.

The DFSS team should focus on the key customers to optimize decisions around them and try to include as many additional customers as possible. The team should establish customer environmental conditions and customer usage and operating conditions, study customer demographics and profiles, conduct customer performance evaluations, and understand the performance of the competition. In addition, the team should

- Establish a rough definition of an ideal service
- Listen to the customer and capture wants and needs through interviews, focus groups, customer councils, field trials, field observations, surveys, etc.
- Analyze customer complaints and assign satisfaction performance ratings to attributes
- Acquire and rank these ratings with the QFD process

- Study all available information about the service including marketing plans
- Create innovative ideas, delights, and new wants by investigating improved functions and cost of ownership and matching service functions with needs, experience, and customer beliefs
- Innovate to avoid compromising for bottlenecks, conflicts, and constraints
- Benchmark the competition to improve weak areas

The following Whats are used.

Direction of improvement
Available products
Professional staff
Flexible processes
Knowledgeable staff
Easy-to-use products
Speedy processes
Cost-effective products
Accuracy

Step 2: Identify the Hows and Relationship Matrix

The purpose of this step is to define a "good" product or process in terms of customer expectations, benchmark projections, institutional knowledge, and interface requirements, and to translate this information into CTS requirements. These will then be used to plan an effective and efficient DFSS project.

One of the major reasons for customer dissatisfaction and warranty costs is that the design specifications do not adequately reflect customer use of the product or process. Too many times the specification is written after the design is completed, or it is simply a reflection of an old specification that was also inadequate. In addition, a poorly planned design commonly does not allocate activities or resources in areas of importance to customers and wastes engineering resources by spending too much time in activities that provide marginal value. Because missed customer requirements are not targeted or checked in the design process, procedures to handle field complaints for these items are likely to be incomplete. Spending time on overdesigning and overtesting items, not important to customers, is wasteful. Similarly, not spending development time in areas important to customers is not only a missed opportunity, but significant warranty costs are sure to follow.

In DFSS, time is spent upfront understanding customer wants, needs, and delights together with corporate and regulatory requirements. This understanding is then translated into CTS requirements which then drive product and process design. The CTS attributes (Hows) are given in the following diagram as well as the relationship matrix to the Whats. A mapping begins by considering the high-level requirements for the product or process. These are the true CTS requirements that define what the customer would like if the product or process were ideal. This consideration of a product or process from a customer perspective must address the requirements from higher-level systems, internal customers (such as manufacturing, assembly, service, packaging, and safety), external customers, and regulatory legislation. Customer Whats are not easily operational in the world of the Black Belt. For this reason it is necessary to relate true quality characteristics to CTS requirements—design characteristics that may be readily measured and, when properly targeted, will substitute or assure performance to the Whats. This diagram, which relates true quality characteristics to substitute quality characteristics, is called a relationship matrix.

Importance to the customer
Meet time expectations
Know my business and offers
Save money and enhance productivity
Do it right the first time
Consultative
Know our products and processes
Talk to one person
Answer questions
Courteous
Adequate follow-up

The logic of a matrix is several levels deep. A tree diagram, one of the new seven management tools, is commonly used to create the logic associated with the customer. The mapping of customer characteristics to CTS attribute characteristics is extremely valuable when done by the DFSS team. A team typically begins differing in opinion and sharing stories and experiences when the logic is only a few levels deep. An experiment may even be conducted to better understand the relationships. When completed, the entire team understands how product and process characteristics that are detailed on drawings relate to functions that are important to customers.

Quality Function Deployment

Direction or improvement		
Maximize	↑	1.0
Target	●	0.0
Minimize	↓	−1.0

			Importance to the customer	Meet time expectations	Know my business and offers	Save money and enhance productivity	Do it right the first time	Consultative	Know our products and processes	Talk to one person	Answer questions	Courteous	Adequate follow-up
			1	1	2	3	4	5	6	7	8	9	10
Direction of improvement	1		↑	↑	↑	↑	↑	↑	↑	↑	↑	↑	↑
Available products	1	2.0	○		○		▽			○			
Professional staff	2	3.0		▽		▽		○	●	▽		●	
Flexible processes	3	4.0	○						●				
Knowledgeable staff	4	4.0	○	●	○	●	●	●	●	●	●		▽
Easy-to-use products	5	4.0	○		○	○			▽	○			▽
Speedy processes	6	5.0	●		●	○			○	○	○		▽
Cost-effective products	7	5.0	○	●	●	○		●	○				
Accuracy	8	5.0		●		●							
	9												

Figure 6.6 The Whats, the Hows, and the Relationship Matrix

The full QFD phases 1 and 2 are given in Figs. 6.6 to 6.9. The following analysis applies to phase 1. The readers are encouraged to analyze the phase 2 as an exercise.

The Hows Importance Calculation

Importance ratings are a relative comparison of the importance of each What or How to the quality of the design. The 9-3-1 relationship matrix strength rating is used. These values are multiplied by the customer importance rating obtained from customer engagement activities (like surveys) resulting in a numerical value. The Hows importance rating is summed by adding all values of all relationships. For example, the importance rating of the first How of Fig. 6.7 is calculated as $2.0 \times 3.0 + 4.0 \times 3.0 + 4.0 \times 3.0 + 4.0 \times 3.0 + 5.0 \times 9.0 + 5.0 \times 3.0 = 102$. Other How importance ratings can be calculated accordingly.

Table 5.6 Market-Perceived Quality Profile of Two Printers

Quality Attributes	Customer's Weight of Attributes (Total = 100)	Industry Comparison		Quality Scores		
		Printer A (Our Printer) (1 = Lowest, 10 = Highest)	Printer B (Competitor's Printer)	Ratio (Ours/Competitor) (ratio > 1.0 means "better than competitor")		Customer Weight × Ratio
Machine uptime	25	8	7	1.14 = 8/7		28.5 = 1.14 × 25
Print speed	15	9	8	1.13		17.0
Image quality	15	7	8	0.88		13.2
Ease of use	5	4	7	0.57		2.85
Service response time	15	5	7	0.71		10.65
Repair time	15	5	6	0.83		12.45
Quality of service	10	7	7	1.0		10.0
Total	100					94.65 = market-perceived quality score

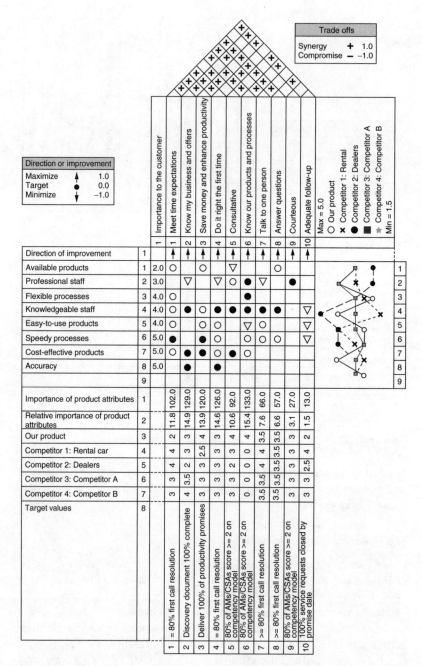

Figure 6.7 Phase 1 QFD

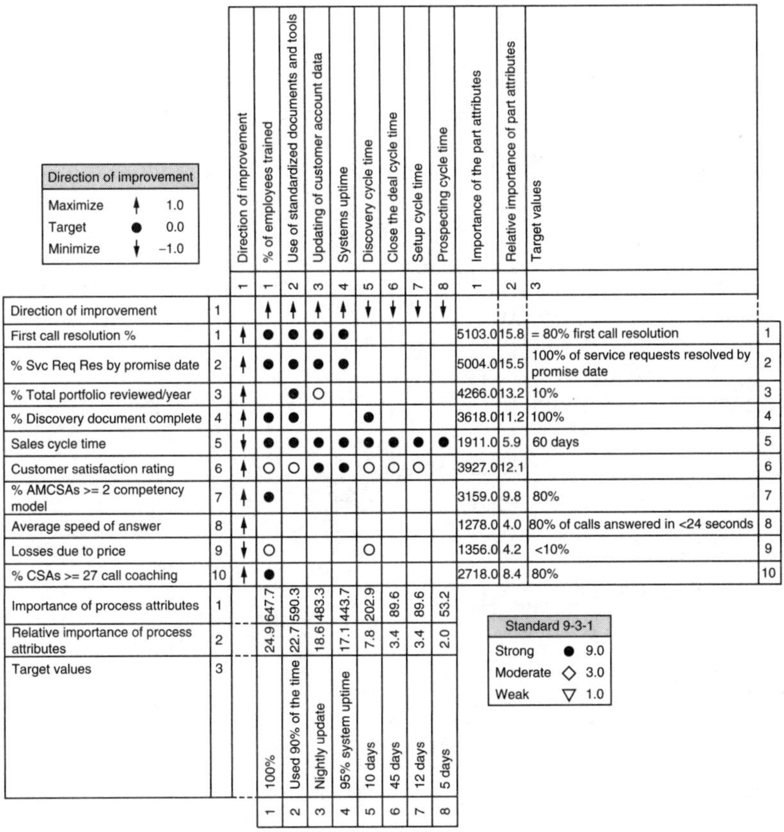

Quality Function Deployment

		Direction of improvement	% of employees trained	Use of standardized documents and tools	Updating of customer account data	Systems uptime	Discovery cycle time	Close the deal cycle time	Setup cycle time	Prospecting cycle time	Importance of the part attributes	Relative importance of part attributes	Target values	
		1	1	2	3	4	5	6	7	8	1	2	3	
Direction of improvement	1		↑	↑	↑	↑	↓	↓	↓	↓				
First call resolution %	1	↑	●	●	●	●					5103.0	15.8	= 80% first call resolution	1
% Svc Req Res by promise date	2	↑	●	●	●	●					5004.0	15.5	100% of service requests resolved by promise date	2
% Total portfolio reviewed/year	3	↑			●	○					4266.0	13.2	10%	3
% Discovery document complete	4	↑	●	●			●				3618.0	11.2	100%	4
Sales cycle time	5	↓	●	●	●	●	●	●	●	●	1911.0	5.9	60 days	5
Customer satisfaction rating	6	↑	○	○	●	●	○	○	○		3927.0	12.1		6
% AMCSAs >= 2 competency model	7	↑	●								3159.0	9.8	80%	7
Average speed of answer	8	↑									1278.0	4.0	80% of calls answered in <24 seconds	8
Losses due to price	9	↓	○				○				1356.0	4.2	<10%	9
% CSAs >= 27 call coaching	10	↑	●								2718.0	8.4	80%	10
Importance of process attributes	1		647.7	590.3	483.3	443.7	202.9	89.6	89.6	53.2				
Relative importance of process attributes	2		24.9	22.7	18.6	17.1	7.8	3.4	3.4	2.0				
Target values	3		100%	Used 90% of the time	Nightly update	95% system uptime	10 days	45 days	12 days	5 days				
			1	2	3	4	5	6	7	8				

Direction of improvement: Maximize ↑ 1.0; Target ● 0.0; Minimize ↓ −1.0

Standard 9-3-1: Strong ● 9.0; Moderate ◇ 3.0; Weak ▽ 1.0

Figure 6.8 Phase 2 QFD

Phase 1 QFD Diagnostics

Weak Whats

The Black Belt needs to identify Whats with only weak or no relationships. Such situations represent a failure to address a customer attribute. When this occurs, the company should try to develop CTS(requirements) to address this What. Sometimes the team may discover that present technology can not satisfy the What. The DFSS team should resort to customer surveys and assessment for review and further understanding.

No such What exists in our example. The closest to this situation is "Available products" in row 1 and "Easy-to-use products" in row 5 (see Figs. 6.6 and 6.7). It was highlighted as the weakest What but not weak enough to warrant the preceding analysis. However, the team is encouraged to strengthen this situation by a CTS requirement with a strong relationship.

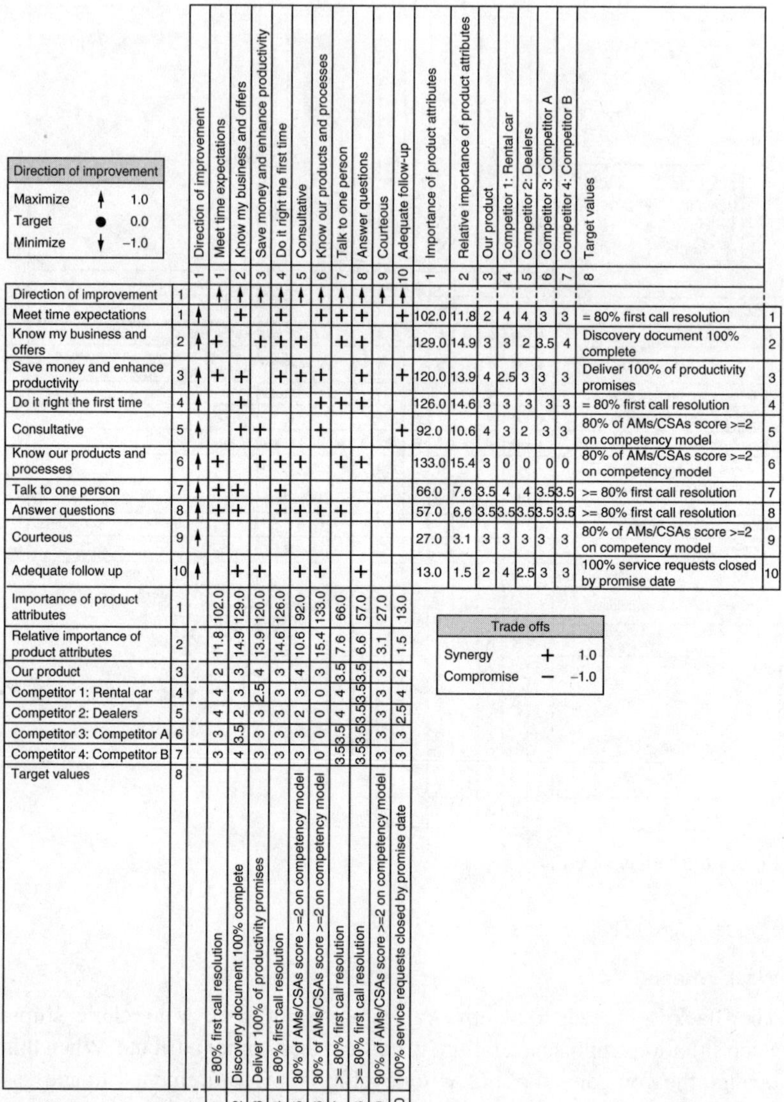

Figure 6.9 Correlation.

Weak Hows

The team needs to look for blank or weak Hows (all entries are inverted deltas). This situation occurs when CTS requirements are included that do not really reflect the customer attributes being addressed by the QFD. The Black Belt and his or her team may consider eliminating CTS requirements from further deployment if they do not relate to basic quality or performance attributes in the Kano model. The theme of DFSS is to be customer driven and to work on the right items; otherwise, we are creating a design "hidden factory."

In our example, the CTS requirement "Adequate follow-up" is weak (rated 13 on the importance rating). However, the What "Easy to use products" has no strong relationship with any CTS requirements and eliminating "Adequate follow-up" may weaken the delivery of this What even further.

Conflicts

The DFSS team needs to look for cases where technical benchmarking rates their product or service high but the customer assessment is low. Misconceptions of customer attributes is the major root cause of these cases. The team together with marketing can remedy these situations.

In our example, "Cost-effective products," a What, is addressed by many CTS requirements including "Save money and enhance productivity." The customer rates our product as weak (rating 2), while the technical assessment is rated the highest (rating 4). Who is right? Conflicts may be a result of a failure to understand the customer and must be resolved prior to further progress.

Strengths

By identifying the CTS requirements that contain the most 9 ratings, the DFSS team pinpoints which CTS requirements have a significant impact on the total design. Changes in these characteristics will greatly affect the design, and such effects propagate via the correlation matrix to other CTS requirements causing positive and negative implications. The following CTS requirements are significant as implied by their importance ratings and number of 9 ratings in their relationships to Whats: "Meet the expectations," "Know my business and offers," "Save money and enhance Productivity," "Do it right the first time," and "Know our products and processes." Examining the correlation matrix (Fig. 6.9), we have positive correlation all over except in the cell "Do it right the first time" and "Meet time expectations."

Eye-Openers

The DFSS team should look at customer attributes where

1. Their design as well as their competitors are performing poorly
2. The Whats are performing poorly compared to their competitors for benchmarking
3. CTS requirements need further development in phase 2

We can pencil in "Flexible processes" in the first category and "Accuracy" and "Easy-to-use products" in the second category. The CTS requirements that deliver these Whats should receive the greatest attention as they represent potential payoffs. Benchmarking areas represent the Whats where competitors are highly rated and it is highly desirable to incorporate their designs. This saves design and research time.

The highest CTS requirements with the largest importance ratings are the most important. For example, "Know our products and processes" has the highest rating at 133. This rating is so high because it has three strong relationships to the Whats. The degree of difficulty is medium (rating 3) in the technical benchmarking. In addition any CTS requirement that has a negative or strong relationship with this CTS requirement in the correlation matrix should be addressed in phase 2.

6.8 QFD Case Study: Yaesu Book Center

This case study is from Akao (1990). The Yaesu Book Center is a bookstore in Japan. When it first opened, the store had few employees experienced in bookselling. Most of the business was conducted by employees who had recently graduated from school. In spite of that, the Yaesu Book Center attracted a great deal of attention and was highly regarded by book lovers.

The Yaesu Book Center has its own quality control (QC) circle. In the QC circle, the area managers are also group leaders. The QC circle determined that the following three things are essential to satisfy customers' needs:

1. Have enough books available
2. Have enough product information
3. Provide enough service

The QC circle members also found that they did not have enough information to figure out how to accomplish the preceding three objectives due to the following:

1. Specific customer demands were not clear.
2. There were no specific quantitative measurements for customer demands.
3. The relationship between the customers' demands and the Yaesu Book Center's service product was not clear.

To solve these problems, the voices of customers were collected and a two-phase QFD was conducted by the Yaesu Book Center to improve the bookstore operation. This QFD study was conducted by the following steps:

Step 1: Determine Customer Attributes (Whats)

First, a lot of customer demands were collected by customer surveys and interviews. The "raw" customer demands are in the customers' own words. In a brainstorming session, this raw customer information was translated into a set of better-defined customer attributes. The following procedures were used in this translation process:

- Vague comments were changed into precise expressions.
- Comments expressed in negative conditions were changed into positive comments.
- Comments were grouped into subcategories, and similar subcategories were combined.
- Customer attributes were fitted into tree diagrams.

Figure 6.10 illustrates a partial tree diagram that organizes the customer attributes in this case.

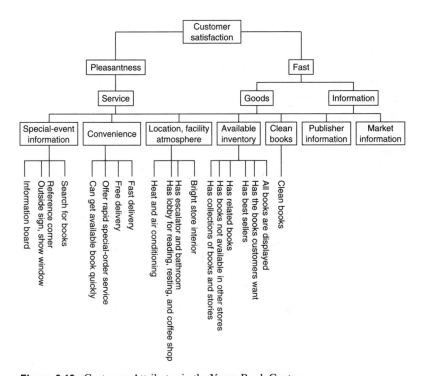

Figure 6.10 Customer Attributes in the Yaesu Book Center

Step 2: Determine Quality Characteristics (Hows)

Quality characteristics are extremely important, so the Yaesu Book Center QC circle made a great effort to identify them. A partial list of quality characteristics is given in Fig. 6.11. A three-level hierarchy of quality characteristics was used.

Step 3: Assign Degree of Importance to Customer Attributes

The survey questionnaire given in Table 6.1 is used to collect customer importance ratings for each customer attribute. It serves as a basis for determining importance ratings in the QFD study.

Step 4: Determine Operation Items

In this case, operation items are what bookstore management and employees are actually doing in their work. Eventually, the QFD study should provide guidelines as to which operation, items were not done enough before and how much effort should go into doing them now. These operation items are organized in a tree diagram and illustrated in Fig. 6.12.

Step 5: Two-phase QFD Analysis for the Yaesu Book Center

Based on the work in the first four steps, two QFD house of quality charts were developed for the Yaesu Book Center. The first house of quality chart relates customer attributes to quality characteristics; the second house of quality chart relates quality characteristics to operations items. The partial listings of these two houses of quality are illustrated in Figs. 6.13 and 6.14.

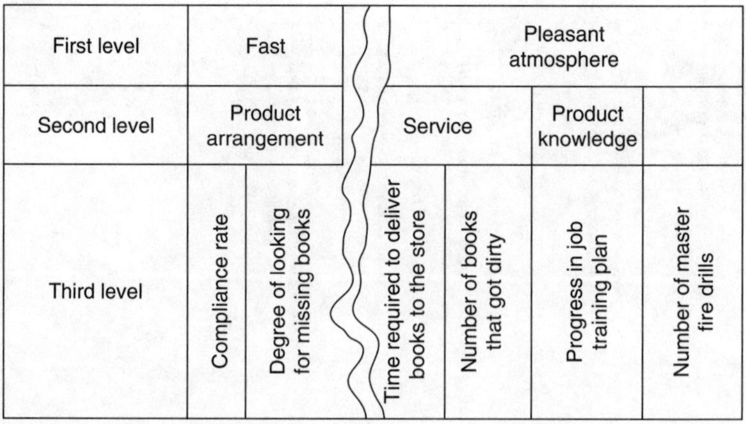

First level	Fast			Pleasant atmosphere		
Second level	Product arrangement		Service	Product knowledge		
Third level	Compliance rate	Degree of looking for missing books	Time required to deliver books to the store	Number of books that got dirty	Progress in job training plan	Number of master fire drills

Figure 6.11 Quality Characteristics (Hows)

Table 6.1 Survey Questionnaire

How important are the following items in a bookstore. Please rank from 1 to 5.

	1 Not Important at All	2 Not Important	3 Neither	4 Somewhat Important	5 Important
1. Has a good variety of best sellers	1	2	3	4	5
2. Scheduled date for availability of out-of-stock books is clear	1	2	3	4	5
3. Has a good variety of art books	1	2	3	4	5
4. Has a variety of books on sociology, literature, science, and history	1	2	3	4	5
...					
13. The store clerks look hard for books for the customers	1	2	3	4	5
14. Can easily find books you want	1	2	3	4	5
15. Book classifications are easy to understand	1	2	3	4	5
16. Attractive, easy-to-find book displays	1	2	3	4	5
17. Books are always clean	1	2	3	4	5

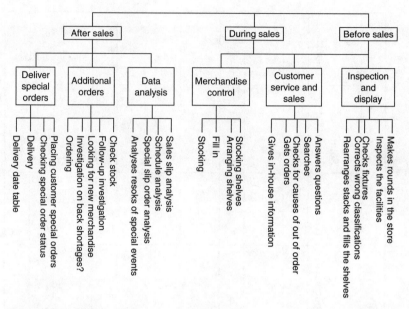

Figure 6.12 Operation Items

After constructing two QFD charts, the QC circle found that there was one major problem in the book center: The customer attribute "book classification is easy to understand" was rated very high in customer survey results, but by QFD analysis, the rating for this is not high with current operation items. The following corrections were made:

1. If a book could fall into more than one category, then it would be displayed in all these categories.
2. Point-of-purchase clerks are placed in the boundary areas between book sections.

6.9 Summary

Quality function deployment (QFD) is a planning tool used to translate customer needs and expectations into the appropriate design actions. This tool stresses problem prevention with an emphasis on results in customer satisfaction, reduced design cycle time, optimum allocation of resources, and fewer changes. Together with other DFSS tools and concepts, it also makes it possible to release the designed entity at the six sigma level. Since

		Quality characteristics															
		Service														Goods	
		Availability					Reception and response to customers				Orders			Environ-ment			
Customer attributes	Importance rating	Clear classification	Immediate response	Rate of availability	Degree of satisfaction	Easy to find	Service satisfaction	Waiting time	Product knowledge	Customer reception	Order lead time	Delivery sale	Procurement time	Cleanliness	Lighting	Damage rate	
Can tell the book is in stock	2.9	●	●		●												
Can tell why not in stock	6.0	●		●													
Can tell if book is available		○															
Can give date of availability	0.1	○							○								
Can tell detailed book description									●	○							
Can find related book	1.8								●	○							
Offer information on the contents									●	○							
Information list of books available							○		○								
Has large variety and volume of books	4.0		●	○	●												
Has books not available in other stores	1.3		●	○	●						●						
Has newly published books	0.2		●														
Has many specialty books			●														
Easy to find books	4.3						●										
Classification is easy to understand	5.8					●											
Signs are easy to see	5.0					●	●										
Display is easy to see	2.5						●										
Has clean books	2.5																●
Has product knowledge	3.4							○		●	○						
Kind and polite	3.8							●			○						
	0.1								●		○						
Assistance in looking for books	3.4									●	○						

Figure 6.13 House of Quality Phase 1

the customer defines quality, QFD develops customer and technical measures to identify areas for improvement.

QFD translates customer needs and expectations into appropriate design requirements by incorporating the voice of the customer into all phases of the DFSS algorithm, through production and into the marketplace. In the context of DFSS, the real value of QFD is its ability to direct the application of other DFSS tools to those entities that will have the greatest impact on the team's ability to design a product, service, or process that satisfies the needs and expectations of the customers, both internal and external.

		Before sales					During sales					After sales									
		Make records in the store	Inspects the facilities	Checks fixtures	Correcting wrong classifications	Rearrange stocks/fill shelves	Answer questions	Searches	Checks for cause of out of stock	Gets orders	Gives in-house information	Sales slip analysis	Schedule analysis	Special order slip analysis	Analyzes results of special events	Check stock	Add on orders	Looks for missing books	Places special orders	Check status of special orders	Delivers
Quality characteristics	Clear classification												O								
	Immediate response											O	●			O					
	Rate of availability											O									
	Degree of satisfaction	●			●											O	O				
	Easy to find	●				●					O										
	Service satisfaction						●														
	Waiting time						●	●													
	Product knowledge						●	●													
	Customer reception																				
	Order lead time											O	O	O			O				
	Delivery rate																				O
	Procurement time								●					●					●	●	
	Cleanliness																				
	Lighting																				
	Damage rate	●	●																		

Figure 6.14 House of Quality Phase 2

The following items are a review of the different parts of the house of quality. The Whats represent customer needs and expectations. The Hows are CTS requirements, or substitute quality characteristics for customer requirements that the company can design and control. Relationships are identified between what the customer wants and how those wants are to be realized. Qualitative correlations are identified between the various Hows. Competitive assessment and importance ratings are developed as a basis for risk assessment when making decisions relative to tradeoffs and compromises.

Chapter 7

Value Engineering

7.1 Introduction

For any service product, customer value and satisfaction can be improved by increasing customer benefits and reducing cost. Among the customer benefits, functional benefits are of key importance. People pay for functions, not for hardware, not for paperwork. For example, people go to a fast-food restaurant to buy such functions as relieving hunger, getting nutrition, and getting taste. People go to hospitals, not to buy doctor's time, surgery, or hospital beds, but to buy such functions as curing a disease and relieving symptoms. Value engineering is a systematic, team-oriented, creative approach that seeks to deliver customer-desired functions with lower cost.

The Society of American Value Engineers (SAVE) defines the term *value engineering* as follows:

> Value engineering is the systematic application of recognized techniques which identify the functions of a product or service, establish a monetary value for that function and provide the function at the lowest cost.

However, value engineering is not merely a cost-cutting program; it only cuts unnecessary cost. *Unnecessary cost* is the cost that can be removed without affecting the functional performance of the product or service. In the value-engineering approach, it is important to maintain a high level of functional performance while cutting cost. That is, the new design coming out of a value-engineering project should have the same or better functional performance than the old design. It has been estimated that 30 percent of the cost of an average product or service is unnecessary. This unintentional cost is the result of habits, attitudes, and all other human factors.

Value engineering originated at the General Electric Company in 1947. Mr. Harry Erlicher, vice president of purchases, noted that during wartime, it was frequently necessary to make substitutions for the critical materials that not only satisfied the required functions but also gave better performance and lower cost. He reasoned that if it was possible to do this in wartime, it might be possible to develop a system that could be applied as a standard procedure to normal operations to increase a company's efficiency and profit. Mr. L. D. Miles was assigned to study the possibility, and the result was a systematic approach to problem solving based on functional performance that he called value analysis.

Value analysis, value engineering, value management, value assurance, and value control are all the same in that they make use of the same set of techniques developed by Mr. Miles in 1947. In many cases, the title tends to describe how the system is being applied. Value analysis is generally considered to apply to removing cost from a product. Value engineering and value assurance are applied in the development phase to keep cost out of a product. Value management and value control are overall programs that recognize that value techniques can be applied in business operations.

Value engineering was first applied in product development, manufacturing, and the construction industry. Since the 1970s, value engineering started to be applied in the service industry. David Reeve's (D. Reeve 1975) case study on the youth service bureau was among the first successful case studies in service organizations. Since then, successful value engineering service case studies are reported in retail, finance, health care, photo shops, and many others.

Value engineering achieves results by following a well-organized, planned approach. It identifies unnecessary cost and applies creative problem-solving techniques to remove it. The three basic steps in this planned approach are

1. Identify the functions. (What does the product or service do for customers?)
2. Evaluate the functions. (What is the lowest cost to create these functions?)
3. Develop alternatives. (What else will do the job?)

Identify the Functions

Function is the very foundation of value engineering. The concern is not with the part or act itself but with what it does; what is its function? It may be said that function is the objective of the action being performed by the product or system. Function is the property that makes something work or sell. We pay for a function, not hardware, not paperwork. Hardware has no value; only function has value. We pay to retrieve information, not file papers.

Defining functions is not always easy. It takes practice and experience to properly define a function. It must be defined in the broadest possible manner so that the greatest number of potential alternatives can be developed to satisfy the function. A function must also be defined using two words, a verb and a noun. If the function has not been defined in two words, the problem has probably not been properly defined.

Function is a forcing technique, which tends to break down barriers to visualization by concentrating on what must be accomplished, rather than the present way a task is being done. Concentrating on function opens the way to new innovative approaches through creativity. Some examples of simple functions are as follows: create design, evaluate information, determine needs, grow wealth, and enclose space.

There are two types of functions, basic and secondary. The basic function describes the most important action performed. The secondary function supports the basic function and almost always adds cost.

Evaluate the Functions

After the functions have been defined and identified as basic or secondary, we must evaluate them to determine if they are worth their cost. This step is usually done by comparison with something that is known to be a best cost. Best cost is the lowest overall cost to reliably provide a function.

Develop Alternatives

Function has been defined as the property that makes something work or sell, and the best cost is the lowest overall cost to reliably provide the function. In value engineering analysis, if we find that the current cost to provide a function is significantly higher than the best cost, then we need to ask: What else will do the job? That is, we will try to develop alternative ways to perform this function.

In order to develop alternatives, we make maximum use of imagination and creativity. This is where team action makes a major contribution. The basic tool is brainstorming. In brainstorming, we follow a rigid procedure in which alternatives are developed and tabulated with no attempt to evaluate them. Evaluation comes later. At this stage, the important thing is to develop the revolutionary solution to the problem.

Free use of imagination means free from the constraints of past habits and attitudes. A seemingly wild idea may trigger the best solution to the problem in someone else. Without a free exchange of ideas, the best solution may never be developed. A skilled leader can produce outstanding results by brainstorming and by providing simple thought stimulation at the proper time.

7.1.1 Evaluation, Planning, Reporting, and Implementation

The creative phase does not usually result in concrete ideas that can be directly developed into outstanding products. The creative phase is an attempt to develop the maximum number of possible alternatives to satisfy a function. These ideas or concepts must be screened, evaluated, combined, and developed to finally produce a practical recommendation. It requires flexibility, tenacity, visualization, and frequently the application of special methods designed to aid in the selection process. The process is carried out during the evaluation and planning phases of the job plan and is covered in detail in those sections of the text.

The recommendations must be accepted as part of a design or plan to be successful. In short, they must be sold. They must show the benefits to be gained, how these benefits will be obtained, and finally, proof that the ideas will work. This takes time, persistence, and enthusiasm, and details of a recommended procedure are covered in Sec. 7.6 of the text.

7.1.2 The Job Plan

These are the basic features that make value engineering an effective tool. All are applied in a step-by-step approach to a value study. The approach is called the job plan and is broken down into six steps:

1. Information phase
2. Creative phase
3. Evaluation phase
4. Planning phase
5. Reporting phase
6. Implementation phase

Each step is designed to lead to a systematic solution to the problem after consideration of all the factors involved.

We are going to discuss these six steps in the value engineering job plan in Secs. 7.2 to 7.7. Section 7.8 will discuss a value engineering project in the service industry.

7.2 Information Phase

The first phase of the value engineering job plan is the information phase. It is the most time-consuming yet most important phase. In this phase, we collect all the necessary raw information for the project, including relevant product descriptions, process flowcharts and layouts, and all relevant cost information. Based on the information collected in this phase, we will produce three important documents for the project: (1) the functions list, (2) the cost-function work sheet, and (3) the FAST (function analysis system technique) diagram. The function list is a complete list of all functions needed in order for the product to work properly; each function is defined and classified. The cost-function work sheet is a complete cost breakdown calculation for all the product elements (subtasks, items, or components) as well as for all the functions; the cost-function work sheet also lists the actual cost and best cost for each function. The FAST diagram is very important and provides an exact logical linkage among all functions. The actual cost and the best cost for each function are also recorded in the FAST diagram.

The information phase is broken down into three distinctly separate parts.

1. Information development
   a. Information collection
   b. Cost visibility
   c. Set goal for achievement
2. Function determination
   a. Define functions
   b. Eliminate duplication
3. Function analysis and evaluation
   a. Construct FAST diagram
   b. Function and cost analysis
   c. Function evaluation
   d. Identify problem areas
   e. Compare potential benefit to goal for achievement

The work done in the information phase is the basis for the development of alternative low-cost methods to perform the required functions. If the functions have not been properly defined and evaluated, then the correct analysis will not be performed and the most satisfactory problem solution is not likely to be developed; if the cost figures are incorrect and/or incomplete, then the low-cost solution will not be identified correctly.

7.2.1 Information Development

The first part of the information phase is the development of all available information concerning the project. This includes drawings, process sheets, flowcharts, procedures, and any other available material. It is important to discuss the project with people who are in a position to provide reliable information and to verify that honest wrong impressions are not being collected; that is, it may have been fact at one time, but is no longer valid.

It is very important that good human relations be used during this data- and information-collecting phase. Get the person responsible for the project or development in the first place to help, by showing the person how he or she will be able to profit from successful results of the completed study.

The project identification checklist, illustrated by Table 7.1, details all the information required for study. If the data or information are not on hand, it will be necessary to obtain them. Table 7.1 is a basic information data sheet that should be filled out as a first step to identify the project. A brief description of the project should be written under operations and performance to be certain all the team members are in at least basic agreement as to the product or process operation.

Cost Visibility

The next step toward a problem solution is to complete the cost visibility section of the cost-function work sheet as illustrated in Table 7.2. This cost-function work sheet is one of the important documents that should be produced in the information phase. The left portion of this work sheet is the cost visibility portion. By cost visibility we mean that we make all costs visible in a very detailed fashion, no ambiguity, no misunderstanding.

Table 7.1 Project Identification Checklist (Service Industry)

1. Flowcharts, organization charts
2. Detailed transaction data
3. Facility layout
4. Service product profile,
5. Cost data (labor, overheads, material)
6. Work instructions

Table 7.2 Cost-Function Work Sheet

Cost Visibility					Cost-Function Analysis			
Total Cost $		Cost Elements			F1 Function 1	F2	.	.
Item No.	Name	Material ($)	Labor ($)	Burden ($)				
1.								
2.								
3.								
...								
				Cost total				
				Best cost				

Cost visibility is required in order to identify the areas of high and unnecessary costs and to find ways to reduce or eliminate these costs.

Cost-visibility techniques are well-ordered and range from very simple to highly complex. These techniques do not tell us where unnecessary costs are; they tell us where high costs are. This is important because they identify a starting point.

The following is a list of definitions commonly used in cost visibility analysis.

Cost The amount of money, time, labor, etc., required to obtain anything. In business, the cost of making or producing a product or providing a service.

Fixed Cost Cost elements that do not vary with the level of activity (insurance, taxes, plant, and depreciation).

Actual Cost Costs actually incurred during the performance of a process. They include labor, material, and burden applied in accordance with local ground rules.

Incremental Cost Not all variable costs vary in direct proportion to the change in the level of activity. Some costs remain the same over a given number of production units or transactions, but rise sharply to new plateaus at certain incremental changes. The costs thus effected are incremental costs.

Material All hardware, raw material, and purchased items consumed in producing a product item.

Labor Work force needed to produce a product or perform a service.

Burden (Overhead) Includes all costs incurred by the company that cannot be traced directly to specific products. The accounting department determines burden rates. These are assigned to individual operations on a formula basis. Burden consists of both fixed and variable categories, and separate rates are often established for each. The method of assigning burden differs from industry to industry and even from one company to another within an industry. Any quantifiable product factor may serve as a basis for assignment of burden, as long as consistent use of the factor across the entire product line results in a full and equitable burden distribution.

Fixed Burden Includes all continuing costs regardless of the production volume for a given item, such as salaries, building rent, real estate taxes, and insurance.

Variable Burden Includes costs that increase or decrease as the volume rises or falls. Indirect materials, indirect labor, electricity used to operate equipment, water, and certain perishable tooling are also included in this classification.

Allowance All costs other than material, labor, and burden that must be included in the total cost of a product, such as packaging materials, scrap, inventory losses, and inventory costs.

Total Cost Includes production cost plus profit and other expenses.

The following expenses are usually added to production cost by the sales and/or accounting departments to make up the total cost.

Administrative and Commercial Costs Costs incurred in the administration of the company, research, and selling of the product. They are usually a factor represented as a percentage of production cost.

Freight Cost Shipping and handling costs.

Profit Amount earned in producing a product or a service. It is usually applied as a percentage of production cost.

Sources of Cost Information

The application of cost-visibility techniques begins with an analysis of total cost, progresses through an analysis of cost elements, and finally ends with an analysis of component or process costs. To perform these steps the best cost information available is required. This information will be available from sources such as

Accounting: Current and historical costs (actual costs)
Purchasing: Cost of purchased items
Suppliers: Estimates and/or quotations, costs, process information, and material prices

In the service industry, labor usually accounts for a big portion of cost. In order to figure out the exact labor cost component in each item, some traditional motion-time study has to be performed. For example, in the health-care industry, the doctor's time is an important source of cost because it is very expensive. If we conduct a value-engineering study on emergency care, we may have to use a stopwatch to track the doctor's time usage for patient visits. After recording the time for a sufficient number of patient visits, we can calculate the average doctor time and use that as a basis in computing the doctor's cost.

Review this cost data in accordance with the process outlined in the text on cost visibility, and make a preliminary judgement of the potential profit improvement. Consider the factors involved, and set a goal for achievement that will provide a profitable position. The target should indicate a 30 to 100 percent cost reduction to be practical. It may seem improbable that this can be achieved; however, it is a target to work toward. A check against this target will be made at the completion of the information phase.

> **Example 7.1: A Cost-Visibility Work Sheet of a Youth Assistance Program**
> David Reeve (1975) did a value-engineering study on the youth assistance program for Oakland County, Michigan. This was one of the very first case studies of value engineering for a government or service organization. The purpose of the youth assistance program was to help troubled teenagers so they would not become problems for society. There are two major activities in the youth assistance program, prevention and rehabilitation. Each activity was to be accomplished through various meetings, contacts, field visits, and office activities.
>
> Tables 7.3 and 7.4 provide cost-visibility sections of the cost-function work sheets for rehabilitation and prevention, respectively. In these cost-visibility calculations, the labor cost is computed based on labor hour times labor rate. The labor hours are determined based on the historical records of meeting length, interview time duration, and so on.
>
> **Example 7.2: A Cost-Visibility Work Sheet of an Automobile Hood Latch**
> Table 7.5 gives a hardware cost-visibility worksheet for an automobile hood latch.

Project Scope

It is now possible to make a preliminary determination of the project scope. By considering the new project as outlined on the project identification sheet, the present cost and target for improvement, and the time available for the study, we can define the scope of the project. Limiting or expanding the scope of a study depends on the objective and the time allowed for the study. In project work, the analysis of function should first be performed upon the total process. If the objectives of the value-engineering study are not achieved at that level, the next lower level should be studied and so on down to the lowest level of indenture. The lower the level of indenture, the more detailed and complex the study might become. This may require additional time in the present study or future studies to consider segments identified by function analysis.

7.2.2 Function Determination

The information on hand, together with an analysis of costs, can be used to define the initial scope of the project. The product or process has been defined

Table 7.3 Cost Visibility of Rehabilitation per Case

Total Cost = $109.64/case		Cost Elements			
Item No.	Name	Material	Labor	Burden	Total Cost ($)
1.	Client contact		27.19		27.19
2.	Organization contact		11.04		11.04
3.	Secretarial center office		6.07		6.07
4.	Secretarial field office		31.12		31.12
5.	Case management		11.59		11.59
6.	General administration		2.21		2.21
7.	Grant administration		1.38		1.38
8.	Others		4.42		4.42
9.	Travel time		5.11		5.11
10.	Administration meetings		1.17		1.17
11.	Supervisory meetings		4.69		4.69
12.	Training meetings		2.35		2.35
13.	Statistical meetings		0.97		0.97
14.	Evaluation meetings		0.34		0.34

and its cost evaluated by the cost-visibility study. It is now possible to start to define the functions to be performed or that are being performed by the system.

What Is a Function?

The usual definition of *function* is the property that makes something work or sell. Miles defines function as a want to satisfy a requirement. Function is the end result desired by the consumer. Function is what is paid for. Function is a requirement, a goal, or an objective.

A function is not an action; it is the objective of an action. For example, "file paper" is an action. But what is the objective or purpose of the action? We file papers not because we enjoy putting papers in folders or cabinets,

Table 7.4 Cost Visibility of Prevention per Case

Total Cost = $41.75/case		Cost Elements			
Item No.	Name	Material	Labor	Burden	Total Cost ($)
1.	Client contact		1.17		1.17
2.	Organization contact		2.90		2.90
3.	Secretarial center office		3.45		3.45
4.	Secretarial field office		8.56		8.56
5.	Case management		0.28		0.28
6.	General administration		2.07		2.07
7.	Grant administration		0.28		0.28
8.	Others		2.55		2.55
9.	Travel time		2.90		2.90
10.	Administration meetings		1.86		1.86
11.	Supervisory meetings		4.90		4.90
12.	Training meetings		0.41		0.41
13.	Statistical meetings		0.14		0.14
14.	Evaluation meetings		0.21		0.21
15.	Advisory council meetings		2.76		2.76
16.	Citizen committee meetings		1.24		1.24
17.	Citizen subcommittee meetings		6.07		6.07

but because we want to keep a record so that we can use it later. Therefore, the objective of filing papers is actually to store information. In this case, the correct function name is "store information." So the function is the desirable result to be accomplished by an action. The action is one method that can be used to accomplish the objective.

Table 7.5 Cost Visibility of Automobile Hood Latch

Total Cost = $2.616		Cost Elements ($)			
Item No.	Part Name	Material (Total 1.545)	Labor (Total 0.713)	Burden (Total 0.358)	Total Cost (Total 2.616)
1.	Primary spring	0.219	0	0.035	0.254
2.	Detent spring	0.09	0	0.015	0.1046
3.	Hook spring	0.09	0	0.015	0.1046
4.	Pivot rivets	0.09	0.005	0.015	0.1104
5.	Hook pivot	0.08	0.005	0.015	0.0988
6.	Fork bolt	0.04	0.096	0.014	0.158
7.	Mounting bracket	0.426	0.198	0.101	0.7253
8.	Back plate	0.08	0.149	0.037	0.2661
9.	Secondary hook	0.26	0.151	0.067	0.4776
10.	Detent level	0.11	0.099	0.034	0.2429
11.	Grease	0.02	0.005	0.004	0.0291
12.	Sleeve	0.04	0.005	0.007	0.0523

The basic rule is to define functions using two words, a verb and a noun. The resultant definition should be such that it is not restrictive in that it defines a method for performance. An abstract definition will offer an opportunity for creative questions that may produce a number of alternatives. For example, using "file papers" as the definition of a function may limit our thought on using papers, folders, and cabinets. By using the more generic definition of "store information," we can open our thoughts to other ideas such as computers, and CDs to read, retrieve and catalog information.

It is also important that the function be measurable in some unit term such as weight, cost, volume, time, or space. In some cases, the measure may be satisfaction, desire, or some other abstract measure that will require more subjective analysis but can still be measured by comparative techniques. The following are some examples of function definitions:

Verb	Noun	Unit
Create	Design	Time
Confirm	Design	Time
Authorize	Program	Cost
Measure	Performance	Workerhours

In his book *Techniques of Value Analysis*, Mr. Miles recognized the difficulty of applying this technically simple concept. He said, "While the naming of functions may appear simple, the exact opposite is the rule. In fact, naming them articulately is so difficult, and requires such precision in thinking, that real care must be taken to prevent the abandonment of the task before it is accomplished." He also said, "Intense concentration, even what appears to be over concentration of mental work on these functions, forms the basis for unexpected steps of advancement of value in the product or service."

The two-word definition of functions is the most difficult naming method. There is some feeling among value-engineering practitioners that it is unnecessary to struggle for two-word definitions; three-word definitions or short-statement definitions should do the job. However, the value-engineering practitioners also found that if the goal of value-engineering study is to generate creative design solutions, then two-word functions are imperative. If the function cannot be defined in two words, more understanding is required. It is a struggle to define good functions, but the result is worth the struggle. The two-word function definition is a forcing technique that requires consensus among team members, eliminates confusion, creates in-depth understanding of the requirement, clarifies overall knowledge of the project, and ultimately breaks down barriers to visualization so necessary to help define the creative questions that will lead to new, outstanding solutions to the project.

Types of Functions

In value engineering a function is defined as something that makes a product work or sell. There are two types of functions, work and sell. *Work* is the function that does the job that customers want. *Sell* is the function that adds appeal for customers to buy.

Different categories of verbs and nouns are used to express the work and sell functions. A work function is always expressed by an action verb and a measurable noun that establish a quantitative statement. A sell function is always expressed by a passive verb and a nonmeasurable noun that establish a qualitative measurement.

The following tables provide examples of work functions and sell functions:

Work Functions

Verb	Noun	Unit
Collect	Payment	Dollar
Remove	Kidney	Time and/or cost
Transfer	Fund	Dollar/time
Sell	Assets	Dollar/cost

Sell Functions

Verb	Noun
Increase	Beauty
Improve	Style
Increase	Prestige

The separation of work and sell functions helps us to define a function more precisely. In defining the work function, the use of measurable nouns provides us with a quantitative means of measuring the work functions. Work functions provide use value. In the case of sell functions, since they are in most cases subjective in nature, their measurement is extremely difficult. Sell functions usually provide prestige value to customers. The separation of work and sell functions can also help us to identify the proportion of cost allocated to use value and prestige value.

All functions can also be divided into two levels of importance, basic functions and secondary functions. The basic functions are those functions that fulfill the primary purpose for a product or service. Secondary functions are those functions that do not directly fulfill the primary purpose of the product or service but support the primary purpose. The result of function determination should be a completed function list as illustrated in Table 7.6.

Table 7.6 Function List

Project Name:	Scope Includes:				
	Scope Does Not Include:				
List All Functions	Function Types				
Verb	Noun	Basic	Second	Work	Sell	Remarks
1.						
2.						
3.						
4.						

Example 7.3: Pencil Function Determination

Figure 7.1 shows a pencil. A pencil has five parts: lead, body, paint, band and eraser. In value engineering, it is a rule of thumb that even a very simple product such as a pencil will have quite a few functions. It is easier to figure out a pencil's function by looking into its five parts. Table 7.7 gives a summary of functions performed by the parts of a pencil. Table 7.8 gives a function list for the whole pencil.

Example 7.4: Function List of Oakland County Youth Assistance Program

This example is a continuation of Example 7.1 (Reeve 1975). In that example we listed all the organizational activities that support a youth assistance program and their cost calculation. Reeve (1975) determined 41 functions in this program. Table 7.9 lists a portion of these 41 functions.

Reeve also provided a *glossary of functions*, which gives a detailed definition for each function. In a value-engineering project, it is highly recommended that such a glossary be developed and that consensus be secured from group members on the definition of each function. Therefore, in later discussions, every team member will be on the same page when each function is discussed. Here is a portion of the glossary of functions for this youth assistance program project:

Identify Need Time spent in written and oral communication, that is, conferences, letters, interviews, etc., with school personnel and/or other

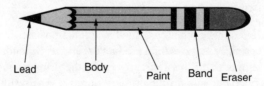

Figure 7.1 A Pencil

Table 7.7 Function List of Pencil Parts

Part	Function	
	Verb	**Noun**
Lead	Make	Marks
Eraser	Remove	Marks
Band	Secure	Eraser
	Improve	Appearance
Body	Support	Lead
	Transmit	Force
	Accommodate	Grip
	Display	Information
Paint	Protect	Wood
	Improve	Appearance

referral sources regarding potential referrals, in order to determine the need for the referral process.

Assist Client Includes counseling; offering alternatives; providing a referral service; indicating community programs; helping kids get to camp; talking to teachers, police, or other authorities on client's behalf; and aiding parents and children.

Eliminate Deviancy The client returns to homeostatic position and development of modification technique to reach normative behavior patterns (measured by time spent).

Define Problems All communication with client, parents, and referral sources for the purpose of describing the client's problem behavior.

Creativity and Function Definition

The ultimate objective of value engineering is to create a better product or service design. Creativity is very important in creating a new and better design. What makes people more creative? The consensus seems to be that to be creative, one must be able to see beyond the conscious, the existing.

Table 7.8 Function List of a Pencil

Project Name: Pencil		Scope Includes:				
		Scope Does Not Include:				
List All Functions		Function Types				
Verb	Noun	Basic	Second	Work	Sell	Remarks
1. Make	Marks	√		√		
2. Remove	Marks		√	√		
3. Secure	Eraser		√	√		
4. Improve	Appearance		√		√	
5. Support	Lead		√	√		
6. Transmit	Force		√	√		
7. Accommodate	Grip		√	√		
8. Display	Information		√		√	
9. Protect	Wood		√	√		
10. Improve	Appearance		√		√	

Table 7.9 A Partial List of Functions for a Youth Assistance Program

List All Functions		Basic	Second	Remarks
Verb	**Noun**			
1. Identify	Need	√		
2. Define	Problem		√	
3. Plan	Treatment		√	
4. Diagnose	Problems		√	
5. Obtain	Information		√	
6. Involve	Client		√	
7. Identify	Client		√	
8. Utilize	Resource		√	
9. Assist	Client	√		
10. Improve	Process		√	
11. Indicate	Trend		√	
12. Maintain	Record		√	
13. Establish	Standard		√	
14. Analyze	Data		√	
15. Terminate	Contact		√	
16. Evaluate	Process		√	
17. Eliminate	Deviancy	√		
18. Plan	Activities		√	
19. Determine	Needs		√	
20. Set	Goals		√	
21. Secure	Action		√	
22. Provide	Alternatives		√	
23. Develop	Programs		√	
24. Establish	Trust		√	
25. Exhibit	Concern		√	
26. Improve	Programs		√	
27. Evaluate	Programs		√	

What is the ingredient that some people have that makes it possible for them to break the barriers to visualization, to be able to look at something and immediately think of new and exciting possibilities for products, services, methods, or other useful or satisfying subjects? This is a provocative question that produces many and varied opinions but no clear-cut formula for producing creative people. It is known that a creative person is somewhat different. It is also known that a creative person exhibits certain characteristics. However, given the same characteristics, another person may not prove to be creative. Many people feel that the seeds for creativity exist in every person. If this is true, it would be exciting to discover the means to release these seeds to foster their growth for the benefit of humankind.

In all probability, at least one of the ingredients of creative people is the ability to visualize, to detach themselves from reality, and to see beyond the stated problem, the object, or the material facts. A creative person must be able to create concepts, broaden and develop them, analyze and examine them, and out of it all select a new idea, new approach, or new solution to a requirement or problem.

According to L. Miles:

1. Creative thinking is constrained by the physical shape or concept of existing products and services.
2. Concentrating on function helps to break down the barriers to visualization and offers outstanding opportunities for creativity.

The conventional approach to product or process improvement is to try to make the existing product work better, cost less, or meet some other objective. Creativity is stifled because the existing form constrains thinking.

The function approach is truly different. It breaks the project into requirements called functions. The process of defining function becomes a method to break the barriers to visualization to make entirely new solutions possible.

The concept is disarmingly simple. It is easy to understand without learning complex systems or studying complex technology. However, the ability to use the system comes only from a thorough understanding of the principles and the determination and discipline to use them.

Function analysis is basic to the system and starts with a need to understand the term *function* and how to define functions that will offer creative opportunities. Function definition and function analysis provide a major discipline for helping a person or group of persons visualize beyond their normally accepted standards. In fact, the forcing and struggle necessary to properly define a function make it possible for someone to look at what has been seen many times before and see new and different things, to see the problem in a new light. It can help someone achieve the ability to visualize beyond the stated problem, as outstanding people have been able to do, throughout the ages.

This means that not only is function the basic ingredient of value engineering, but it provides the opportunity for a person to break down barriers to seeing new things, to eliminate prejudices, and to come up with insights never before thought possible.

Functions for Creativity

In the definition of function, it is important that several key questions be kept in mind at all times. These questions are

- What are we really trying to do when we perform this action? Why is it necessary to do this?
- Why is this part or action necessary?

Specific answers to these questions will aid you in zeroing in on a useful definition.

It is also necessary to be aware that the functions of a product will be different depending on who (for example, the plant manager, the product, or the customer) is using it. Role playing the parts of these different people will assist you in determining all these functions. This role playing may be difficult at first, but it becomes easier with practice. The idea is to "let the job be the boss," as Kettering said. Be the crankshaft. What do you do? How do you feel? Act the part of the customer. What do you see? What does it do for you? If you were the plant manager, what would you want? How would you get it? This system helps to eliminate bias in that functions can be defined from all viewpoints and sorted out in the FAST diagram through cause-and-effect relationships for maximum understanding and subject evaluation.

Start at the Top

In defining functions, first start with the assembly, complete process, program, organization, or whatever the total project may be. Define the functions. Do not haggle over whether the function has been properly defined at this stage; it can be redefined later. Write every thought down so it will not be forgotten.

After it is believed that all functions of the assembly have been defined, take each part or segment of the system and define the function of each. There will be some duplication, but this will be screened out later.

After all functions have been defined, screen the list to eliminate duplicate functions and redefine functions for clarity of understanding. Now, screen the list again to define the basic function. The basic function is the function upon which all other system functions depend. If the basic function is not needed, none of the other functions will be needed.

In many cases, a number of functions beyond the system scope will be defined. These are called *high-order functions* and are those functions that cause the basic function to be performed. A detailed discussion of the scope and high- and low-order functions is beyond the intent of this orientation and are mentioned here only to note that it is not necessary to struggle over this step as the scope will become clear during the construction of a FAST diagram. The team may even reconsider the original scope and redefine the scope because of the new understanding of the overall project.

By application of the function definition principles cited here the end result will be clearly understandable, be measurable for use in cost-function analysis and function evaluation, and lead to outstanding opportunities in the creative phase.

7.2.3 Function Analysis and Evaluation

After the functions have been determined, identify the basic function or functions, as well as all the supporting functions. It is time to create a functional analysis system technique (FAST) diagram. The functional analysis system technique was developed by Charles Bethway in 1964, and first presented and published as a paper at the Society of American Value Engineers Conference in 1965. FAST contributed significantly to the most important activity in the value-engineering project, the function analysis and evaluation.

A FAST diagram is a logic chart that organizes the functions of a project and arranges them in a cause-and-effect relationship. Construction of a FAST diagram is necessary to ensure that the functions have been properly defined and that nothing has been overlooked. Although it follows a simple concept, the process of creating a FAST diagram is often difficult and frustrating, and it forces people to think out their project in a detailed and precise manner. Construction of a FAST diagram creates a focal point for the entire project, because eventually, all important information on the project is precisely defined and displayed in the FAST diagram. The FAST diagram is especially useful in dealing with projects where there might be widely different opinions, fuzzy understandings, and cloggy definitions among team members, all of which are very common in the analysis of organizations, operations, and the service industry. Construction of a FAST diagram tends to pull together the thinking process of a group to create a dynamic, enthusiastic team.

Determining the basic function is the first step in the construction of a FAST diagram. The basic function is the function that cannot be eliminated unless the product is eliminated. There may be more than one, but an effort should be made to determine the one most likely basic function. We will use Example 7.5 as a starting point to discuss the FAST diagram.

Example 7.5: Portion of FAST Diagram for Youth Assistance Program
This example is a continuation of Example 7.4. Of all the functions defined in Example 7.4, eliminate deviancy, identify needs, and assist clients are considered to be basic functions, because if any of these functions are not performed, the whole youth assistance program will not perform as intended. We can easily notice that the functions are not working in isolation; they are related to each other depending on the overall mission of the system. We can ask the question, Why do we need these three basic functions, that is, eliminate deviancy, identify needs, and assist clients? If we think really hard, we may get to the conclusion: Ah ha, because this is a youth assistance program, all we want to do is change the life of these troubled youth so they can become better kids. Then we may come to another function: modify behavior. That is, the three basic functions are needed because we want to modify behavior. Figure 7.2 illustrates this relationship.

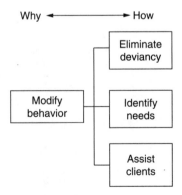

Figure 7.2 Relationships between Basic Functions and Higher-Order Functions

At the top of Fig. 7.2, a why-how arrow tells the relationships among these four functions. To modify behavior, you have to do all three functions, eliminate deviancy, identify needs, and assist clients. Why do we need to perform these three functions? Because we want to modify behavior.

In a similar manner, we can expand Fig. 7.2 by adding more functions, as illustrated by Fig. 7.3. Again, the function on the left gives the reason why the functions on the right should be performed, and the functions on the right tell how the function on the left can be accomplished. For example, the function to the right of "eliminate deviancy" is "plan activities"; the function to the right of "plan activities" is "determine needs." Why do we plan activities? Because we want to eliminate deviancy. Why do we determine needs? Because we want to plan activities. How do we eliminate deviancy? By planning activities! How do we plan activities? By determining needs!

Now we are ready to go over the details about establishing the FAST diagram for a value-engineering project. The general format of a FAST diagram is illustrated in Fig. 7.4.

We now define the terminologies used in the FAST diagram.

Scope of the Project The scope of the project is depicted as two vertical dotted lines. The scope lines bound the project under study or a portion of the problem with which the study team is concerned.

Highest-Order Function(s) The objective or output of the basic function(s) and subject under study is referred to as the highest-order function(s); it appears outside the left scope line and to the left of the basic function(s). Any function to the left of another on the primary path is a higher-order function.

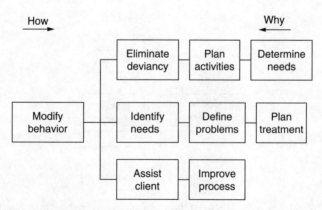

Figure 7.3 Part of FAST Diagram for Youth Assistance Program

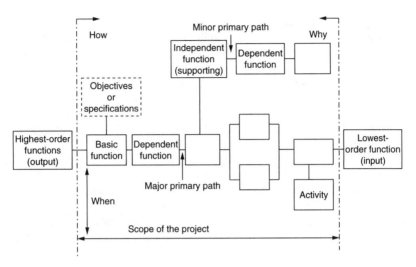

Figure 7.4 FAST Diagram Format

Lowest-Order Function(s) The functions to the right and outside of the right scope line represent the input side that "turn on" or initiate the subject under study and are known as lowest-order functions. Any function to the right of another function on the critical path is a lower-order function.

The terms *higher-* and *lower-order functions* should not be interpreted as relative importance, but rather the input and output side of the process. As an example, "receiving objectives" could be a lowest-order function, with "satisfying those objectives" being the highest-order function. How best to accomplish the satisfy objectives (highest-order function) is therefore the scope of the problem under study.

Basic Function(s) Those function(s) to the immediate right of the left scope line represent the purpose of the mission or the subject under study. By definition, basic functions cannot change. Secondary functions can be changed, combined, or eliminated.

Concept All functions to the right of the basic function(s) describe the approach to achieve the basic function(s). The concept represents either the existing conditions (as is) or proposed approach (should be). Which approach to use (current or proposed) is determined by the task team and the nature of the problem under study.

Objectives or Specifications Objectives or specifications are particular parameters or requirements that must be achieved to satisfy the highest-order function in its operating environment. Although objectives or specifications are not in themselves functions, they may influence the method selected to best achieve the basic function(s) and satisfy the user's requirements. Note: The use of objectives or specifications in the FAST process is optional.

Primary Path Functions Any function on the How or Why logic is a primary path function. If the function along the Why direction enters the basic function(s), it is a major primary path; otherwise it will be identified as an independent (supporting) function and be a minor critical path. Supporting functions are usually secondary. They exist to achieve the performance levels specified in the objectives or specifications of the basic functions or because a particular approach was chosen to implement the basic function(s). Independent functions (above the critical path) and activities (below the critical path) are the result of satisfying the When question.

Dependent Functions Starting with the first function to the right of the basic function, each successive function is dependent on the one to its immediate left (higher-order function) for its existence. That dependency becomes more evident when the How question and direction is followed.

Independent (or supporting) Function(s) Independent (or supporting) functions do not depend on another function or method selected to perform that function. Independent functions are located above the critical path function(s) and are considered secondary with respect to the scope, nature, and level of the problem, and its critical path.

Activity The method selected to perform a function (or a group of functions) is an activity.

7.2.4 Symbols and Graphs Used in FAST Diagram Construction

Why, How, and When

Figure 7.5 show the directions in a FAST diagram. The How and Why directions are always along the primary path, whether it is a major or minor primary path. The When direction indicates an independent or supporting function (up) or an activity (down). We have already discussed the How and Why directions in Example 7.5. The lower-order function on the How direction (immediate right) always explains how a particular function can be accomplished; the higher-order function on the Why direction (immediate left) always tells the reason why a particular function should be performed.

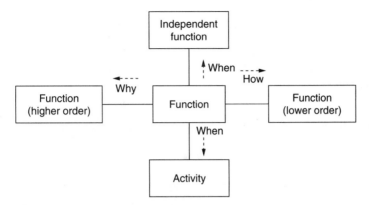

Figure 7.5 Directions in FAST Diagram

All the functions and/or activities along the When direction with a particular function will happen at the same time. We detect these functions or activities by asking the question, When a function occurs, what else happens? The independent functions and supporting functions are listed above the particular function; the activities will be listed under the particular function.

Common Symbols Along a Primary Path

In a primary path of a FAST diagram, it is possible that several functions have to be performed simultaneously as the precondition for lower-order function(s). Sometimes, these functions are related by logical AND, and sometimes they are related by logical OR. Figs. 7.6 to 7.9 illustrate such cases.

In both Figs. 7.6 and 7.7, the fork is read as "and." In Fig. 7.6, how do you build the swim club? By constructing the pool *and* constructing the club house. "Construct pool" and "construct club house" are equally important. In Fig. 7.7, how do you determine compliance deviations? By analyzing the design *and* reviewing proposals. However, "analyze design" is more important than "review proposals."

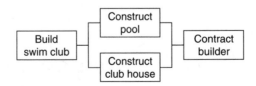

Figure 7.6 Two Equally Important Functions in AND Relation

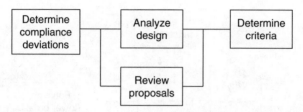

Figure 7.7 Two Unequally Important Functions in AND Relation

In both Figs. 7.8 and 7.9, the multiple exit lines represent an OR. In Fig. 7.8, how do you convert books (to delivery)? By extending bookings *or* forecasting orders, not both. "Extend bookings" and "forecast orders" are equally important. In Fig. 7.9, how do you identify discrepancies? By monitoring performance *or* evaluating the design. However, "evaluate design" is less important than monitor performance.

Symbols Along the When Direction

In the FAST diagram, the When direction is the vertical direction. When several functions are located along the same vertical line, it means that these functions will be performed at the same time. In addition, when these functions are connected by lines, it means that there is an AND relationship among these functions. Figure 7.10 illustrates such an example.

In Fig. 7.10, when you influence the customer, you inform the customer *and* apply skills. If it is necessary to rank the AND functions, those closest to the primary path should be the most important.

Now that we have discussed symbols and notation used in the FAST diagram, we are ready to go over the step-by-step procedure to complete the FAST diagram.

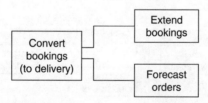

Figure 7.8 Two Equally Important Functions in OR Relation

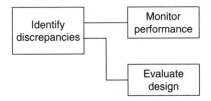

Figure 7.9 Two Unequally Important Functions in OR Relation

7.2.5 Step-by-Step Procedure to Establish FAST Diagram

Step 1: List all functions by using the function list illustrated by Table 7.6. Be sure to identify each function by a verb and noun. Identify basic functions and secondary functions.

Step 2: Prepare a 1" × 2" card for each function. Take a close look at all functions and try to identify the relationships among all functions. We can use the following logical questions for this purpose:

How is this function accomplished?
Why is this function performed?
When is this function performed?

Select the function that you think is the basic function, and apply the logic questions to the right and left of the basic function. To determine the function to the right ask, How is this function performed? To determine the function to the left ask, Why is this function performed? Repeat this process until the lowest-order function is included. The path of functions thus created is called a primary path. We may get multiple primary paths.

Step 3: When the primary path has been selected and positioned on the chart, position all secondary functions that did not fit into the primary path by applying the When question and add them above or below the primary path depending on whether they are supporting functions, independent functions, or actions. If the secondary functions are actually objectives or specifications, put them into the upper-left corner of the FAST diagram.

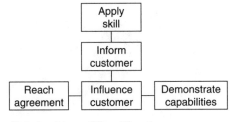

Figure 7.10 AND Relationship on When Direction

Example 7.6 illustrates this step-by-step process.

Example 7.6: Cigarette Lighter FAST Diagram

Figure 7.11 shows a typical cigarette lighter. Before constructing the FAST diagram, assume that we have compiled the function list illustrated in Table 7.10. First, we pick up the basic function "produce flame" and ask the Why and How questions as illustrated by Fig. 7.12. The basic function of a cigarette lighter is to produce a flame. By answering the question, Why produce a flame? we get the higher-order function "ignite cigarette." By answering the question, how do we produce a flame? we get the lower-order function "ignite fuel." We then can ask a further question, How do we ignite the fuel? By answering this question, we find that we need two lower-order functions to be performed, "produce spark" *and* "release fuel." These two functions are of equal importance; therefore, we add these two functions in the FAST diagram, as illustrated in Fig. 7.13.

Now we could continue to ask Why and How questions to find lower-order functions for the "release fuel" and "produce spark" functions, and continue this process. We would end up with the diagram illustrated by Fig. 7.14. The functions picked in the figure form the primary path of the FAST diagram of the cigarette lighter.

There are still many functions in the function list that cannot be fitted into the primary path. By asking the When question, we can fit the rest of the functions into the FAST diagram. The final FAST diagram is illustrated by Fig. 7.15.

7.2.6 Cost-Function Relationship

The completion of the FAST diagram makes it possible to complete the cost-function work sheet. The cost-function work sheet lists all functions versus all parts of a product or actions of a system, procedure, or administrative activity. The objective is to convert product cost to function cost.

Figure 7.11 A Cigarette Lighter

Table 7.10 Function List of Cigarette Lighter

Verb	Noun	Basic	Second	Remarks
1. Produce	Flame	√		
2. Protect	Flame		√	
3. Manage	Flame		√	
4. Ignite	Fuel		√	
5. Release	Fuel		√	
6. Produce	Spark		√	
7. Control	Flow		√	
8. Restrict	Exit		√	
9. Energize	Particles		√	
10. Strike	Flint		√	
11. Generate	Heat		√	
12. Contain	Fuel		√	
13. Open	Valve		√	
14. Depress	Lever		√	
15. Enclose	Fuel		√	
16. Rub	Material		√	
17. Rotate	Wheel		√	
18. Apply	Force		√	
19. Activate	Thumb		√	
20. Accommodate	Hand		√	
21. Stimulate	Muscle		√	

The cost of each piece of hardware or service activity is redistributed to the function performed. This proportional redistribution of cost to function requires information, experience, and judgment, and all team members must contribute their expertise.

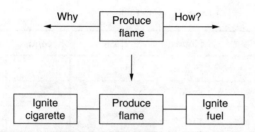

Figure 7.12 Start-up of FAST Diagram Construction

After the cost of each part or action has been redistributed to the functions performed, the cost columns are totaled to obtain the function cost. This cost is then placed on the FAST diagram. The FAST diagram then becomes a very valuable tool. It tells what is happening, why, how, when, and what it costs to perform the function. It is now possible to evaluate the functions to determine if they are worth what is being paid for them. In other words, a value must be set on each function.

Determining the value of each function is a subjective process. However, it is a key element in the value process. Comparing the function cost to function value provides an immediate indication of the benefit being obtained for expended funds. The ratio of value cost to function cost is the *performance index*. The sum of all values is the *value of the system* or the lowest cost to reliably provide the basic function. It should be compared to the preliminary goal set earlier.

It may be that the new goal is considerably higher than the original. If this is the case, an evaluation of the diagram will indicate what must be done to achieve the original goal. It may indicate an entirely new concept is required, or it may be that it will be acceptable to settle for less. It is often the case that the original goal and the new value are close. An analysis of the function costs will again indicate necessary action.

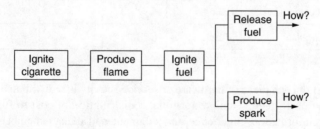

Figure 7.13 Partial FAST Diagram for Cigarette Lighter

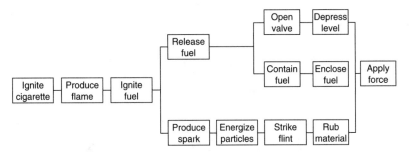

Figure 7.14 Primary Path of FAST Diagram for Cigarette Lighter

This analysis clearly defines the task for product improvement. It breaks the problem down into functions that must be improved, revised, or eliminated to achieve the goal. The FAST diagram clearly identifies functions and their relationship to each other. Cost visibility analysis can identify high-cost areas. We now are ready to identify the relationship between cost and function. Specifically, we are ready to identify the cost for each function. Also, after clearly defining each function, we are able to identify the best cost for each function. The difference between current cost and the best cost is the profit improvement target. This provides us with an estimate of profit improvement potential. Table 7.11 provides an example of cost-function work sheet based on Example 7.3.

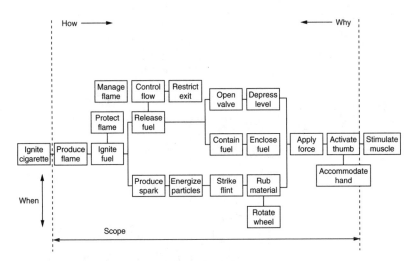

Figure 7.15 FAST Diagram for Cigarette Lighter

Table 7.11 Cost-Function Work Sheet for Pencil

Pencil Components	Cost (Cents)	Functions														
		Remove Marks		Secure Eraser		Improve Appearance		Make Marks		Transmit Force		Accommodate Grip		Display Information		Support Lead		Protect Wood	
		%	Cost	%	Cost	%	Cost	%	Cost	%	Cost	%	Cost	%	Cost	%	Cost	%	Cost
Eraser	.43	100	0.43																
Metal band	.25			50	0.13	25	.06			25	.06								
Lead	1.2							100	1.2										
Body	.94									50	.47			10	.09	40	.38		
Paint	0.10					50	.05											50	.05
Total cost	2.92	16	.43	5	.13	4	.11	40	1.2	17	.53			3	.09	13	.38	2	.05
Best cost			.34		.10		.10		.8		.30				.09		.28		.04
Profit improvement potential			.09		.03		.01		.4		0.03				.0		0.1		0.01

We now need to determine the cost of each function by distributing the cost of each part to its related function. For example, the cost for the pencil body is 0.94 cent, 50 percent of the pencil body cost is used to perform the function "transmit force," 40 percent of its cost is used to perform the function "support lead," 10 percent of its cost is used to perform the function "display information." This breakdown of cost is based on qualified judgment from the whole team. It is subjective; hopefully it is not too biased because it is based on the consensus of the team. By adding all the cost portions from all relevant parts for a function, we can get the cost for performing that function. For example, in Table 7.11, the cost of the "transmit force" function consists of 25 percent of the metal band cost, which is 0.06 cent, and 50 percent of the pencil body cost, which is 0.47 cent; therefore the cost of the "transmit force" function is 0.53 cent.

We also need to determine the best cost for each function. By definition, the *best cost* is the lowest cost to adequately and reliably provide the function. The best way to determine the best cost of a function is by comparison to another function that we know there is a 'best deal.' For example, if a function is "tell time," then we need to know what is the time precision requirement. The required precision might be ± 30 seconds after a month of use. Next, we will find a watch that has just enough to provide time with this precision reliably. This watch should not provide any other functions, such as decoration or brand-name recognition. In this way, a cheap, no-brand, plain, 99-cent electronic watch might be adequate. Then the best cost for the "tell time" function is 99 cents. To make sure we determine the best value, we can ask the following questions:

1. Can we do without it? (If yes, the best cost is zero.)
2. Does it need all its features? (If no, get rid of all unnecessary features and then figure out the best cost.)
3. Is anyone buying it for less?
4. Is there something better that can do the job?
5. Can it be made by a less costly method?
6. Can a standard item be used?
7. Can another dependable supplier provide it for less?
8. Would you pay the price if you were spending your own money?

The best cost is not always lower than the current cost. As stated before, by definition, the best cost is the lowest cost to adequately and reliably provide this function. It is possible that in the current system this function is not adequately and reliably provided. In this case, we may have to increase the cost for this function. This is also why question 4 (Is there something better that can do the job?) is asked.

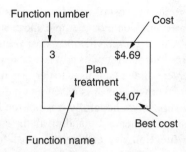

Figure 7.16 A Fully Marked Function Block in a FAST Diagram

The cost and the best cost for functions are also often marked in a FAST diagram. A fully marked function block in a FAST diagram has the format indicated by Fig. 7.16. Figure 7.17 gives a portion of the FAST diagram for the youth assistance program with fully marked function blocks.

After the FAST diagram is fully developed and the cost-function work sheet is fully filled, we will get into the next stage of the value-engineering job plan, the creation stage.

7.3 Creative Phase

At the end of the information phase, we have listed all relevant functions for the project, filled out the cost-function work sheet, and developed the FAST diagram. The difference between the cost and the best cost of each function is the profit improvement potential. The functions that have high profit improvement potentials are the perfect candidates for cost saving. The creative phase of the value-engineering project is to use team members' creativity to develop alternative solutions to perform the functions that have high profit improvement potentials. The creative phase is where free

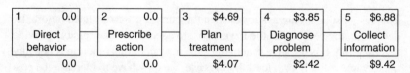

Figure 7.17 A Portion of the FAST Diagram with Fully Marked Function Blocks for the Youth Assistance Program

development of ideas is fostered. These ideas will form the basis for concepts that will lead to recommendations for improvement.

Brainstorming is extremely helpful in the creative phase of the value-engineering project. It helps to loosen up the mental barriers to creativity for the group, in order to create a great volume of ideas. In the beginning of brainstorming, the quantity of ideas is important. Though many obvious wrong ideas will be generated, the more ideas generated, the better the chance that some really brilliant but nonobvious idea will be among them. This large number of ideas will be screened and evaluated in the next stage.

In the brainstorming process, an atmosphere is generated that permits each person to freely depart from his or her mental barriers. It is in this process that you may hear yourself differently and feel uncomfortable; this is a necessary stretch that pushes you to think more freely and more creatively. The brainstorming session should follow a specific set of ground rules. These are necessary to ensure the proper environment for idea development.

The first step in brainstorming is to select questions for discussion. These questions are often selected based on the functions that have high profit improvement potentials. The question is often in the form, What else can perform this function? The questions are then presented to the group, and the group will toss out ideas regarding the question. Any idea is acceptable, and no discussion is allowed. For each question, it is desirable that at least one new idea should be generated that has never been thought of before.

The following ground rules for brainstorming must be followed to ensure success:

1. No criticism allowed during the session.
2. A peer group is desired. Never have members of high-level management or their assistants attend.
3. Quantity is desired. The more ideas there are, the more likelihood of at least one outstanding item.
4. Six to ten participants is best.
5. No publicity on the session after its completion.
6. Combine ideas.
7. Wild ideas wanted. Usually the first 90 percent of all ideas will be those that have come up before.
8. Record all ideas; on paper is best.

Table 7.12 gives a template for idea generation. The function under discussion is "enhance appearance" for a decoration.

Table 7.12 Idea Generation Form

	What Else Will Do the Job?
	Function: Enhance Appearance
1.	Use laminate instead of paint
2.	Use stainless-steel parts
3.	Use plastic material
4.	Use curves instead of sharp edges
5.	Delete complicated features
6.	Paint parts individually before assembly
7.	Use chrome plating
8.	Use multicolor paint
9.	Use gold material

7.4 Evaluation Phase

Evaluating the ideas developed during the creative phase is a critical step in the value-engineering job plan. The ideas generated will include practical suggestions as well as wild ideas. Each and every idea must be evaluated without prejudice to determine if it can be used or what characteristics of the idea may be useful.

Proper evaluation of the ideas is a critical step. Remember, if an idea is discarded without thorough evaluation, the key to a successful solution may be lost. The time to create ideas is in the creative phase. If an idea is discarded, there may not be another opportunity to develop it again.

During the screening process, it must be kept in mind that the objective is not to discard ideas but to look for the good in them. All too frequently, a new idea will create a negative reaction, for example, "That's a great idea but let me tell you what is wrong with it." We should say, "That's a great idea. What can we do to make it work?" There never seems to be any problem thinking of reasons why something will not work. However, developing ways to make an idea work takes ingenuity. How can we make it work or what is there about this idea we can use should be the state of mind during the screening process.

Evaluation processes can range from the simple to the complex. The method selected depends to some degree on the quantity and quality of ideas generated. The number of ideas can run from less than a hundred to over a thousand depending on the scope of the project. The first screening of the list should be to eliminate the ones that obviously are of no use to the project. However, each idea must be reviewed with a positive attitude. Look for the good rather than the bad and do not be too critical.

The following process is suggested for initial screening.

Step 1: Each item on the idea generation form will be read. Each team member will vote whether to keep the idea for future evaluation or drop it. This is an impulse decision. Each person will decide by his or her initial reaction as to whether to keep the idea or drop it. However, if one person on the team wants to keep the idea, it must be kept on the list without question. During this initial screening, there should be no discussion of the idea; only a yes or no vote is acceptable. The result will be elimination of the obviously impractical ideas for this project.

Step 2: Then each of the remaining items on the list will be read, and the group will discuss each idea. Table 7.13 can be used as a template for this step. The intent is to determine what there is about each idea that may be useful and decide whether to keep it on the list or to drop it. It will be found that at this stage many ideas will combine with

Table 7.13 Idea Screening Work Sheet 1

Ideas	Implementation Cost	Development Cost	Total Cost
1.			
2.			
3.			
4.			
5.			
6.			
7.			
8.			
9.			

other ideas to form basic groups or categories such as materials, methods, and organization. The discussion may also result in new ideas that can be added to the list.

Step 3: After the initial screening process has been completed, it will be necessary to resort to systems designed to aid in identifying the best choice and an alternative, or to rank and weigh alternatives. It is always important to have a second choice to fall back on, just in case the first choice cannot be implemented for reasons that may not become apparent until detailed development is under way. When the initial list of ideas has been reduced to a choice of only a few alternatives, the simple system illustrated in Table 7.14 may be used. This sheet identifies the advantages and disadvantages of each alternative concept. In most cases, an idea with more advantages than disadvantages listed will be the first choice. However, there may be an overpowering disadvantage that creates a serious roadblock. Can it be eliminated? If it can, the choice may be clear. If it cannot, the second alternative may be the best choice.

Table 7.14 Idea Screening Work Sheet 2

Idea 1		Idea 2	
Advantage	Disadvantage	Advantage	Disadvantage
Idea 3		Idea 4	
Advantage	Disadvantage	Advantage	Disadvantage

There may also be situations where the choice of alternatives will require more complex systems to aid in the evaluation process. Two systems favored because of their convenience, simplicity, and effectiveness are Pareto voting and paired comparisons. They may be used separately, or in sequence depending on the situation. Each of these systems is described in detail in the following text. They have been found applicable in a large number of cases and are extremely useful.

There are also cases involving high risk or a substantial amount of money where even more detailed analysis is required. These may be situations where risk is critical and alternatives and tradeoffs are necessary. In these cases, a matrix analysis may be necessary.

Experience has shown that this evaluation process is a difficult task. The impulse to quickly screen through the list to zero in on the best ideas must be controlled. The mass of data must be handled systematically to obtain maximum benefit from the creative phase. Careful screening is essential to isolating the best concept to carry over into the planning phase where the idea will be developed into a practical recommendation for action.

7.4.1 Selection and Screening Techniques

A difficult problem that frequently confronts decision makers is the need to organize a large amount of data, so that one or several of the most important items may be identified. It may be required to determine which of several alternatives appears to be the best, or it may be necessary to select a number of items so that they may be ranked and weighted by order of importance or some other criteria.

Experience has shown that most people are not able to handle this task quickly and effectively. For this reason, it was decided to develop a simple method that would be applicable in most cases. More complex situations may require more sophisticated methods. However, experience has shown that a combination of two simple methods, Pareto voting and paired comparisons, will satisfy a majority of requirements.

A literature search by the author identified 13 methods for evaluating data to aid in decision making. The methods are listed in Appendix 2 along with references, so that further in-depth study by the reader can be made.

Pareto Voting

Pareto voting is based on Pareto's law of maldistribution. Alfredo Pareto (1846–1923), a political economist, observed a common tendency of wealth and power to be unequally distributed. This observation has been refined to the degree that it can be said that there is an 80/20 percent relationship between similar elements.

For example, 20 percent of the parts in an assembly contain 80 percent of the cost. This is very useful information in cost estimating; however, the relationship holds for many diverse examples such as the following:

> Twenty percent of the states use 80 percent of the fuel oil.
> Twenty percent of the activities create 80 percent of the budgeted expense.
> Twenty percent of the items sold generate 80 percent of the profit.

In value engineering it is frequently necessary to select the best ideas, the highest value functions, the highest potential projects, or any of a number of other requirements. It has been found that the application of Pareto voting can help to simplify the list and will, in most cases, ensure that the most important items have been selected. It also produces results quickly and can be incorporated into the value-engineering process to allow continuous operations without undue disruptions.

Pareto voting is conducted by requesting each team member to select what they believe are the items or elements that have the greatest effect on the system. This list of items is limited to 20 percent of the total number of items. For example, each team member would be allowed to select 6 items out of a list of 30. The vote is taken on an individual basis to obtain as much objectivity as possible.

The resultant lists are then compared and arranged into a new consolidated list, in descending order, by the number of votes each item received. Usually, several items will have been selected by two or more team members. The top 10 to 15 items are then ranked and weighted in a second step by using paired comparisons.

Example 7.7: Pareto Voting

This example refers to the idea generation form used in Table 7.10. A team of six people will conduct a Pareto voting on the nine ideas; each member can only vote for two ideas, so a total of 12 votes will be received. The number of votes for each idea will be tallied; the result is summarized in Table 7.15.

Paired Comparisons

Paired comparisons, or numerical evaluation as it is sometimes called, compares a list of items to rank and weights them in order of importance or

Table 7.15 Pareto Voting

Rank	What Else Will Do the Job? Function: Enhance Appearance	Votes Received
1.	Use laminate instead of paint	5
2.	Paint parts individually before assembly	4
3.	Use curves instead of sharp edges	2
4.	Use plastic material	1
5.	Use stainless-steel parts	0
6.	Delete complicated features	0
7.	Use chrome plating	0
8.	Use multicolor paint	0
9.	Use gold material	0

some other criteria. *Ranking* is the assignment of a preferred order of importance to a list of items. *Weighting* is the determination of the relative degree of difference between items.

In paired comparisons each item is compared to every other item on the list in turn, using a simple matrix. It is most convenient for up to 15 items. A comparative decision is made between any two items on a two- or three-level basis. In a two-level comparison, 2 = major difference and 1 = minor difference. In a three-level comparison, 3 = significant difference, 2 = moderate difference, and 1 = minimal difference.

Example 7.8 shows how paired comparison works.

Example 7.8: Paired Comparison for Pencil Improvement
This example refers to the case presented in Example 7.3. After some discussion by the team about how to improve the pencil, several ideas about cost reduction for the pencil are proposed (see Table 7.16).

The next step will be to evaluate idea A with respect to B, idea A versus C, and so on, for all possible pairs. Is A or B a better idea based on cost, benefit, customer satisfaction, etc.? Table 7.17 summarizes the comparisons for all possible pairs. When comparing A and B, a B-2 result indicates the team thinks that idea B is moderately better than A. Similarly, when comparing A and C, an A-1 result indicates that the team thinks idea A is minimally better than C.

Table 7.16 Pencil Improvement Ideas

Key Letter	Idea
A	Eliminate paint
B	Reduce the length of lead
C	Remove eraser
D	Stain wood in lieu of paint
E	Make body out of paper

After the team compare all pairs, all the boxes in Table 7.15 will be filled. The values for each idea are then added up, $A = 1$, $B = 2 + 3 + 2 + 1 = 8$, etc. Table 7.18 summarizes the values for each idea in this paired-comparison study. Clearly, ideas B and E are the top choices.

The whole evaluation stage may go through several screening steps. Table 7.19 is a convenient template to use to record the whole evaluation stage.

7.5 Planning Phase

After the evaluation phase, we have a final list of ideas at hand that are ready to be recommended to the management for implementation. Now is the time to develop the best ideas in detail so recommendations can be made

Table 7.17 Paired Comparison of Pencil Ideas

	B	C	D	E
A	B-2	A-1	D-2	E-2
	B	B-3	B-2	B-1
		C	D-1	E-3
			D	E-2

3 Significant
2 Moderate
1 Minimal

Table 7.18 Final Evaluation Results for Paired Comparison

Key Letter	Idea	Value
A	Eliminate paint	1
B	Reduce the length of lead	8
C	Remove eraser	0
D	Stain wood in lieu of paint	3
E	Make body out of paper	7

convincingly. At this stage, we need to determine costs more accurately and discuss proposed solutions with relevant people. We need to get the latest material, labor, process, and cost data. We shall develop a cooperative atmosphere with everyone able to contribute to a successful problem solution, refine the cost of each solution, and determine the best and alternate recommendations for the performance of basic functions.

Table 7.19 Idea Screening Result

	What Else Will Do the Job?	First Screening	Second Screening	Final Screening
	Function: Enhance Appearance			
1.	Use laminate instead of paint	√		
2.	Use stainless-steel parts	√		
3.	Use plastic material	√		
4.	Use curves instead of sharp edges	√	√	
5.	Delete complicated features	√		
6.	Paint parts individually before assembly	√	√	√
7.	Use chrome plating			
8.	Use multicolor paint			
9.	Use gold material			

Table 7.20 Identify Roadblocks

Best Idea: Reduce the Length of Lead		
Roadblock	Where/Why	Action Required
Differ from traditional design practice	Design/out of specification Marketing/bad customer image	1. Show that people seldom use the full pencil length 2. Show that good style and low price is more important to customers
Alternative Idea: Use Body out of Paper		
Roadblock	Where/Why	Action Required
Effect on strength and durability unknown	Design/no previous experience with this design	Show strength/durability test results
Perceived as a risk idea	Marketing/no idea if customer will buy in	Show this new design can make pencil body self-peeling to expose lead, no need for pencil sharpeners

For a successful project completion, we need to determine potential roadblocks, where they may come up, and how they may be eliminated. Table 7.20 can be used as a template.

In the planning stage, it is also very important to discuss how this project can be sold and implemented. Table 7.21 is a planning form that lists the names of all persons who will be involved in accepting and implementing the proposal. We need to figure out possible problem areas and decide how they can be eliminated.

7.6 Reporting Phase

The object of the study is to develop a successful recommendation for improvement in products, systems, organizations, etc., and therefore, in turn, profits. For the study to be worth anything to you and your company, it must be presented so it will be accepted and implemented. Your best recommendation is prepared for presentation to the responsible organization. Before and after costs and potential savings must be shown and clearly

Table 7.21 Action Plan for Selling Value Engineering (VE) Ideas

Department	Supervisors	Action Required	Problem	How to Solve Problems
1.				
2.				
3.				
4.				
5.				
6.				
7.				

defined. Sketches should show the basic changes in whatever detail is necessary to prove results. It may be necessary to provide simple models in some cases. You should list all advantages and disadvantages and show how the disadvantages were considered in your decision. If the procedure has been followed, all necessary data should be available in your notes and records.

It is important to realize that it is necessary to present the recommendation in a manner that will clearly demonstrate the advantages of its acceptance from the standpoint of the organization required to implement it. The importance of the reporting phase should not be overlooked. If the recommendations are not presented properly and effectively, the result may be the loss of a good idea or an excellent recommendation.

The work sheets provided in this chapter have been developed to provide all the information necessary to prepare an effective recommendation. They are complete and concise. The next step is to arrange the material so that it will sell your idea. One of the most important considerations here is to provide complete information. Failure to provide complete information has been proven to be a major cause for rejection of a proposal. Persons who are required to review or approve proposals of one type or another will verify that it is rare when complete information is provided. In fact, the government made an analysis of 90 rejected contractor proposals submitted to them for approval and found that 40 percent of the rejections were the result of incomplete or inaccurate technical or cost information.

Some of the factors covered in the preparation of the final recommendation and report are listed here:

1. Plan the proposal to cover all the facts. Do not skip an important consideration on the basis that it can be considered later. Do not plan surprises.
2. Justify the recommendation on both technical and economic grounds. Show the risk involved, as well as the rewards, and the cost to verify the idea, as well as total lifetime program costs such as design and developmental expenses, capital investments necessary for buildings, and tools.
3. Indicate the effect on corporate profit, competitive position, or other important factors.
4. Discuss the proposal with people who will be affected by the idea.

7.7 Implementation Phase

7.7.1 Introduction

The objective of a value-engineering study is the successful incorporation of recommendations into the product or operations. However, a successful project often starts back at the beginning. Each project must be thoroughly analyzed to determine its potential for benefit and the probability of implementation. This is as important as the knowledge and skill required to apply the system to attain successful results.

An excellent idea is worthless unless it can be properly implemented. If it is not implemented, no one will obtain the benefit. It must also be implemented in the manner intended. Unfortunately, there have been many cases on record where the idea could not be implemented because of the high cost to make the change. There are other cases where the recommendations were not properly understood and implementation resulted in increased cost. This often results in disillusionment or the feeling that value engineering does not work for our problems. Actually in most cases, the real problem was that the problem was not properly diagnosed. It was not that value engineering does not work; it was inefficient preliminary analysis and preparation.

It does not seem reasonable to expend the effort and funds required to make a value study without first having done the necessary work to assure that the project is practical, that it can be implemented, and that the necessary funds and work force will be available.

Selection of projects is a part of the entire value-engineering implementation process. Many times management will assume that any project will prove profitable. This is not always the case. The project must be practical in relationship to its effect on the organization.

To aid in the selection of projects, development of people, implementation of projects, and all the other aspects necessary to successfully achieve the stated objective, we have prepared some guidelines. They are guidelines, not rules, since every organization is different and successful value-engineering operations must be integrated into operations to become part of the day-to-day decision-making process of the company.

To begin with, we will look at the overall organization and implementation of value-engineering operations. Then we will look at some of the details that make for success.

7.7.2 Goal for Achievement

What do we want to get from value engineering? What will be the objective? This is the first question to answer.

Value engineering can increase productivity, reduce product cost, improve quality, reduce administrative costs, and provide a number of other benefits that may be critical to operations. Whatever the goal, it should be defined in specific terms: increase productivity by a specific percent, reduce product cost by a specific number of dollars per unit, etc. Whatever the initial goal may be, it can be revised and broadened as skill in application and implementation of the process develops, and understanding and credibility increase.

Value engineering is a people-oriented program, designed to help people do a better job by aiding them in breaking down constraints to understanding. It provides some very specific methods and systems to achieve results.

Since people perform a wide range of jobs in an organization, it is certainly logical to expect that if they can be provided with a system that can help them do a better job, anything that they are expected to do can be improved. In the end it is people who do the thinking. If they can improve their performance, everyone will benefit. This has been the experience of value engineered professionals. Many people, highly skilled in their jobs, have developed new insights that have created breakthroughs in technology as well as major organizational and operational improvements.

The goal for achievement should be known to everyone. It can be product-oriented, or directed toward manufacturing or administrative operations. It need not be companywide. However, the scope can be broadened at any time. Once the goal has been determined, the means to achieve the objective can be developed.

7.7.3 Develop a Plan

There are five steps to incorporating value engineering into operations:

1. Evaluate the system.
2. Define an objective.
3. Develop a plan and organization to achieve the objective.
4. Understand the principles.
5. Implement the plan.

Each step can be approached in a number of different ways. However, there are certain specific problems to be considered and pitfalls to be avoided in each. Understanding the problems and pitfalls rather than outlining a specific method or procedure should provide the necessary guidelines for an effective operation. In many cases, a consultant can aid in the initial stages and support each step of the process, by providing the broad range of his or her experience for the client to build upon. However, it is important that the consultant have the type and quality of experience needed to ensure success.

The coordinator should be required to select a consultant, develop an educational plan, aid in organizing and conducting workshops, and identify people who may be developed into value specialists. The extent of these programs will depend upon the size and scope of the company.

From what we have noted here, it is obvious that the problem is complex from the standpoint of options. However, successful operations do not have to be extensive. Starting small and developing successfully is preferred to a lot of noise and a big crash because of poor planning.

Attitude

One of the most important factors in value engineering is attitude; attitude of management and people on task teams. A positive, cooperative, supportive attitude is required. In many cases value engineering actually requires a new management style. It cuts across organizational lines, looks at taboo aspects of a problem, and recommends drastic changes compared to the past. To accept these disruptions to the old way of doing business requires faith, understanding, and a positive attitude.

In most cases whenever a new idea is presented to an American group, the initial reaction is negative. The first remarks are, "It's interesting but let me tell you what's wrong with it." The best approach to this reaction is to listen carefully. They may have some ideas you overlooked. After all negative reaction has run out, be prepared to ask some specific positive questions of the group that will develop positive responses. For example, "I understand your difficulty in producing this in the plant. What do you think we would have to do to make this practical? Do you see any changes we might make to satisfy our methods?" This will usually work to achieve a positive result.

Never argue. In many cases it is beneficial to solicit negative ideas, but be prepared to develop positive questions. Our attitude is that we must begin to ask, What's good about this idea? How will it help us to do a better job?

Changing people's attitudes is difficult and may never happen, but understanding the reasons behind the negative reaction should make it possible to persuade most people that they can benefit from success. Remember, there is a risk of failure in new ideas. New ideas require change, and they may not work. People want proof. It has to work before they will support it. However, maybe you can show them that the benefits are greater than the risks. The best way to change people's attitudes is to show that top management is interested in value engineering and expects participation and results in achieving the stated goals.

Value Council

The value council is a small group of high-level executives who oversee operations. In a small company, it might be chaired by the president, or in a large company, by a division manager.

The council should be staffed with people who have the authority to make decisions relative to acceptance and/or rejection of proposals and authorization of funds and work force changes. They set the attitude, develop the environment, break bottlenecks, and by their interest and visibility create credibility to participation and provide authority to operations.

It is important that members of the council make every effort to attend council meetings except in cases of dire emergency. When a member is unable to attend, he or she should authorize a key assistant to attend. If council members' attendance degenerates, the message sent is that they are losing interest.

7.8 Value-Engineering Case Studies

7.8.1 Automobile Dealership Construction

This case study is from Park (1999). A large real-estate company built and maintained many automobile dealership facilities all over the country. One major problem faced by this company was the long duration required from dealership project authorization to dealer occupancy. History showed that this long duration in construction cycle would cause a tremendous dollar amount in lost sales, so the company wanted to use value-engineering techniques to shorten this cycle.

At the start of this project, it was found that the average duration from site selection and land purchasing to construction and leasing was 502 days, or about $1\frac{1}{2}$ years. A review of the project process flowchart identified the activities that were eating up these 502 days. They included selecting and obtaining the options on the land, topographical surveys, soil borings, facility layouts, bid estimates and analysis, budget reviews, design, construction, and many others.

In this value-engineering project, a FAST diagram was developed, as illustrated in Fig. 7.18. The twenty or so functions illustrated consumed these 502 days. Out of these 30 functions, three functions, "resolve restrictions," "obtain data," and "construct facility," took 85 percent of the time. This evaluation was obtained by using time, instead of cost, as a measure in the FAST diagram.

As a result of project recommendations, the project process procedure was revised to make it possible to conduct several of the long-term activities in parallel with other activities. For example, approval for early site work was obtained from property owners before ownership was transferred so that topographic surveys and soil boring could be made as soon as possible. Standard designs were developed for several parts of the facility to reduce overall design and development time, and a single source contracting procedure was developed to reduce contractor project interface.

The result of these recommendations was a potential average saving of 262 days, or a 47 percent average saving in time per project. Based on the average annual construction program the yearly benefit in increased rent would be over $1,250,000 per year. The additional increased vehicle sales were not included in the benefit.

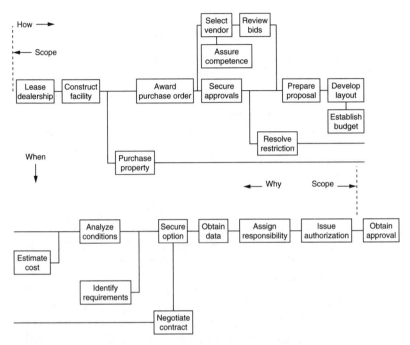

Figure 7.18 FAST Diagram for Automobile Dealership Construction

7.8.2 Engineering Department Organization Analysis

This case study is from Park (1999). A leading automobile company spent about $200 million dollars per year and employed 4000 people, including engineers, designers, technicians, technical specialists, and financial analysts. An economical downturn forced the company to cut costs. A painful lesson had been learned from across-the-board budget cuts, for which an equal share of the budget was cut across all departments. The result of this kind of budget cut was that some vital operations were seriously damaged; others simply slid by. This time, the situation was critical, the budget had already been cut several times, and no one knew where to look next. A value-engineering project was initiated in order to identify hidden, unnecessary costs.

In this project, after 72 hours of total effort by a team of six people, a FAST diagram was developed that had 72 functions. The chart was then thoroughly discussed to ensure that it covered all aspects of the operation, and a glossary of the functions was made to ensure future understanding.

Table 7.22 gives a partial list of functions for this engineering operation. Here is a sample term from the glossary of functions:

Create Design To generate a new system, assembly, or component, measured by time, which include time to come up with design ideas, design and layout time, engineers' working time, programming time and etc.

The FAST diagram provided some interesting information. Most importantly, it showed that many functions were performed to satisfy functions outside the scope of engineering responsibilities. Many of these functions contributed to higher-order functions to support other company operations, such as the purchasing and legal departments.

The next step was to determine how much each function cost and how funds were distributed among all the functions. To do that, departmental managers were asked to distribute their departmental cost by function. Cost-function work sheets were filled out. One portion of a cost-function work sheet is illustrated in Table 7.23.

A partial FAST diagram is illustrated in Fig. 7.19. As we can see in the FAST diagram, 43 percent of the available funds went to confirming the design and only 14 percent went to creating the design. This is considered a poor distribution of funds. Team members thought that this lopsided fund distribution was a major source of the problem and across-the-board budget cuts would likely create a big problem in new product design.

It was recognized that confirming the design was a required function. However, changing the way that this function was performed would offer opportunities to make a major improvement in productivity and could improve the overall engineering operation and obtain more efficient use of

Table 7.22 Partial List of Functions in Engineering Operation

Verb	Noun	Verb	Noun
Create	Design	Prepare	Plan
Transmit	Information	Negotiate	Alternatives
Evaluate	Information	Evaluate	Capabilities
Confirm	Design	Allocate	Resources
Model	Concept	Appropriate	Fund

Table 7.23 Cost-Function Work Sheet

Cost (Hours)			Functions						
Item No.	Activity	Hours	Trans. Info.	Create Design	Auth. Prog.	Conf. Design	Evaluate Info.	Collect Data	Make Model
1.	Manager	1,300	100	60	40	150	40		
2.	Secretary	1,736	1,000			60	60		
3.	Design supervisor	1,438	40	40		60	40	40	
4.	Engineering supervisor	1,344	20	100	40	80	80	40	
5.	Development supervisor	2,270				160	280		
6.	Technical specialist	1,078		200	40				
7.	Sr. design engineers	2,790	40	200	40	160	200	120	
8.	Sr. development engineers	2,790				320	1,290	560	120

(*Continued*)

Table 7.23 Cost-Function Work Sheet (*Continued*)

Cost (Hours)

Item No.	Activity	Hours	Functions							
			Trans. Info.	Create Design	Auth. Prog.	Conf. Design	Evaluate Info.	Collect Data	Make Model
9	Design engr	11,109	280	3,360	140	420	560	560	
10	Development engr	22,560				1,200	4,200	3,000	750
11	Design leader	4,909	190	480		180	180		
12	Technician	22,204		8,320		1040	780		
13	Modeler	17,580		4,800					
14	Clerk	3,392	880			800	400		
15	Mechanics	26,528						4,800	8,600
16	Material	$198,000							
Total	Material								
Total	Hours								

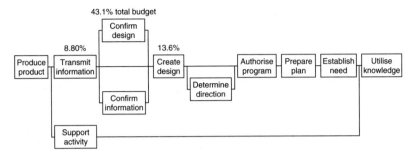

Figure 7.19 FAST Diagram for Engineering Operation

engineering funds. The immediate recommendation was to review all areas involved in the "confirm design" function. By following the value-engineering job plan, major changes were made in several areas that substantially increased output, cost saving, and avoidance of major capital investment. A substantial cost reduction was achieved without affecting the vital design functions.

Chapter 8

Brand Development and Brand Strategy

8.1 Introduction

Famous brand names make a big difference in the marketplace. Figure 8.1 shows that several T-shirts are made of exactly the same fabric, the same style, and the same quality, but because they have different brand names, the retailing prices of these T-shirts are vastly different.

A good brand name brings extra value to the product and the company that makes the product. McDonald's, Coca-Cola, Disney, Kodak, and Sony are among the most globally recognized names in the world (Kochan et al., 1997). The name recognition of these brands brings tremendous marketplace successes and high profitability. In modern history, brand development is one of the key sources of competitive advantage for companies worldwide. Brands are regarded among the most valuable assets owned by a company (Batra 1993, Davis 2000). Some brands are valued so highly that companies have paid huge amounts of money to acquire the rights to them. For example, in 1988 Philip Morris bought Kraft, the maker of cheese products, for $12.9 billion, a sum that was four times the value of the assets of the company (Murphy 1989). Sometimes companies that have good brand names can defend their market positions for a long period of time (Arnold 1992), as illustrated by Table 8.1.

What is a brand? Why do brand names have such magical power? What is the importance of brand names in developing a service product? These are some of the questions that we try to answer in this chapter.

The Merriam-Webster Dictionary, a brand is defined as "a mark made by burning with a hot iron to attest manufacture or quality or to designate ownership," or "a characteristic or distinctive kind." Peter Kotler (1984) defines

Figure 8.1 The Power of Brand Names

a brand as "a name, term, symbol, or design, or a combination of them, which is intended to signify the goods or services of one seller or group of sellers and to differentiate them from those of their competitors." Scott Davis (2000) defines a brand as "an intangible but critical component of what a company stands for," and "a brand is a set of promises, it implies trust, consistency, and a defined set of expectations. The strongest brands in the world own a place in the consumer's mind, and when they are mentioned almost everyone thinks of the same things." Mercedes Benz stands for prestige and the ultimate driving experience; Ralph Lauren stands for classic looks, high status, and pride.

The strongest brands usually stand for superior functions, benefits, and quality; without these, you cannot be the strongest brand in the world. However, superior functions, benefits, and quality alone will not make the strongest brands. Does

Table 8.1 Leading U.S. Brands from 1933 to 1990

Brand	Market
Eastman Kodak	Cameras/film
Del Monte	Canned Fruit
Wrigley	Chewing gum
Nabisco	Baked goods
Gillette	Razors
Coca-Cola	Soft drinks
Campbells	Soup
Ivory	Soap
Goodyear	Tires

Source: Arnold (1992).

McDonald's offer much better food than Burger King? Is Starbuck's much better than Caribou Coffee in terms of the coffee it offers? Probably not. The difference between the top brand and second-tier brands is mostly psychological. Research in psychology has shown that the name recognition alone can result in more positive feelings toward nearly everything, whether it is music, people, words, or brands. In a study, respondents were asked to taste each of three samples of peanut butter. One of these samples contained an unnamed superior (preferred in blind tests 70 percent of the time) peanut butter. Another contained an inferior (not preferred in taste test) peanut butter labeled with a brand name known to the respondents but neither purchased nor used by them before. Surprisingly, 73 percent of respondents selected the brand name (inferior) option as being the best-tasting peanut butter. This test result clearly shows the power of brand-name recognition. A mere name recognition will make people feel an inferior peanut butter tastes better than an actually better-tasting peanut butter. From this example, we can see clearly that consumers' psychology plays a very important role in brand-name strength. Davis (2000) calls this consumer psychological reaction to brand names *PATH*; it is the acronym for promise, acceptance, trust, and hope. A strong brand makes the intangible feelings of promise, acceptance, trust, and hope tangible.

The benefits from strong brands are numerous; Davis (2000) listed the following benefits:

1. Seventy-two percent of customers say that they will pay a 20 percent premium for their brand of choice, relative to the closest competitive brand; 50 percent of customers will pay a 25 percent premium; 40 percent of customers will pay up to a 30 percent premium.
2. Twenty-five percent of customers state that price does not matter if they are buying a brand that owns their loyalty.
3. Over 70 percent of customers want to use a brand to guide their purchase decision, and over 50 percent of purchases are actually brand driven.
4. Peer recommendation influences almost 30 percent of all purchases made today, so a good experience by one customer with your brand may influence another's purchase decision.
5. More than 50 percent of consumers believe a strong brand allows for more successful new product introductions, and they are more willing to try a new form of a preferred brand because of the implied endorsement.

These benefits clearly indicate that strong brands do create tremendous values for the companies who own them, so the making of strong brands should be an integral part of product development strategy. The making of a strong brand, usually called *brand development*, is a very elaborate process;

it involves the coordinated efforts of product development, marketing, promotion, customer service, and corporate leadership. Since the strength of the brand is directly related to the value of service products, the people who work on product development must understand the basics of brand development.

This chapter covers the important aspects of brand development. Section 8.2 makes a deep dive into the question, What is a brand? Section 8.3 discusses the brand development process. Section 8.4 discusses the role of brand development in the Design for Six Sigma practice.

8.2 The Anatomy of Brands

Strong brands have magical power to add value to products and bring customer loyalty. It is really important to understand how strong brands influence consumers' minds and what are essential components of a strong brand. In this section we are going to discuss all important aspects of brands.

8.2.1 People's Buying Behavior and Brands

According to Arnold (1992), the power of brands can be explained by some truths about how people make buying decisions:

1. Most customers, especially consumers in the mass market, will not understand a product or service as well as the company selling it. Most customers only have some superficial knowledge about the product or service, and many are not even interested in product or service details.
2. Customers will perceive a product or service in their own terms. Since the customers usually only have imperfect knowledge about the product or service, customers have to select some attributes that are most obvious to them and will develop their perceptions based on their opinion of these attributes. For example, airline customers may rate the airline based on the things they see; if they see there are stains on the flip-down table or see an imperfect bathroom, they may doubt the whole operation of the airplane, including the airplane engine maintenance. Customers may judge a detergent by its smell, not by how well it washes. Different customers may choose different attributes; every customer has a personalized view.
3. Customers' perception often focuses on benefits of the product or service. The benefits are what a product or service can do for a customer. Each customer may see different benefits; some of them may see some functional benefits, while others are more interested in

emotional benefit. For example, some kids would like to buy cereal with a sports star on the box for the sake of emotional benefit, not the taste of the cereal itself.
4. Customer perception is not always at a conscious level. If we ask a customer why a product or service is chosen, sometimes we may get a rational answer, and sometimes we may not. Even if there is a rational answer, it may not be the whole story. Feelings about a product or service may not be easily articulated, because these feelings are complex, hard to explain, sometimes subconscious, and may not be rational.

Because the relationship between customers and the things they buy is complex, brand names become a short cut for customers to choose the products or services. When customers gradually develop a positive perception about a product or service, the thing that they remember about the product or service is its brand name. The feelings and perceptions are often contagious; these customers will spread their feelings and perceptions to friends, family members and other people with similar opinions and this will create a snowball effect. Watkins (1986) used Fig. 8.2 to illustrate the model of customer choice:

Customers derive their buying decision making by a complex set of perceptions and demands. Therefore, a successful brand should also address many elements of customer perception and demand. The following criteria for a successful brand is adapted from Arnold (1992):

1. On the product or service level, it must deliver the functional benefits to meet the market need at least as well as the competition. No product or service will survive in the long run if it does not perform. A brand is not merely the creation of advertising and packaging.
2. A first-of-its-kind product or service in a particular area is a strong basis to build a brand. However, the brand will not be successful in the long run if it cannot make customers satisfied. When competing in a crowded market with many existing brands of similar products, a newcomer has to provide a significant advantage in some area of benefits (functionality, price, emotional) in order to compete effectively with incumbent brands.

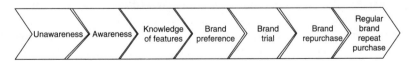

Figure 8.2 Model of Customer Choice (*Watkins 1986*)

3. Besides the functional benefits, a brand will have to offer intangible benefits, such as emotional, belonging, prestige, or style, in order to shine. Figure 8.1 is a perfect example of this. People are willing to pay a high price for a top-brand T-shirt for the sake of pride, belonging, and prestige.
4. The benefits offered by a brand should be consistent with each other and present a unified character or personality. If the benefits offered from the same brand are too confusing, or change from time to time, it will drive customers away. This is because customers will often come to a quick and superficial conclusion when purchasing a product. Customers form a stereotype about a brand quickly; if they like the brand, they will stick with it. For example, both McDonald's and Chinese restaurants are providing foods, and both will have loyal customers based on customers' perception about the food. If a McDonald's store starts offering some Chinese food, though it is a good food, it will really send a very confusing signal to customers, and finally many customers may be turned away. To maintain a brand, a company must actively manage the personality of the brand to make it clear and consistent over time.
5. The benefits offered by a brand must be wanted by the customer. No brand image, however clear and consistent, is of any use unless it meets customer wants. If people's wants have changed, the benefits offered by the brand will have to change.

Brands have a magical power, and brand building is an important element in value creation. To build a strong brand, we first need to know what are the essential elements of a brand and how these elements are related to each other really well. In the next few subsections, we will discuss some important concepts about the essential elements of a brand, such as brand identity and brand equity.

8.2.2 Brand Identity

What Is Brand Identity?

Customers' perceptions about a brand are very much similar to people's perceptions about a person. A person's name is simply a symbol. People form an opinion about a person based on their perceptions. They may ask themselves, What is this person good at? What is his or her personality? What does this person look like? What does he or she stand for? What are his or her core values? and so on. The answers to these questions allow people to form a perception of the person's identity. According to David Aaker (1996), brand identity "provides direction, purpose and meaning for the brand." Specifically, he said:

> Brand identity is a unique set of brand associations that the brand strategist aspires to create or maintain. These associations represent

what the brand stands for and imply a promise to customers from the organization members. Brand identity should help establish a relationship between the brand and the customer by generating a value proposition involving functional, emotional, or self-expressive benefits.

Brand Identity Models

There are several models that describe what brand identity is. Aaker (1996) proposed a brand identity model based on four perspectives: (1) brand as product, (2) brand as organization, (3) brand as person, and (4) brand as symbol. Davis (2000) used the brand image model, which has two components: brand association and brand persona.

Aaker's Brand Identity Model

Figure 8.3 illustrates the framework of Aaker's brand identity model. As stated, this model describes brands from four perspectives, but a brand may not actually employ all of them. It may employ only a subset of these perspectives. For brands that relate to a larger corporation and its products, it is very likely

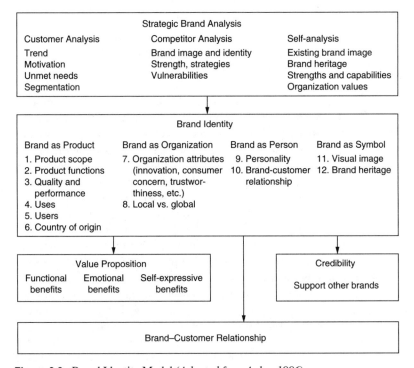

Figure 8.3 Brand Identity Model (*Adapted from Aaker 1996*)

that all four perspectives are employed. Brand identity is a perception in customers' minds that results from their entire experience with the products and services of that brand. What perception customers get depends on the products and services themselves, on how customers are treated by the company, as well as on advertisements and promotions. Ultimately, however, it depends on the company's business strategy and business operation. To build a strong brand, it is desirable that the company should carefully design a good brand identity for their products or services and make this ideal brand identity a reality. On the top of Fig. 8.3, a strategic brand analysis is performed first. Strategic brand analysis consists of three components: customer analysis, competitor analysis, and self-analysis. The purpose of strategic brand analysis is to provide a basis on which to design an appropriate brand identity for the product or service offered by the company. The details of strategic brand analysis will be discussed in Sec. 8.4. To design a good brand identity, we need to consider all four perspectives though we may not deploy all of them. We describe these four perspectives of brand identity in detail.

The Brand as a Product: Product-Related Associations

Product-related association is always an important part of brand identity, because customers are buying the product. The product-related association has the following aspects:

1. *The product scope:* This aspect deals with what product class the brand is associated with. For example, McDonald's is associated with the product class of fast food; Visa is associated with credit cards; Hertz is associated with rental cars. For an unsuccessful brand, when the brand name is mentioned, most people do not know what product class it is related to. For a nondominant brand, when the brand name is mentioned, people know what product class it is related to. However, for a dominant brand, only the product class needs to be mentioned and most people will recall the name of the brand. For example, when "soft drink" is mentioned, the name Coca-Cola will at least be thought about once. On the other hand, when Faygo, a nondominant brand is mentioned, some people will recognize it is a soft drink brand; however, many people will not recall Faygo as a soft drink choice.
2. *Product functions:* This aspect deals with what functional benefits, as well as some emotional benefits, the product or product class that is related to the brand can provide to customers. How well this aspect will perform depends on how well the customers' needs (told and untold) are met. For example, McDonald's functional benefits include all their breakfast and ordinary meal items, hamburgers, fries, soft drinks; fast purchase cycle time (time from ordering to getting the food); Happy Meals and toys; playgrounds; unrivaled worldwide

product consistency; and clean restrooms. McDonald's emotional benefits include friendly service and being kid-friendly.
3. *Quality and performance:* This aspect deals with how well and how consistently the functional and emotional benefits are provided. For example, fast purchase cycle time is a key functional benefit for McDonald's, but how fast and how consistent is a matter of quality and performance. McDonald's is famous for its unrivaled worldwide product consistency, which is also a matter of quality. For products with similar functional benefits, the performance level is dealing with how well these functional benefits are delivered. For example, Mercedes, Buick, and Kia all produce cars, but the performance levels of the cars are different.
4. *Uses:* This aspect deals with the particular use or application associated with a brand. For example, Gatorade specializes in providing a drink for athletes to maintain a high level of performance.
5. *Users:* This aspect deals with the type of users targeted by a particular brand. For example, Motherhood Maternity targets pregnant women, and its products are for pregnant women.
6. *Country of origin:* Association of a brand with a country of origin will add credibility if the country that the brand relates to is good in this product area. For example, French fashion is more highly regarded than French electronic goods.

A product-related association is an important part of brand identity. After all, people are buying products and the benefits related to product functions. However, if the brand identity is only associated with product attributes, it will have serious limitations. Specifically some of these limitations are as follows (Aaker 1996):

1. *Failure in brand differentiation:* A product attribute can be extremely important to customers, but if all brands are perceived to be adequate on this attribute, it does not differentiate the brand. For example, in the hotel business, cleanliness is always rated as one of the most important attributes to customers. Thus it would be appropriate for cleanliness to be a part of Hilton's brand identity. However, because all hotels are expected to be clean, it will not be a brand differentiator. Without brand differentiation, the brand name will not stand out in customers' minds when the product is needed.
2. *Easy to copy:* Product functional benefits are easy to copy. A brand that relies on the superior performance of functional attributes will eventually be beaten, because functional attributes are transparent, a fixed target. If the brand name does not have a psychological dimension, a low-cost competitor could easily nudge the brand out of the marketplace.

3. *Limitation on brand extension:* An overly strong association to particular product attributes may limit the ability for a brand name to extend to other fields. For example, both GM and Ford have strong financial arms, and they make a profit out of them. However, they are mostly automobile-related financial operations. Because of the overwhelming brand association with the auto industry, if would be hard for them to stretch into non-auto-related financial operations in a big way.
4. *Limitation on business strategy change:* An overly strong association to particular product attributes will limit a brand's ability to respond to changing markets. The Atkins brand is closely associated with the Atkins diet theory. It is doing fine today. However, if the Atkins diet theory becomes out of favor, this brand will have a big problem.

Therefore, it is important for a brand name to address other perspectives of brand identity.

The Brand as Organization

The brand as organization perspective focuses on attributes of the organization rather than on those of the product or service. Such organizational attributes as innovation, a drive to quality, and concern for the environment are created by the people, culture, values, and programs of the company. Some organization-related attributes can also be related to product; for example, innovation and quality could also be related to product design. However, when these attributes are related to the organization, they usually mean different things such as culture and values. For example, Toyota's lean manufacturing principles are easy to copy from a procedural point of view; however, it is Toyota's culture that is really difficult to imitate. And this is why so many companies want to implement lean manufacturing but fail to reach its full benefits (Liker 2004).

Organizational attributes are more enduring and more resistant to competitive claims than are product attributes. First, it is much easier to copy a product than to duplicate an organization with unique people, values, and culture. Second, organizational attributes usually apply to a set of product classes, and a competitor with a single product class is difficult to match. Third, the organizational attributes such as an innovative and quality culture are difficult to measure and communicate, so it is difficult for a competitor to convince consumers that it has closed a conceived gap.

The Brand as Person

The brand as person perspective suggests a brand identity that is richer and more interesting than one based on product attributes. Like a person, a brand can be perceived as being upscale, competent, impressive, trustworthy, fun, casual, youthful, and so on. The most important concept here is the brand personality. A *brand personality* can be defined as the set of human characteristics associated with a given brand. Thus it includes such characteristics as gender, age, and socioeconomic class, as well as such classic human personality traits as warmth, concern, and sentimentality.

A brand personality can create a strong brand in several ways. First, it can help to create a self-expressive benefit that becomes a vehicle for the customer to express his or her own personality. For example, a rich man may want to drive a Mercedes Benz to show his affluence and pride. Second, brand personality can be the basis of a relationship between the customer and the brand. For example, the Harley Davidson brand has the personality of a rugged, free-wheeling, outdoors guy. It helps the buyer use the product as an identifier for his or her own personality. Third, a brand personality may help communicate product attributes.

A customer's perceived brand personality is created by many factors. Some of the factors are product-related, and some are non product related. Table 8.2 summarizes these factors:

Product-related characteristics could be the primary drivers of a brand personality. Even the product class can affect the personality. For example, a bank

Table 8.2 Brand Personality Drivers

Product-Related Characteristics	Nonproduct-Related Characteristics
Product class	User image
Package	Sponsorship
Price	Symbol
Attributes	Age
	Ad style
	Country of origin
	Company image
	CEO
	Celebrity endorsers

Source: Aaker (1996).

or insurance company tends to assume a "banker" personality (competent, serious, male, older, upper class). Huggies' packaging always features healthy, happy kids.

Price is a complex factor in brand personality; if the price is low, it may attract low-end buyers and thus increase sales, but on the other hand, it also gives the brand a "cheapo" image. If the products related to a brand have top performance and top quality, a higher price actually may psychologically give an image of a prestigious brand. Product attributes also affect brand personality; for example, the strong flavor in Marlboro brand cigarettes suggests a rugged male personality.

Nonproduct-related characteristics could also affect the brand personality. Important nonproduct-related characteristics are listed in Table 8.2. The user image refers to either the profile of typical users (the people who use the brand) or idealized users (as portrayed in advertising and elsewhere). The user image can be a powerful driver for brand image; for example, the Marlboro man is the defining image of Marlboro's brand personality, a free-spirited, rugged man. Sponsorship of particular events can influence a brand's personality; for example, Nautica sponsorship of Olympic swimming events gives its brand of swimming wear a personality of a world class swimmer. How long a brand has been in the market (age) can affect its personality. An old brand usually gives a "traditional" or "reliable, but old fashioned" brand personality. A newcomer tends to have a younger brand personality. A brand symbol can have a powerful influence on the brand personality because it is visible every time a consumer sees the advertisement or product. The "Intel inside" symbol created a very strong psychological impact on buyers that a computer without Intel's CPU would not be as good. Marlboro country and the Maytag repairman are among the most successful brand symbols that provide desirable stereotypes in customers' minds. The country of origin is also a very powerful opinion-shaping factor of brand personality; a German brand might capture some perceived characteristics of German people (precise, serious, hardworking, and so on). A CEO's personality, such as Bill Gates of Microsoft, influences people's perception of a company and its products. Celebrity endorsements can also be influential. For example, Michael Jordan's endorsement of Gatorade gave the brand a personality of a strong, thirsty athlete. Advertisement style and company image affect brand personality as well.

Value Proposition

The purpose of having products or a product class under a brand name is to provide customers with benefits. There are many kinds of benefits.

In Aaker's brand identity model illustrated in Fig. 8.3, three kinds of benefits are listed: functional, emotional, and self-expression. For each brand, the benefits that are offered will be different. Aaker (1996) calls this the value proposition. Specifically, a brand value proposition is a statement of the functional, emotional, and self-expressive benefits delivered by the brand that provide value to the customer. An effective value proposition should lead to a brand-customer relationship and drive purchase decisions.

The concepts of functional, emotional, and self-expressive benefits are explained as follows:

1. *Functional benefits:* Functional benefits are the aggregated product functions that a product provides to customers. The functional benefits of a car include movement from point A to B, change of directions and speed, a nice driving environment, and styling. Besides some must-have functions, a brand often provides some functional benefits that are special features. For example, Volvo is featured by its safety and durability, 7-Eleven is featured by its convenience, and Nordstrom is featured by its customer service. Functional benefit is important; if a brand can dominate a key functional benefit for which the customers really care, it can dominate its product class. The challenge is to select functional benefits that will "ring the bell" with customers. Just delivering this functional benefit is not enough though; customers buy products based on *perceived* quality and *perceived* functional superiority. Convincing customers that the brand is truly the leader in a key functional area might be more challenging than delivering these key functional benefits.
2. *Emotional benefits:* When purchasing or using a particular brand gives customers a positive feeling, that brand is providing them with an emotional benefit. For example, you feel safe when you drive a Volvo and you feel important when you shop at Nordstrom. The strong brand value proposition often includes an emotional benefit, on top of functional benefits. If a brand only has functional benefits, it is vulnerable, because if a low-cost producer can duplicate the same functional benefits, the price of the brand product must be lowered or it will be priced out of the market. Emotional benefits are more complex and much more difficult to copy. They are intertwined with functional benefits. Therefore it is important to study the relationship between functional attributes and emotional benefits.
3. *Self-expressive benefit:* Some customers use brands to show themselves off. We call this the self-expressive benefit. For example, some youngsters buy fashions from the Gap to show off themselves; likewise, a successful businessperson might drive a Lincoln, Lexus, or Mercedes Benz.

Example 8.1: McDonald's Brand Identity

McDonald's is one of the most successful global brands. Its brand identity can be summarized as follows:

Brand as Product

Product scope: Fast food, children's entertainment, eating spaces
Product functions:
    Variety of fast-food items: hamburgers, Big Mac, Happy Meals, Egg McMuffin, etc.
    Service: Fast, accurate, friendly, and hassle-free
    Cleanliness: Spotless in eating spaces, restrooms, and counter;
    Low prices
Quality and performance: Consistent temperature, taste, portion, layout, decoration, cleanliness all over the world
User: Family and kids are the focus, but company serves a wide clientele
Country of origin: United States

Brand as Organization

Convenience: McDonald's is the most convenient quick-service restaurant. It is located close to where people live, work, and travel; features efficient, time-saving service, and serves easy to eat food.

Brand Personality

Family-oriented, all-American, genuine, wholesome, cheerful, fun

Brand as Symbol

Logo: Golden arches
Characters: Ronald McDonald; McDonald's doll and toys

Value Proposition

Functional benefits: Good-tasting burgers, fries, and drinks; extras such as playgrounds, prizes, and games
Emotional benefits: Kids fun via excitement of birthday parties; joy from toys and playgrounds; the feeling of special family times

Davis' Brand Image Model

Scott Davis (2000) developed a brand image model. The brand image has two components: brand associations and brand persona. Brand associations describe what kinds of benefits the brand delivers to customers and the role it plays in their lives. Brand persona is a description of the brand in terms of human characteristics. The brand image model is a concise model for the brand identity. We will discuss brand associations and brand persona in detail.

Brand Associations

Brand associations relate to the product, service, and organization aspects of the brand. They describe a hierarchy of benefits that a brand provides to its customers. Brands associations are best described by the brand value pyramid, which is illustrated in Fig. 8.4. The features and attributes layer is at the bottom of this pyramid. Here the features and attributes are the most essential product functions, performances, and quality levels that must be delivered to customers in order for the brand to survive in the marketplace. The benefits layer is at the middle of pyramid. Here the benefits are additional functional and/or emotional benefits that the brand provides to its customers, given that the features and attributes have been satisfactorily provided by the brand. The beliefs and values layer is at the top of the pyramid; this layer represents the emotional, spiritual, and cultural values that are addressed by the brand, given that all the benefits from the benefits layer and features and attributes layer have been provided by the brand.

Many brands may not be able to fill all the layers of the brand value pyramid. If a brand cannot fill the bottom layer, then it cannot even deliver the most basic benefits to its customers for this kind of product and this brand will fail in the long run. If a brand can only fill the bottom layer, then it is a very marginal brand, nothing special. It is an essential commodity, such as raw cotton, raw sugar, or it is the leftmost no-brand T-shirt in Fig. 8.1, Its market

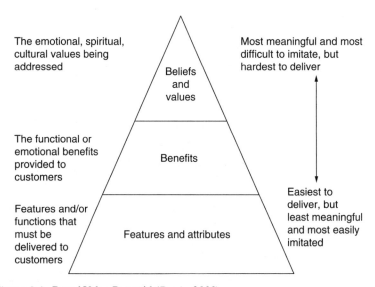

Figure 8.4 Brand Value Pyramid (*Davis, 2000*)

survival is mostly dependent upon having a low price. If a brand fills or somewhat fills the middle layer, it becomes a surviving brand. It is better than the commodity. The most powerful brands fill all the layers in the brand value pyramid.

Figure 8.5 shows Ralph Lauren's brand value pyramid. Ralph Lauren has achieved the strongest brand status in its product class. Many brands can deliver the features and attributes illustrated in Fig. 8.5, that is, offering high quality, durable, and classic-looking clothes. But few of them can say their clothes allow their customers to make a statement. Wearing Ralph Lauren clothes is like driving a Mercedes Benz in its appeal to social status. The psychological benefits of this kind of brand name usually take years to evolve; they are difficult to explain and even more difficult to duplicate.

Brand Persona

According to Davis (2000), brand persona is the set of human characteristics that consumers associate with the brand, such as personality, appearance, values, likes and dislikes, gender, size, shape, ethnicity, intelligence, socioeconomic class, and education. Brand persona brings the brand to life, and

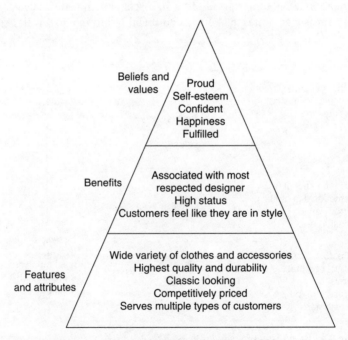

Figure 8.5 Ralph Lauren's Brand Value Pyramid (*Davis, 2000*)

customers subconsciously decide if they want to be associated with this brand, just like they decide if they want to be associated with other people. If a brand persona is unpopular and unattractive, then it will affect the sale. The brand persona here is very similar to the brand personality in Aaker's brand identity model.

Example 8.2: Brand Personas for Mail Services
This example is from Davis (2000). Table 8.3 lists the brand personas of three major mail carriers: Federal Express, U.S. Post Office, and UPS.

8.2.3 Brand Equity

Brand equity is the set of assets (and liabilities) that is linked to a brand name and symbol. The brand equity adds (or subtracts) the value provided by a product or service to a firm and/or that firm's customers (Aaker 1996).

During the 1980s, a lot of research was done to define and estimate the true value of brands to the competitive position of enterprises (Keller 1993, Aaker 1991, Farquhar 1989, Tauber 1988). There were two reasons for this

Table 8.3 Brand Personas for Three Mail Carriers

FedEx	U.S. Post Office	UPS
Male or Female	Male	Male
Young	Old	Middle-aged
Athletic	Grumpy	Evolving
Friendly	Not reliable	Inconsistent
Prompt	Low technology	Friendly
Dependable	Unsophisticated	Brown uniforms
Energetic	Overweight	Unionized
High technology	Complacent	Okay service
Problem solvers	Slow	Professional
Motivated	Rigid	International
Professional	Problem makers	Problem solvers

(Keller 1993). The first reason was an accounting one, and it was to better estimate the value of brands more precisely for the balance sheet especially in cases of mergers, acquisitions, and divestitures. The second reason was a strategy-based motivation to improve marketing productivity (Keller 1993).

Brand equity provides a mechanism for capturing the marketing effects uniquely attributable to the brand (Keller 1993). Aaker's brand equity model (Aaker 1991, p. 269) is one of the best-known models of brand equity. It is a tool for understanding the linkage between the brand and the value it provides the firm and its customers beyond what is inherent in the functional attributes of the products and services. The brand equity model defines five dimensions of value that the brand provides the firm: brand loyalty, name awareness, perceived quality, brand associations, and other proprietary brand assets (Fig. 8.6). Each of these dimensions is important in influencing customers' purchasing decisions, and thus, is a contributor toward the viability of the enterprise. The strategic role of each of these brand equity dimensions will be described in more detail.

Brand Loyalty

In a competitive environment the ability of a company to retain its existing customers is of key importance. Brand loyalty is a key factor influencing the repeat-buying behavior of customers (Keller 1993), and it reduces vulnerability to competitive actions in the marketplace. Secondly, it reduces the cost of doing business for a company because it is more expensive for a business to try to acquire new customers than to retain existing ones, especially when the existing ones are satisfied with the brand (Aaker 1991). Thirdly, brand loyalty can be powerful leverage for negotiating more favorable terms in the distribution channels (Aaker 1991).

Name Awareness

Brand-name awareness relates to the likelihood that a brand name will come to mind and the ease with which it does so. Brand-name awareness consists of two dimensions: brand recall and brand recognition. Brand recognition reflects a familiarity gained from past experience with the brand (Aaker 1996). Studies have shown that people often buy a brand because they are familiar with it (Aaker 1991). Brand recall refers to how strongly the brand comes to mind when the consumer thinks about that product category or the needs fulfilled by that product category (Keller 1993). Brand-name awareness plays an important role in consumer decision making because it allows the brand to be included in the consideration set, which is a prerequisite for its eventual choice.

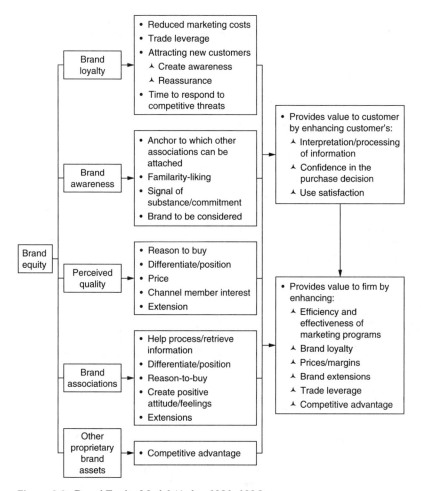

Figure 8.6 Brand Equity Model (*Aaker 1991, 1996*)

Perceived Quality

Perceived quality is part of the human experience and is developed entirely from the perspective of the consumer, based on those product attributes that are important to them. Perceived quality may be different from the actual quality of the product. Studies have shown that the customer's perception of quality has one of the greatest impacts on the financial performance of a company (Buzzell et al. 1987, p. 7; Jacobson and Aaker 1987; Anderson et al. 1994). A study of 33 publicly traded stocks over a 4-year period demonstrated that perceived quality had an impact on stock return (Aaker 1996).

Perceived quality is a key strategic variable for many companies (Aaker 1996), and it is a key-positioning dimension for corporate brands.

Brand Associations

Brand associations can be anything that connects the customer to the brand (Aaker 2000). These associations help determine the brand image with the customer and marketplace. Brand associations can be *hard*—related to specific perceptions of tangible functional attributes, such as, speed, user-friendliness, taste, and price. Brand associations can also be *soft*—emotional attributes like excitement, fun, trustworthiness, and ingenuity (Biel 1993). Apple is an example of a brand with values that have resonated with those of its target audience. Emphasizing values such as fun, excitement, innovation, and humor (Kochan 1997), the company has succeeded in carving out a niche for itself in the highly competitive personal computer marketplace (Levine 2003).

Other Proprietary Brand Assets

Beyond their use as a tool in achieving a competitive advantage, brands are also a financial asset to a company. Successful brands can be traded or used to increase the valuation of a company during a corporate acquisition.

8.3 Brand Development

Strong brands can create tremendous values for the companies that own them. Developing strong brands that lead the market is always one of the most important goals for companies, that strive to excel in the marketplace. The brand development process develops strong brands that fit the owner companies' business goals and their comparative advantages.

The objective of a brand development process is to create a brand that achieves and maintains the intended position in the minds of customers within the targeted market group. In other words, the brand development process must create a brand image in the minds of customers that reflects the brand identity defined by the company. The creation of this brand position in the minds of customers involves the creation of the brand identity, the transmission of the brand image to customers, and the receipt and acceptance of this image by customers, as illustrated by Fig. 8.7. This process is influenced by a variety of factors, which include the nature of the brand identity, organization factors, the communications media, and market forces.

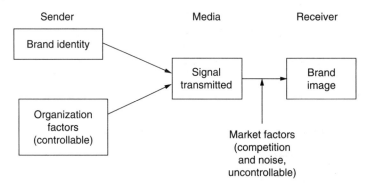

Figure 8.7 Brand Identity Transmission Process

In this section, we will first identify and discuss the factors that influence the brand development and then will discuss several key steps in the brand development process.

8.3.1 Key Factors in Brand Development

There are many factors that affect the brand development process. They can be categorized into two classes: controllable and uncontrollable. The controllable factors are those over which the company that owns the brand will have some degree of control. The uncontrollable factors are those over which the company will have little or no control. These controllable and uncontrollable factors are also illustrated in Fig. 8.7.

Controllable Factors

There are three classes of controllable factors. The first is brand identity. The second is called marketing mix factor, or the 4 Ps: product, price, promotion, and place. The third is time to market. In brand development, these factors can be used to shape a desirable brand image by the company and transmit this brand image to customers.

Brand Identity

Brand identity was thoroughly discussed in Sec. 8.2.2. If we use Aaker's brand identity model, then there are four perspectives: brand as product, brand as organization, brand personality, and brand symbol. Clearly, the company that owns the brand has full control of the brand symbol, a good degree of control over product development, and a relatively good degree of control over organization behavior. Brand personality takes years to form; it is more difficult to change.

In brand development, the company should design a desirable brand identity based on a thorough analysis of the marketplace, competitors, and the relative strengths and weaknesses of the company itself, in order to achieve the best customer brand image possible.

Product

Customers do not buy brand symbols; they purchase products. Initially, customers may be influenced by advertisements or their friends' advice to try a product with a certain brand name, but the product has to perform up to customers' expectations. If the product performs equal to or better than customers' expectations, the perceived brand image will be confirmed by customers' experience; that will trigger word-of-mouth recommendations, sales will grow, and the positive brand image will spread among more and more customers. In order to accomplish this, it is very important that the actual product characteristics, such as functions, performances, and quality levels, be consistent with the brand identity. Therefore, product development has to go hand in hand with brand identity design; the product development process has to be in tune with brand development.

Price

The role of price in brand development is quite interesting. The brand price is related to the benefits that the brand provides, as illustrated by Fig. 8.8. If a price is too high relative to the benefits that the brand provides, the perceived value in customers' minds will be low. Customers will think this brand is overpriced. However, if the benefits that the brand provides are high, but the price is low relative to the benefits, the customers' reaction can be quite complex. Theoretically, customers will be happy to get more and spend less, but the perception of a "cheapo" product may creep in, which may undercut the brand image. Usually, benefits are the main focus for brand identity creation. If you have more benefits and customers are happy

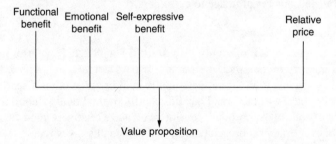

Figure 8.8 The Relationship Between Price and Benefits of a Brand (*Aaker 1996*)

with them, it is always easy to raise the price to match the benefits. If the benefits are really low in comparison with other brands, the price will have nowhere to go but lower.

Promotion (Communication)

The effectiveness of the communications campaign is a critical factor in creating the desired brand position and image (Aaker 1996). The purpose of marketing programs, such as advertising, is to transmit the brand image in order to increase brand awareness and create a strong, favorable, and unique brand identity in the customer's mind. The strength of the brand image is greatly influenced by the communications in the marketing programs, particularly by the effectiveness with which the brand identity is integrated into the marketing programs (Keller 1993). Given the many choices of media available, selecting the correct media mix to reach the targeted audience and ensuring that the message is integrated is a key factor in creating a distinct brand image.

Place (Distribution)

Distribution channels used to deliver products to customers are also very important factors for brand development. It is important that the distribution channels deliver the goods to the right customers. The key variables in the selection of distribution channels include the type of channels, the number of outlets, the locations of outlets, and stock levels. The distribution process of the products should be synchronized with the promotion activities so that the desired customers will know what and where the products are. An efficient distribution process will give customers service benefits and reduce the hassles in obtaining the products, thus enhancing the brand image.

Time to Market

Brands that are able to position themselves first in the minds of their customers have the best chance of achieving the highest brand awareness (Reis 1981). So the ability to be first in the marketplace will affect the ability to create a strong brand. However, this factor is not totally controllable, because it also depends on the competition from competitors.

Uncontrollable Factors

Uncontrollable factors in brand development are mostly marketing factors, as illustrated in Fig. 8.7. The demographics of the target population for a brand is an uncontrollable factor. It includes age group, income level, sex, marital status, area of residence, location, social group membership, and stage in life. The variation in demographics makes the targeted customers very inhomogeneous, so the brand image that the company tries to communicate to customers will be perceived differently. For example, a particular brand image might be very attractive to one age group, but unattractive to all other age groups.

Culture is also an important uncontrollable factor. People from different cultural backgrounds and business clients from different corporate cultures will see things in different ways, so cultural factors may affect how customers perceive the brand image.

Competitive activity is clearly a very important, uncontrollable factor. The number of competitors and the relative strength of competitors will make a tremendous impact on brand image. Therefore, it is critical to conduct a competitive analysis in the early stages of the brand development process.

The customer is another uncontrollable factor. Clearly, the company that owns the brand cannot control customers. Customers are at the receiving end of the brand and its associated products. The products and service must meet the needs and wants of customers. The company cannot control the customers, but it can try to understand the customers' needs. The ability of the product or service to meet the needs of its customers is a critical factor in creating the brand loyalty that will determine the success or failure of the brand.

8.3.2 Overview of the Brand Development Process

There have been different processes proposed for brand development. These processes are fundamentally based on two paradigms. The first paradigm views the brand development process as being closely associated with the development and marketing of new products (Watkins 1986). In this paradigm, the brand development process is called the classical brand management process (Aaker 2000), and it consists of the following steps:

1. Market exploration
2. Preliminary financial analysis and screening
3. Formal business analysis and planning
4. Product and brand development
5. Product testing
6. Product launch

Recent trends in brand development have elevated brand management to a more strategic position within the organization, as defined by the brand leadership model (Aaker 2000) and brand asset management process

Table 8.4 Paradigms of the Brand Development Process

Features	Classical Brand Management Model	Brand Leadership Model
Perspective	Tactical and reactive	Strategic and visionary
Brand manager status	Less experienced, short time horizon	Higher in the organization, longer time horizon
Conceptual model	Brand image	Brand equity
Focus	Short-term financials	Brand equity measures
Product-market scope	Single products and markets	Multiple products and markets
Brand structures	Simple	Complex brand structures
Number of brands	Focus on single brands	Category focus—multiple brands
Country scope	Single country	Global perspectives
Brand manager's communication role	Coordinator of limited options	Team leader of multiple communication options
Communication focus	External/customer	External and internal
Driver of strategy	Sales and share	Brand identity

Source: Adapted from Aaker (2000).

(Davis 2000). The differences between the two brand development paradigms are shown in Table 8.4.

The new paradigm for the brand development and management is more focused on the management of the brand as a strategic asset of the company (Arnold 1992; Davis 2000; and Aaker 1996). The brand development process defined from this paradigm generally consists of the following phases (Fig. 8.9):

1. Brand strategy analysis
2. Brand strategy development
3. Brand implementation
4. Brand evaluation

We discuss these four phases in detail.

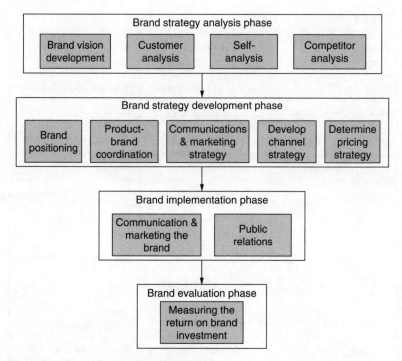

Figure 8.9 Brand Development Process

8.3.3 Brand Strategy Analysis

The strategy analysis phase in brand development focuses on understanding the brand's competitive position in the marketplace. It also looks at the ability of the company to influence this positioning through its capabilities and the attributes of its products and services. The steps of the strategy analysis phase are described in the following:

Brand Vision Development

The brand vision is a short, succinct statement of what the brand is intended to become and what is to be achieved at some point in the future, often stated in competitive terms. Brand vision refers to the category of intentions that are broad, all-intrusive, and forward thinking. It is the image that a business must have of its goals before it sets out to reach them. It describes aspirations for the future, without specifying the means that will be used to achieve those desired ends.

IBM's brand vision is stated as follows:

> At IBM, we strive to lead in the creation, development, and manufacturing of the industry's most advanced information technologies, including computer systems, software, networking systems, storage devices, and microelectronics. We translate these advanced technologies into value for our customers through professional solutions and service businesses throughout the world.

According to Davis (2000), a good brand vision should have four components: (1) a statement of the overall goal of the brand, (2) the target market that the brand will pursue, (3) the points of differentiation that the brand will strive for, and (4) the overall financial goals for which the brand will be accountable. Davis (2000) gives an example of such a brand vision statement:

> Around the world, our eye care brand will stand for leadership in visual care. Consumers and the professional channel will recognize us as the industry leader in visual care solutions, including the best service, follow-up, expertise, and product innovation. Our brand will help us fill one-third of our stated financial growth gap through price premiums, better relationships with the channel, and close-in brand extensions.

Development of a brand vision links the brand development process to the strategic objectives of the company. This linking is an important step to ensure the necessary top management and financial commitment to the brand (Davis 2000). During this step the strategic and financial goals of the brand are defined, and the commitment of senior management to the goals and objectives of the brand are obtained.

Customer Analysis

The activities in this step are focused on understanding the trends, motivation, and unmet needs of the various segments of the customer market (Aaker 1996). This step creates an understanding of how the "customer thinks and acts and why and how they make a purchasing decision" (Davis 2000). The objectives of this step are fourfold (Aaker 1996):

1. *To determine the functional, emotional, and self-expressive benefits that customers seek when they buy and use the brand.* Customer surveys are usually needed to determine these benefits as well as the relative importance of these benefits. The following set of questions is used to assess the functional benefits:
   - What functional benefits are relevant to customers?
   - What is the relative importance of each functional benefit?
   - Can benefit segments be identified?

Emotional and self-expressive benefits are more subtle than functional benefits and are thus more difficult to determine.

2. *To understand the different segments of the customer market and their different needs, wants, and behaviors.* We need to find out how the market segments. Because different market segments may have different functional, emotional and self-expressive benefit needs, they may respond differently to a brand promotion program. There are many possible segmentation schemes; however, in the brand development process, the major task is to find out which segments are the most attractive target for the brand and most relevant to the brand identity development. Therefore, the commonly used segmentation schemes include segmentation by benefits sought and segmentation by price sensitivity.

3. *To understand trends occurring in the customer markets so that the current and future positioning of the brand can be better assessed.* By trends we mean the dynamics of the market and how the demand pattern will change. Analysis of market data, such as sales volume trends and profitability prospects of the submarkets, may help to understand market trends. Understanding market trends provides insight into changing motivations and emerging segments with strategic importance. For example, in the coffee market, the sales of regular supermarket brands declined from 1962 to 1993, but gourmet coffee and coffeehouse sales increased. If a company is in the coffee business, this information will certainly help to develop the position of its brand in future markets.

4. *To identify customer needs that current products do not meet.* Unmet needs are customer needs that are not met by existing products in the market. They are strategically important because they can represent opportunities for a company to make beneficial moves in the market. For example, Black & Decker organized a focus group of 50 power tool owners. The executives of Black & Decker visited the focus group members' homes and found out several major unmet needs. One of the problems with cordless drills is that they run out of battery power before the job is done. Black & Decker responded by offering detachable battery packs that could be recharged quickly. Tool owners can have several battery packs charged and when one battery runs out while they are working, they can replace it with another in no time and recharge the run-out battery. Several of these kinds of design moves really give Black & Decker's Quantum brand a core identity and competitive advantage.

Self-Analysis

The objective of this step is for management to examine the strengths and weaknesses of its brand's current situation, so as to understand how the brand is positioned in the marketplace and what circumstances contributed to the achievement of this current position (Arnold 1992). The areas analyzed during this step include

1. *The current brand image, i.e., What is the perception of the brand in the marketplace?* This analysis can be done by using a customer survey (Chap. 4). The following types of questions should be included in the survey:
   - How is the company's brand perceived?
   - What associations are linked with the brand?
   - Why do customers like the brand? Why do customers not like the brand?
   - How does a company's brand differ from competing brands?
   - What benefits do customers get from the brand?
   - Has the company's brand changed over time? If yes, how?
   - For different market segments, does the company's brand image differ? If yes, then how does it differ?
   - Does the company's brand have a personality? If yes, then what is it?
   - What are the intangible attributes and benefits of the brand?

   In assessing the current brand image, it is important that customer research include a study of not only product-related attributes but also of nonproduct-related attributes, such as organizational association, brand personality, brand-customer relationships, and emotional and self-expressive benefits.

2. *The fundamental values of the brand, i.e., What does the brand stand for? Is it for fun, luxury, or an active lifestyle? What is the heritage of the brand?* Besides studying the current image of the brand, it is important to understand the heritage of the brand. Any surviving brand has some reasons why it survived; it must have done something right. Many brands get into trouble because they deviate from their heritage. Arbitrary changes in brand identity may hurt a brand more than help it. The answers to such questions as, Who were the early pioneers of the brand? How did it originate? What was the brand image when it first started? can help to understand the brand heritage.

3. *Links and associations to other brands.* Some companies offer several brands of products or services. In this case, a change in one brand position may affect other brands offered by the company. So the brand position decision should not be made in isolation. Each brand should have well-defined roles, and all brands offered by the company should work together in a synergistic manner.

4. *The strengths and weaknesses of product and service offerings and the capabilities of the organization, i.e., What is the organization good at?*

For a realistic brand strategy, the desired brand identity should be supported by the organizational strength of the company. It is necessary to find out the company's and products' strengths and weaknesses. We need to find out what the company is good at, what the company is not good at, and how and how much these weaknesses can be changed. If a company is pursuing a goal that cannot be substantiated due to its weakness, then the goal will not be achieved.

Competitor Analysis

The competitor analysis is focused on understanding the current image, positioning, strengths, and weaknesses of competitive brands in the marketplace, as well as the possible future trajectories of these competitive brands (Aaker 1996). This analysis will develop an understanding of the following aspects:

1. *The customer's perception of competitive brands.* This information is a fundamental input for brand identity determination. Because it tells how customers perceive competitors' brands, it provides a basis for determining what needs to be done to differentiate the company's brand from competitors' brands. It is important to find out what are the functional benefits to customers, how these benefits compare with that of the company's brand, the brand-customer relationship, and brand personality.

   There are two sources from which to get this information. One is from a customer survey study on competitors' brands. The other is from competitors' own information; for example, competitors' advertisements and advertisement plans can provide clues as to what kind of brand image they want customers to perceive.

2. *Previous changes in competitive brand positioning and the future market positioning (strategic brand objectives) of competitors.* In the brand development process, it is important to consider not only the current images of competing brands but also past changes and possible future changes in these images. A thorough examination of such changes can provide useful information about the reasons for such changes and the reality of the competitive environment.

3. *Strengths and vulnerabilities of competitors.* Information on strengths and vulnerabilities of competitors provides valuable inputs for a company's brand position. It is difficult and costly to compete head to head with the strong points of competitors. It is much easier to compete in the areas in which your competitors are not strong.

8.3.4 Brand Strategy Development

The purpose of the development phase is to develop the brand strategy. During this phase the positioning of the brand is developed. Also channel strategy, pricing strategy, and future extensions to the brand are developed or aligned. The following describes the steps in the brand strategy development phase.

Brand Positioning

"The brand position is part of the brand identity and value proposition that is to be actively communicated to the target audience and that demonstrates an advantage over competing brands" (Aaker 1996). The purpose of brand positioning activities is to create an identity that provides the brand with a unique, credible, sustainable, and valued place in customers' minds (Davis 2000). The positioning of the brand is the place in customers' minds that the brand is intended to own (Davis 2000, Reis 1981). Positioning is the process of determining the impact the message (the brand) will make on the mind of the prospect (Reis 1981). The effect of the positioning is to create the necessary associations customers will think of when they recall the brand. Some examples of associations of some well-positioned brands are given in Table 8.5.

A brand's position should be updated every 3 to 5 years, or as often as needed to update the company's growth strategy. Senior management has to lead in developing, updating, and implementing the brand position.

Table 8.5 Examples of Top Brands and Their Attributes

Brand	Attributes
Disney	Family fun entertainment
Nordstrom	Highest level of retail service
Saturn	Your car company
FedEx	Guaranteed overnight delivery
Wal-Mart	Low prices and good values
Hallmark	Caring shared

Source: Davis (2000).

Brand positioning defines the following aspects of the brand's position in the marketplace:

1. *The target market segment.* The company that owns the brand needs to know who are the intended customers for its brand. Customer surveys and self-study can be used to determine the target market segment. The following types of questions should be asked:
   - Is the target market both identifiable and reachable?
   - Would the current customers be part of our target market segment?
   - Will the target market be attracted to our distinct brand identity?
   - If we never served this market segment before, why do we want to serve it now?
2. *The business it provides to the market segment.* The company that owns the brand needs to know what kinds of businesses it will provide and what kinds of businesses it will not provide in the target market segment. Customer surveys and self-study can be used to determine this. The following types of questions should be asked:
   - What is the category, industry, or business that we compete in?
   - How has this changed over time?
   - Will the marketplace value and believe in our participation in this business?
3. *Key benefits and points of differentiation of the company's products and brand in the marketplace.* The company that owns the brand needs to know what differentiates its brand from other brands in the target market segment, as well as what are the key benefits of its brand to customers. Customer surveys and self-study can be used to determine this. The following types of questions should be asked:
   - Are our key benefits important to customers?
   - Can we deliver these benefits satisfactorily?
   - What are our key points of difference from other brands?
   - Can we own these key points of difference over time?
   - Are we competing at the features and attributes layer, the benefits layer, or the beliefs and values layer? (See Fig. 8.4.)
4. *The contract of the brand with the market, which defines its brand's promises (in terms of its products and service quality) to its customers.* The contract of the brand with the market is also called a brand contract (Davis 2000). A brand contract is a list of all promises the brand makes to customers. Such a contract is executed internally, but it is defined and validated externally by the marketplace.

The brand contract is derived by analyzing
- The current promises the brand makes to the marketplace

- Positive and negative feedbacks regarding current promises from customers
- The results of the brand position analysis

Example 8.3: Starbucks' Implicit Brand Contract
This example is from Davis (2000). Starbucks promises to do the following:

1. Provide the highest-quality coffee available on the market today
2. Offer customers a wide variety of coffee options as well as complementary food and beverage items
3. Have an atmosphere that is warm, friendly, homelike, and appropriate for having a conversation with a good friend or reading a book
4. Recognize that visiting Starbucks is as much about the experience of drinking coffee as it is about coffee itself
5. Have employees who are friendly, courteous, outgoing, helpful, knowledgeable, and quick to fill customer orders
6. Provide customers with the same experience at any one of the several thousand Starbucks worldwide
7. Stay current with the times, meet customer needs, and help customers create the Starbucks experience on their own terms
8. Provide customers with an environmentally friendly establishment
9. Educate customers on the different types of coffee offered

Five Principles of Effective Brand Positioning

Davis (2000) proposed five principles of effective brand positioning: value, uniqueness, credibility, sustainability, and fit.

1. *Value:* The proposed brand position should provide the targeted customers with superior values to that of competitors. It should provide customers with functional, emotional, and self-expressive benefits that a wide range of customers will appreciate and be willing to pay a premium price to get.
2. *Uniqueness:* The proposed brand position should have some unique attributes that no other competitors can deliver. These should be important to and appreciated by customers. The uniqueness should make the company's brand stand out in the crowd.
3. *Credibility:* The proposed brand position should be implementable in a credible manner, and the company's effort should be able to make customers believe that all the promises will be met. The brand position should be in line with customers' perception of the company's ability.
4. *Sustainability:* Once a proposed brand position is implemented, it is desirable that this brand position last as long as it can. Changing brand positioning involves a lot of investment, and frequent change cuts

down on the credibility of the brand. A good brand position should be difficult to copy by competitors and meet the customers' changing needs for a long time.

5. *Fit:* The brand position should fit the company's objectives and culture.

Example 8.4: Several Bookstores' Brand Positions
This example is from Davis (2000). Borders, Crown Books, Barnes & Noble, and Amazon.com are four dominant booksellers in America. The following table lists each of their unique brand positions.

Company	Target Market Segment	Business Provided	Point of Difference
Borders	Individuals looking for a community meeting place	Books, music, multimedia, and on-line	Fun place to go
Crown Books	Price-sensitive individuals, strip-mall shoppers	Bookstores	Discount pricing
Barnes & Noble	Individuals looking for a quiet gathering place	Books, music, multimedia and on-line	Library-like setting
Amazon.com	Individuals who are Internet-active and shop on-line	On-line books, music, and many other items	Personalized on-line service, huge variety

Product-Brand Coordination

After all, customers are buying products or services; a brand is a symbol and a short cut for customers to use in selecting the products they need. Customers' total experience with the products, including purchasing, consuming, and servicing, has to be in tune with the brand position. For example, if a hospital's brand position is to be the premier hospital of choice that provides customers with an attentive team of caring experts working together to provide the highest level of professional care, then every word promised, such as premier, attentive, team, and caring, should have concrete actions behind it. Table 8.6 provides an example of a well-coordinated product-brand combination.

Table 8.6 Product-Brand Coordination of a Hospital

Brand Position	Product (Patient Treatments) Attributes
Premier hospital	• Excellence in all performance metrics • Staffed with first-class doctors, administrators, and nurses • Excellent infrastructure, first-class equipment • Modern appearance, spotless, well organized
Attentive team	• Reduction of patients' waiting time to industry's best • Reduction of unneeded paperwork, and tests • Clearly explained treatment plan, hospital protocols, discharge procedures • Quick feedback to patients' requests
Caring	• Reduction of patients' waiting time to industry's best • Caring nurses • Prompt response to all patients' care issues • Excellent in-patient facility
Expert	• Competent doctors • State-of-the-art medical equipment and first-class technical support
Highest-level of professional care	• Reduction of treatment errors and diagnostic errors to a minimum • Reduction of unneeded treatment to a minimum

The following are the key issues in ensuring product-brand coordination:

1. The brand development team should include product development people.
2. Key product development professionals should learn the basics in brand development and management.
3. Brand positioning and product development should go hand in hand.

Communications and Marketing Strategy

Marketing determines what in the brand's positioning will be communicated and how it will be communicated to the marketplace (Levine 2003). There are numerous vehicles for communicating the brand to the marketplace, and these include

- Advertising
- Internet
- Public relations

- Trade and sales promotions
- Consumer promotions
- Direct marketing
- Event marketing
- Product placement
- Internal employee communications

In order for the brand to achieve its intended positioning in the marketplace, it is very important that the brand image be communicated to the marketplace through various vehicles using an integrated marketing communications strategy (Davis 2000). The message delivered through all these vehicles must be consistent and relate back to the brand image. The communications strategy determines the best mix of vehicles in communicating the brand image.

Develop Channel Strategy

The objective of this step is to determine the appropriate distribution channel strategy that will enhance the brand image, and in the case of existing brands leverage the strength of the brand. The selection of the appropriate distribution channel is very important because of the association that is created between the image of the channel and that of the brand. Also, because a strong brand can create a draw to a distribution channel, it is necessary during this step to leverage the power of the brand to create the best distribution arrangements and ensure more control over the distribution of products and services.

Determine Pricing Strategy

This step focuses on determining the correct pricing policy for the brand. A brand's price must be related to the benefits it provides (Aaker 1996). An overpriced brand will not be rewarded in the marketplace, and an underpriced brand can negate certain associations with the brand's image. Also, the ability to charge premium prices is one of the benefits of developing a strong brand (Davis 2000), so this must be leveraged in determining the pricing strategy.

8.3.5 Brand Implementation

During this phase the plans developed in the brand strategy are executed.

Communicating and Marketing the Brand

The objective in this step is to communicate the brand to the marketplace using the integrated marketing and communication strategy developed in the previous phase. During this step, care is taken to use all the selected communication vehicles effectively to achieve the optimal sales per dollar spent.

Also, in order to improve the effectiveness of the communications, it is important at this stage, to track all marketing expenditures by product, promotional tool, stage of the life cycle, and observed effects in order to establish a baseline for the improvement of the usage of these tools.

Public Relations

The purpose of public relations in the development of the brand image is to encourage the public to have positive feelings about the brand. However, unlike marketing, which is an essential activity that is very visible in the development of the brand, a well-executed public relations campaign is not visible (Levine 2003, p. 17). The public relations activity achieves its objectives by encouraging third parties to deliver positive messages about the brand. These messages are usually delivered through news organizations and print journalists in the form of news and press releases (Levine 2003, p. 17). Because the company does not have any control over the news outlet, a challenge during this step is ensuring that the delivered message is true to the brand image.

8.3.6 Brand Evaluation

Measuring the Return on Brand Investment

This purpose of this step is to measure the performance of the brand in the marketplace. The classical brand management process emphasized two metrics: recall and awareness of the brand (Davis 2000). However, these measures alone are not suited for measuring the brand performance, as determined by the equity value of the brand. In order to provide information for managing the brand as an asset, brand performance measures should (Davis 2000)

- Provide an understanding of how the brand is performing internally and externally
- Provide information about the return on investment of marketing and branding strategies
- Assist the organization in its resource allocation decisions
- Provide information for rewards and incentive systems

Some of the additional brand performance measures include

- Acquired customers
- Lost customers
- Customer satisfaction
- Purchase frequency
- Market share
- Return on advertising
- Price premium

We use the legendary new Marlboro cigarette brand development process as a comprehensive example to illustrate a successful brand development.

Example 8.5: New Marlboro Cigarette Brand Development

This example is from Arnold (1992). The Marlboro cigarette brand is the best-selling packaged cigarette in the world. However, as recently as the late 1950s, it was an old, dying, tobacco brand in the United States. In 1954, after careful analysis of the trends in the tobacco market, the management of Philip Morris made a number of key decisions on the changes of the brand position:

To match the newly designed brand image, the product was totally redesigned. At that time, 90 percent of U.S. smokers used unfiltered cigarettes. The company realized that the coming trend would be filters, and this could also help to

Old Marlboro	New Marlboro
Mild tar blend	Stronger blend
Less flavor	More flavor
Nonfilter	Filter
White pack design	Red and white design
Older image	More modern image
Aimed at women	Aimed at men
Product-based advertising	Imagery advertising

modernize the image of the brand. To change the perception that the Marlboro cigarette is a mild cigarette for women, the flavor of the cigarette was made stronger and the filter was covered in tobacco brown paper, indicating strength and flavor. To shape the new Marlboro brand identity, a new advertisement agency, Leo Burnett, was contracted by Phillip Morris to develop a campaign to relaunch the brand using male role models in tough, rugged jobs, in order to project a new Marlboro brand personality. At the beginning, pilots, deep-sea fishermen, cowboys, and engineers were tried. In 1963, market research indicated that Marlboro needed a more clear-cut identity. The Marlboro Man, symbolized by a cowboy, was established.

Campaign guidelines were laid down as follows:

- The cowboy must symbolize the type of man that other men would prefer to be like and women would like to be with.
- He must be believable.

- Marlboro country must always be magnificent, never ordinary.
- Every ad in the campaign must be candid and have an impact.
- Variety must be achieved by rotation of cowboy portraits, smoking moments, and magnificent country material.

To the present day these guidelines have been maintained throughout all media. To ensure the projection of a consistent brand image, the Marlboro advertisement and campaign style is highly consistent worldwide. After all these efforts, the sales of Marlboro brand cigarettes steadily increased; by 1975, Marlboro had grown to U.S. brand leadership.

Chapter

9

Theory of Inventive Problem Solving (TRIZ)

9.1 Introduction

TRIZ (Teoriya Resheniya Izobreatatelskikh Zadatch) is the theory of inventive problem solving (TIPS) developed in the Soviet Union starting in the late 1940s. TRIZ was developed based on 1500+ person years of research and study of many of the world's most successful solutions of problems from science and engineering, and systematic analysis of successful patents from around the world, as well as the study of the psychological aspects of human creativity (Mann 2002).

Dr. Genrich S. Altshuller, the creator of TRIZ, started the investigation on invention and creativity in 1946. After initially reviewing 200,000 former Soviet Union patent abstracts, Altshuller selected 40,000 as representatives of inventive solutions. He separated the patents' different degrees of inventiveness into five levels, with level 1 being the lowest and level 5 being the highest. He found that almost all invention problems contain at least one contradiction, where a *contradiction* is defined as a situation where an attempt to improve one feature of the system detracts from another feature. He found that the level of invention often depends on how well the contradiction is resolved.

Level 1. Apparent or Conventional Solution: 32 Percent; Solution by Methods Well Known within Specialty

Inventions at level 1 represent 32 percent of the patent inventions and employ obvious solutions drawn from only a few clear options. Actually level 1 inventions are not real inventions but narrow extensions or improvements of existing systems, which are not substantially changed due to the application of the invention. Usually a particular feature is enhanced or strengthened. Examples of level 1 inventions include increasing the thickness of walls to allow for greater insulation in homes or increasing the distance between the front skis on a snowmobile for greater stability. These solutions may represent good engineering, but contradictions are not identified and resolved.

Level 2. Small Invention Inside Paradigm: 45 Percent; Improvement of an Existing System, Usually with Some Compromise

Inventions at level 2 offer small improvements to an existing system by reducing a contradiction inherent in the system while still requiring obvious compromises. These solutions represent 45 percent of inventions. A level 2 solution is usually found through a few hundred trial-and-error attempts and requires knowledge of only a single field of technology. The existing system is slightly changed and includes new features that lead to definite improvements. The new suspension system between the track drive and the frame of a snowmobile is a level 2 invention. The use of an adjustable steering column to increase the range of body types that can comfortably drive an automobile is another example at this level.

Level 3. Substantial Invention Inside Technology: 18 Percent; Essential Improvement of an Existing System

Inventions at level 3 significantly improve the existing system and represent 18 percent of the patents. At this level, an invention contradiction is resolved with the existing system, often through the introduction of some entirely new element. This type of solution may involve 100 ideas, tested by trial and error. Examples include replacing the standard transmission of a car with an automatic transmission, or placing a clutch drive on an electric drill. These inventions usually involve technology that is integral to other industries but is not well known within the industry in which the invention problem arose. The resulting solution causes a paradigm shift within the industry. A level 3 invention is found outside an industry's range of accepted ideas and principles.

Level 4. Invention Outside Technology: 4 Percent; New Generation of Design Using Science, Not Technology

Inventions at level 4 are found in science, not in technology. Such breakthroughs represent about 4 percent of inventions. Tens of thousands of random trials are usually required for these solutions. Level 4 inventions usually lie outside the technology's normal paradigm and involve using a completely different principle for the primary function. In level 4 solutions, the contradiction is eliminated because its existence is impossible within the new system. That is, level 4 breakthroughs use physical effects and phenomena that had previously been little known within the area. A simple example involves using materials with thermal memory (shape-memory metals) for a key ring. Instead of taking a key on or off a steel ring by forcing the ring open, the ring is placed in hot water. The metal memory causes it to open for easy replacement of the key. At room temperature, the ring closes.

Level 5. Discovery: 1 Percent; Major Discovery and New Science

Inventions at level 5 exist outside the confines of contemporary scientific knowledge. Such pioneering works represent less than 1 percent of inventions. These discoveries require lifetimes of dedication for they involve the investigation of tens of thousands of ideas. This type of solution occurs when a new phenomenon is discovered and applied to the invention problem. Level 5 inventions, such as lasers and transistors, create new systems and industries. Once a level 5 discovery becomes known, subsequent applications or inventions occur at one of the four lower levels. For example, the laser, a technological wonder of the 1960s, is now used routinely as a lecturer's pointer and a land surveyor's measuring instrument.

Based on the extensive studies of inventions, other major findings of TRIZ include the following:

1. Through inductive reasoning on millions of patents and inventions, we can find a very small number of inventive principles and strategies that summarize most innovations.
2. Outstanding innovations are often featured by complete resolution of contradictions, not merely a tradeoff and compromise on contradictions.
3. Outstanding innovations are often featured by transforming wasteful, or harmful elements in the system, into useful resources.
4. Technological innovation trends are highly predictable.

9.1.1 What Is TRIZ?

TRIZ is a combination of methods, tools, and a way of thinking (Mann 2002). The ultimate goal of TRIZ is to achieve absolute excellence in design and innovation. In order to achieve absolute excellence, TRIZ has five key philosophical elements.

1. *Ideality:* This is the ultimate criterion for system excellence; it is the maximization of the benefits provided by the system, and minimization of the harmful effects and costs associated with the system.

2. *Functionality:* This is the fundamental building block of system analysis; it builds models about how a system works and how it creates benefits, harm, and costs.
3. *Resource:* Maximum utilization of resources is one of the keys to achieving maximum ideality.
4. *Contradictions:* A contradiction is a common inhibitor for increasing functionality; removing the contradiction usually greatly increases the functionality and raises the system to a totally new performance level.
5. *Evolution:* The evolution trend of the development of technological systems is highly predictable, and it can be used to guide further development.

Based on these five key philosophical elements, TRIZ developed a system of methods. The methods defined here are a complete problem definition and solving process. It is a four-step process, consisting of (1) problem definition, (2) problem classification and tool selection, (3) solution generation, and (4) evaluation.

Problem Definition

This is a very important step in TRIZ. If you can accurately define the right problem, then you have 90 percent of the solution. The problem definition step includes the following tasks:

- *Function analysis:* This includes the function modeling of the system and analysis. This is the most important task in the definition step. TRIZ has highly developed tools for function modeling and analysis.
- *Technological evolution analysis:* This step looks into the relative maturity in technology development of all subsystems and parts. If a subsystem and/or part is technically too mature, it may reach its limit in performance and thus become a bottleneck for the whole system.
- *Ideal final result:* The ideal final result is the virtual limit of a system in TRIZ. It may never be achieved, but it provides us with an ultimate dream and will help us to think "out of the box."

Problem Classification and Tool Selection

TRIZ has a large array of tools for inventive problem solving; however, we must select the right tool for the right problem. In TRIZ, we must first classify the problem type and then select the tools accordingly.

Solution Generation

In this step, we apply TRIZ tools to generate solutions for the problem. Because TRIZ has a rich array of tools, it is possible to generate many solutions.

Evaluation

In any engineering project, we need to evaluate the soundness of the new solution. TRIZ has its own evaluation approach. However, other non-TRIZ methods might also be used at this stage, such as axiomatic design and design vulnerability analysis.

In subsequent sections, we first discuss the philosophical aspects of TRIZ in order to lay a foundation for understanding. Then we discuss the four-step TRIZ problem definition and solving process, together with the tools used in TRIZ.

9.2 TRIZ Fundamentals

Ideality, functionality, contradictions, use of resources, and evolution are the pillars of TRIZ. These elements make TRIZ distinctively different from other innovation and problem-solving strategies. In this section, we describe all five elements.

9.2.1 Function Modeling and Functional Analysis

Function modeling and functional analysis originated in value engineering (Miles 1961). A function is defined as the natural or characteristic action performed by a product or service. Usually, a product or service provides many functions. For example, an automobile provides customers with the ability to get from point A to point B, with a comfortable riding environment, air conditioning, music, and so on.

Among all functions, the most important is called the *main basic function*. It is defined as the primary purpose or the most important action performed by a product or service. The main basic function must always exist, although methods or designs to achieve it may vary. For example, for an automobile, the ability to get from point A to B is the main basic function.

Besides the main basic function, there are other useful functions as well; we can call these *secondary useful functions*. There are several kinds of secondary useful functions:

1. *Secondary basic functions:* These are not main basic functions, but customers definitely need them. For example, providing a comfortable riding environment is a must-have function for automobiles.
2. *Nonbasic but beneficial functions:* These are functions that provide customers with esteem value, comfort, and so on. For example, the paint finish on an automobile provides both basic and nonbasic

functions; it protects the automobile from corrosion and rust, as well as creating a sleek look for the car.

Besides secondary useful functions, there are two other types of functions:

1. *Supporting function:* This function supports the main basic function or another useful function. A supporting function results from the specific design approach to achieve the main basic function or other useful functions. As the design approach to achieve the main basic function and other useful functions changes, supporting functions may also change. There are at least two kinds of supporting functions, assisting functions and correcting functions.
   - *Assisting functions:* These functions assist other useful functions. For example, the engine suspension system provides the function of locking the position of the engine in the automobile so that the engine can provide power without falling off the car.
   - *Correcting functions:* These functions correct the negative effects of another useful function. For example, the main basic function of the water pump in the automobile internal combustion engine is to circulate water in the engine system in order to cool it off; it is a correcting function for the automobile engine. The main basic function of the engine is to provide power for the automobile, but the internal combustion engine also creates negative effects, such as heat. A water pump's function is to correct this negative effect. If we change the design and use electricity as the power source of the automobile, the function of a water pump will no longer be needed.
2. *Harmful function:* This is an unwanted, negative function caused by the method used to achieve useful functions. For example, an internal combustion engine not only provides power; it also generates noise, heat, and pollution, and these are harmful functions.

In summary, the main basic function and secondary useful functions provide benefits for the customer. Supporting functions are useful, at least they are not harmful, but they do not provide benefits directly to the customer and do incur costs. Harmful functions are not useful and provide no benefits at all.

Functional Statement

A function can be usually fully described in three parts: a subject, a verb, and an object. For example, for the automobile, its main basic function can be described as

Car	**moves**	**people**
(Subject)	(Verb)	(Object)

For a toothbrush, its main basic function can be described as

Toothbrush   **brushes**   **teeth**
(Subject)   (Verb)   (Object)

Functional Analysis Diagram

The functional analysis diagram is a graphic tool used to describe and analyze functions. The following graph is a typical template for a functional analysis diagram:

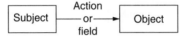

Where the *subject* is the source of action, the *object* is the action receiver. Action is the *verb* in the functional statement, and it is represented by an arrow. In a technical system, the action is often accomplished by applying some kind of *field*, such as a mechanical, electrical, or chemical field. For example, the function "brush teeth" can be described by the following functional analysis diagram:

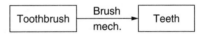

In the diagram, *Mech.* stands for "mechanical field." Clearly, brushing teeth is an application of one kind of mechanical field, force.

In the functional analysis diagram, there are four types of actions. They are represented by four types of arrows as illustrated in Fig. 9.1.

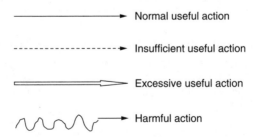

Figure 9.1 Legends for Various Actions in the Functional Analysis Diagram

Example 9.1: Brushing Teeth

If we use a toothbrush correctly, and our teeth get cleaned properly, then we call this brush action a normal useful action. We can illustrate this by the following functional analysis diagram:

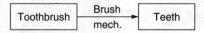

However, if we use the toothbrush too gently and do not brush long enough, or we use a worn toothbrush, then our teeth will not get enough cleaning. In this case, we can use the following functional analysis diagram:

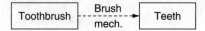

Clearly, this is a case of an insufficient useful action.

If we use a very strong toothbrush and brush our teeth with much force and big strokes, then our gums will get hurt, and so will our teeth. We can use the following functional analysis diagram to describe this situation:

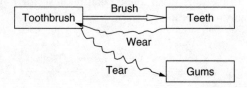

That is, the toothbrush delivers excessive brushing action to the teeth. The excessive toothbrush action is harmful since it tears the gums and makes them bleed. The teeth also may deliver a harmful action, by causing wearing of the toothbrush.

Functional Modeling and Analysis Example

Figure 9.2 is a schematic view of an overhead projector. Figure 9.3 shows a graph of the functional modeling and analysis diagram for the whole system.

In this functional analysis graph, E stands for electric field and M stands for mechanical field. In this example, there are many chains of action; that is, an object can be another object's subject. Then we have a sequence of subject-action-object-action chains. Each chain describes a complete function. We can identify the following functions:

1. From *electric power* to *image* to *screen*, that is, the function of projecting an image in the film onto the screen. We can think of this as the main basic function.

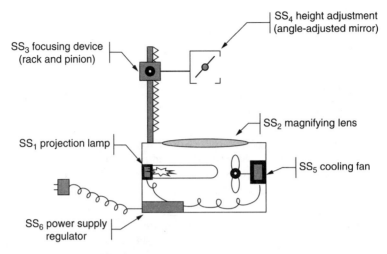

Figure 9.2 Overhead Projector

2. From *hand* to *focusing adjuster* to *mirror*, that is, the function of focusing the image. This is a secondary basic function.
3. From *hand* to *angular adjuster* to *mirror*, that is, the function of projecting the image to the right position on the screen. This is also a secondary basic function.
4. From *electric power* to *projection lamp* to *lens*, etc. There is a harmful function chain, there are several harmful functions that are all related to unwanted heat, without correction, these harmful functions will damage the film and device.
5. Because of the harmful function, we have to add a supporting function, that is the chain from *electric power* to *fan* and end with *lens* and *film*.

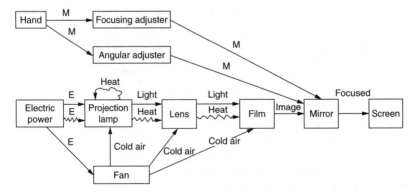

Figure 9.3 Whole System Functional Modeling and Analysis Diagram

This function is a correcting function to compensate for the negative effect of the harmful function.

9.2.2 Resources

The maximum effective use of resources is very important in TRIZ. We also need to think of resources and make use of them in creative ways.

The primary mission for any product or process is to deliver functions. Because substances and fields are basic building blocks of functions, they are important resources from the TRIZ point of view. However, substances and fields are not sufficient by themselves to build and deliver functions. Space and time are also needed; they are also important resources. From the TRIZ point of view, an information and knowledge base about how to use available resources is also an important resource.

We can segment resources into the following categories:

1. Substance resources
    a. Raw materials and products
    b. Waste
    c. By-products
    d. System elements
    e. Substance from surrounding environments
    f. Inexpensive substances
    g. Harmful substances from the system
    h. Altered substances from the system
2. Field resources
    a. Energy in the system
    b. Energy from the environment
    c. Energy or field that can be built upon existing energy platforms
    d. Energy or field that can be derived from system waste
3. Space resources
    a. Empty space
    b. Space at interfaces of different systems
    c. Space created by vertical arrangement
    d. Space created by nesting arrangement
    e. Space created by rearrangement of existing system elements
4. Time resources
    a. Prework period
    b. Time slot created by efficient scheduling
    c. Time slot created by parallel operation
    d. Post work period

5. Information and knowledge resources
   a. Knowledge of all available substances (material properties, transformations, etc.)
   b. Knowledge of all available fields (field properties, utilizations, etc.)
   c. Past knowledge
   d. Other people's knowledge
   e. Knowledge of operation
6. Functional resources
   a. Unutilized or underutilized existing system main functions
   b. Unutilized or underutilized existing system secondary functions
   c. Unutilized or underutilized existing system harmful functions

In TRIZ, it is more important to look into cheap, ready-to-use, abundant resources than expensive, hard-to-use, and scarce resources. Here is an example.

Example 9.2: Cultivating Fish in Farmland
The southeastern part of China is densely populated, so land is a scarce resource. Many pieces of land are used to plant rice. Agricultural experts suggest that farmland can be used to cultivate fish while it is used to grow rice, because in rice paddies, water is a free and ready resource, and the waste from fish can be used as a fertilizer for the rice.

9.2.3 Ideality

Ideality is a measure of excellence. In TRIZ, ideality is defined by the following ratio:

$$\text{Ideality} = \frac{\sum \text{benefits}}{\sum \text{costs} + \sum \text{harm}} \qquad (9.1)$$

where $\sum$ benefits = sum of values of system's useful functions (Here the supporting functions are not considered as useful functions, because they will not bring benefits to customers directly; we consider supporting functions to be part of the costs to make the system work.)

$\sum$ costs = sum of expenses for systems performance

$\sum$ harm = sum of all harm created by harmful functions

In Eq. (9.1), a higher ratio indicates a higher ideality. When a new system is able to achieve a higher ratio than that of the old system, we consider it a real improvement.

In TRIZ, there is a law of increasing ideality, which states that the evolution of all technical systems proceeds in the direction of increasing degree of ideality. The ideality of the system will increase in the following cases:

1. Increasing benefits
2. Reducing costs
3. Reducing harm
4. Benefits increasing faster than costs and harm

From the TRIZ point of view, any technical system or product is not a goal in itself. The real value of the product or system is in its useful functions. Therefore, the better system is the one that consumes fewer resources in both initial construction and maintenance.

When the ratio becomes infinite, we call this the *ideal final result* (IFR). Thus, the IFR system requires no material, consumes no energy on space, needs no maintenance, and will not break.

9.2.4 Contradiction

From the TRIZ standpoint, a challenging problem can be expressed as either a technical contradiction or a physical contradiction.

Technical Contradiction

A technical contradiction is a situation where efforts to improve some technical attributes of a system lead to deterioration of other technical attributes. For example, as a container becomes stronger, it becomes heavier, and faster automobile acceleration reduces fuel efficiency.

A problem associated with a technical contradiction can be resolved either by finding a tradeoff between the contradictory demands or by overcoming the contradiction. Tradeoff or compromise solutions do not eliminate the technical contradictions, but rather soften them, thus retaining the harmful (undesired) action or shortcoming in the system. Analysis of thousands of inventions by Altshuller resulted in formulation of typical technical contradictions, such as productivity versus accuracy, reliability versus complexity, and shape versus speed. It was discovered that despite the immense diversity of technological systems and even greater diversity of inventive problems, there are only about 1250 typical system contradictions. These contradictions can be expressed as a table of contradiction of 39 design parameters (Table 9.1).

From the TRIZ standpoint, overcoming a technical contradiction is very important, because by overcoming a contradiction, both attributes in the contradiction can be improved drastically and the performance of the system

Table 9.1 Thirty-nine Parameters

1. Weight of moving object	21. Power
2. Weight of nonmoving object	22. Waste of energy
3. Length of moving object	23. Waste of substance
4. Length of nonmoving object	24. Loss of information
5. Area of moving object	25. Waste of time
6. Area of nonmoving object	26. Amount of substance
7. Volume of moving object	27. Reliability
8. Volume of nonmoving object	28. Accuracy of measurement
9. Speed	29. Accuracy of manufacturing
10. Force	30. Harmful factors acting on object
11. Tension, pressure	31. Harmful side effects
12. Shape	32. Manufacturability
13. Stability of object	33. Convenience of use
14. Strength	34. Repairability
15. Durability of moving object	35. Adaptability
16. Durability of nonmoving object	36. Complexity of device
17. Temperature	37. Complexity of control
18. Brightness	38. Level of automation
19. Energy spent by moving object	39. Productivity
20. Energy spent by nonmoving object	

will be raised to a whole new level. TRIZ developed many tools for elimination of technical contradictions.

Physical Contradiction

A physical contradiction is a situation where a subject or an object has to be in a mutually exclusive physical state. A physical contradiction has the typical pattern: To perform function F_1, the element must have property P, but to perform function F_2, it must have property $-P$, or the opposite of P. For example, an automobile has to be light in weight (P) to have high fuel economy (F_1), but it also has to be heavy in weight ($-P$) in order to be stable in driving (F_2).

Example 9.3
Problem: Some buildings are supported by piles. The pile should have a sharp tip to facilitate the driving process. However, the sharp piles have reduced support capability. For better support capacity, the piles should have blunt ends. However, it is more difficult to drive a blunt-tipped pile.
Contradiction: A pile should be sharp to facilitate the driving process, and it should be blunt to provide better support of the foundation.
TRIZ Solution: The situation clearly calls for the solution providing separation of contradictory properties in time. The pile is sharp during the driving process, and then its base is expanded, which could be realized by a small explosive charge.

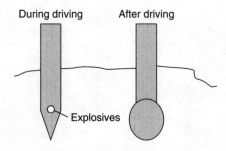

Conventional design philosophy is based on compromises (tradeoffs). Contrary to this approach, TRIZ offers several methods to overcome physical contradictions completely.

9.2.5 S-curve and the Evolution of a Technical System

Based on research of the evolution histories of many technical systems, TRIZ researchers have found that the trends of evolution of many technical systems are very similar and predictable. They found that many technical systems go through five stages in their evolution process. These five stages are pregnancy, infancy, growth, maturity, and decline. If we plot a time line on the horizontal axis (X axis), and plot

1. Performance
2. Level of inventiveness
3. Number of inventions (relating to the system)
4. Profitability of inventions

on the verticle axis (Y axis), we get the four curves shown in Fig. 9.4. Because the shape of the first curve (performance versus evolution stages) (Fig. 9.4a) has an S-shape; it is also called an S-curve.

Pregnancy

For a technical system, its pregnancy stage is the time between an idea's inception and its birth. A new technological system emerges only after the following two conditions are satisfied:

- There is a need for the function of this system
- There are means (technology) to deliver this function

The development of a technical system say, the airplane, can be used as an example. The need for the function of the airplane, that is, "to fly" was there a long time ago in many people's dreams and desires. However, the technical knowledge of aerodynamics and mechanics was not sufficient for the development of human flight until the 1800s. The technologies for the airplane became available after the development of glider flight in 1848 and the gasoline engine in 1859. It was the Wright brothers who successfully integrated both technologies in their aircraft in 1903—and a new technology got off the ground.

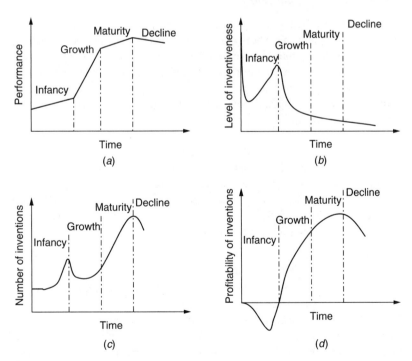

Figure 9.4 Curves of Technical System Evolution

Infancy

The birth of a new technical system is the starting point of the infancy stage; it is the first stage of an S-curve. The new system appears as a result of a high-level invention. Typically, the system is primitive, inefficient, and unreliable and has many unsolved problems. It does, however, provide some new functions or the means to provide the function. System development at this stage is very slow, due to lack of human and financial resources. Many design questions and issues must be answered. For example, most people may not be convinced of the usefulness of the system, but a small number of enthusiasts who believe in the system's future continue to work toward its success.

In the infancy stage, the performance level is low and its improvement is slow (Fig. 9.4a). The level of inventions is usually high, because the initial concept is often very inventive and patentable. It is usually level 3, 4, or even 5 (Fig. 9.4b). But the number of inventions in this system is usually low (Fig. 9.4c), because the system is fairly new. The profit is usually negative (Fig. 9.4d), because at this stage of the technology usually the customers are few but the expense is high.

Growth (Rapid Development)

This stage begins when society realizes the value of the new system. By this time, many problems have been overcome; efficiency and performance have improved in the system, people and organizations invest money in development of the new product or process. This accelerates the system's development, improving the results and, in turn, attracting greater investment. Thus, a positive *feedback* loop is established, which serves to further accelerate the system's evolution.

In the growth stage, the improvement of performance level is quick (Fig. 9.4a) because of the rapid increases in investment and the removal of many technical bottlenecks. The level of inventions is getting lower because most inventions in this stage deal with incremental improvements. They are mostly level 1 or 2 (Fig. 9.4b), but the number of inventions is usually high (Fig. 9.4c). The profit is usually growing fast (Fig. 9.4d).

Maturity

In this stage, system development slows as the initial concept upon which the system was based nears exhaustion of its potential. Large amounts of money and labor may have been expended; however, the results are usually very marginal. At this stage, standards are established. Improvements occur through system optimization and tradeoffs. The performance of the system

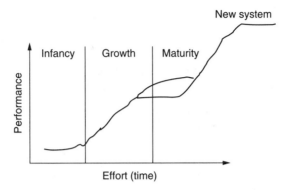

Figure 9.5 S-curve for Two Generations of a System

still grows but at a slower pace (Fig. 9.4a). The level of invention is usually low (Fig. 9.4b), but the the number of inventions in the forms of industrial standards is quite high (Fig. 9.4c). The profitability is usually dropping because of saturation of the market and increased competition (Fig. 9.4d).

Decline

At this stage, the limits of technology have been reached and no fundamental improvement is available. The system may no longer be needed, because the function provided may no longer be needed.

It is really important to start the next generation of the technical system long before the decline stage in order to avoid failure of the company. Figure 9.5 illustrates the S-curves of the succession of two generations of a technical system.

9.3 TRIZ Problem-Solving Process

TRIZ has a four-step problem-solving process. The four steps are (1) problem definition, (2) problem classification and problem tool selection, (3) problem solution, and (4) solution evaluation. We describe each step in detail.

9.3.1 Problem Definition

Problem definition is a very important step. The quality of the solution is highly dependent on the problem definition.

The problem definition starts with several questions:

1. What is the problem?
2. What is the scope of the project?
3. What subsystem, system, and components are involved?
4. Do we have a current solution? Why is the current solution not good?

These are common questions to be asked in any engineering project. By answering them, we are able to define the scope of the project and focus on the right problem area.

Besides answering these common questions, several TRIZ methods are also very helpful in the problem definition stage.

Functional Modeling and Functional Analysis

After identifying the project scope, it is very helpful to establish the functional model of the subsystem involved in this project. Functional modeling and analysis enables us to see the problem more clearly and precisely. We recall the toothbrush example (Example 9.1) to illustrate how functional analysis can help in the problem definition.

Example 9.4: Toothbrush Problem Revisited
Assume that we are a toothbrush manufacturer and the current regular toothbrush does not perform satisfactorily; that is, the teeth cannot be adequately cleaned. We can first draw the following functional diagram:

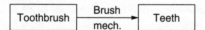

By analyzing the functional diagram, we may come up with the following possibilities:

1. The current lack of performance may be caused by inadequate action; that is, the actual functional diagram is the following:

   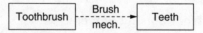

   If that is the case, it belongs to the problem of inadequate functional performance; we can use the TRIZ standard solution technique to resolve this problem.
2. We may find the current functional model is too limiting, because the function statement "toothbrush brush teeth" limits our solution to using only a toothbrush and to use a mechanical action only. We can develop the following alternative functional modeling:

   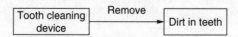

The subject *toothbrush* is replaced by the more general *tooth cleaning device*. The object *teeth* is changed to *dirt in teeth*; it is more precise. The action *brush* is changed to the more general term *remove*. Under this alternative functional modeling, many possible choices of subjects and actions can be open for selection. For example, we can use hydraulic action or chemical action to clean teeth; we can even consider pretreatment of teeth to make them dirt-free and so on. Clearly this alternative functional modeling opens up a lot of gates for problem solving and innovation.

Ideality and Ideal Final Result

After functional modeling and functional analysis, we can evaluate the ideality of the current system by using

$$\text{Ideality} = \frac{\sum \text{benefits}}{\sum \text{costs} + \sum \text{harm}}$$

Ideal final result means the ultimate optimal solution for the current system in which

$$\sum \text{Benefits} \to \infty \quad \text{and} \quad \sum \text{Costs} + \sum \text{harm} \to 0$$

By comparing the ideality of the current system with the ideal final result, we can identify where the system improvement should go and what aspects of the system should be improved. This will definitely help the problem definition and identify what problem should be solved.

S-curve Analysis

It is very beneficial to evaluate the evolution stage of the current technical system involved in any TRIZ project. For example, if our current subsystem is at the growth stage, then we should focus our attention on gradual improvement. If our subsystem is near the maturity stage, then we will know that it is time to develop the next generation of this subsystem.

Contradiction Analysis

By using the method described in Sec. 9.2.4, we can identify if there are any physical or technical contradictions in our current system. TRIZ has many methods to resolve contradictions.

9.3.2 Problem Classification and Tool Selection

After we are finished with the problem definition, we should be able to classify the problem into the following categories. For each category, there are many TRIZ methods available to resolve the problem.

1. *Physical contradiction*
   Methods: Physical contradiction-resolution by using separation principles.
2. *Technical contradiction*
   Methods: Inventive principles
3. *Imperfect functional structures:* This problem occurs when
   - There are inadequate useful functions or lack of needed useful functions
   - There are excessive harmful functions

   Methods: Functional improvement methods and TRIZ standard solutions
4. *Excessive complexity:* This problem occurs when the system is too complex and costly and some of its functions can be eliminated or combined.
   Methods: Trimming and pruning
5. *System improvement:* This problem occurs when the current system is doing its job but enhancement is needed to beat the competition
   Method: Evolution of technological systems
6. *Develop useful functions:* This problem occurs when we can identify what useful functions are needed to improve the system but we do not know how to create these functions.
   Methods: Physical, chemical, and geometric effects database

9.3.3 Solution Generation

After problem classification, there are usually many TRIZ methods available for solving the problem, so many alternative solutions could be found. These solutions will be evaluated in the next step.

9.3.4 Concept Evaluation

There are many concept evaluation methods that can be used to evaluate and select the best solution. These methods are often not TRIZ-related. The frequently used concept evaluation methods include Pugh concept selection, value engineering, and axiomatic design method.

9.4 Technical Contradiction Elimination and Inventive Principles

Genrich Altshuller analyzed more than 40,000 patents and identified about 1250 typical technical contradictions. These contradictions are further expressed by a matrix of 39 by 39 engineering parameters. To resolve these contradictions, Altshuller compiled 40 principles. Each of these principles contains a few subprinciples, totaling up to 86 subprinciples.

It should be noted that the 40 principles are formulated in a general way. If, for example, the contradiction table recommends principle 30, flexible shell and thin films, it means that the solution of the problem relates somehow to changing the degree of flexibility or adaptability of the technical system being modified.

The contradiction table and the 40 principles do not offer the direct solution to the problem; they only suggest the most promising directions for searching for a solution. The problem solver has to interpret these suggestions and find the way in which they can be applied to a particular situation.

Usually people solve problems by analogical thinking. We try to relate the problem confronting us to some familiar standard class of problems (analogs) for which a solution exists. If we draw upon the right analog, we arrive at a useful solution. Our knowledge of analogous problems is the result of educational, professional, and life experiences.

What if we encounter a problem analogous to one we have never faced? This obvious question reveals the shortcomings of our standard approach to invention problems. So, the contradiction table and 40 principles offer us clues to the solution of problems with which we are not familiar.

When using the contradiction table and 40 principles, following this simple procedure will be helpful:

1. Decide the attribute to be improved, and use one of the 39 parameters in the contradiction table to standardize or model this attribute.
2. Answer the following questions:
   a. How can this attribute be improved using conventional means?
   b. Which attribute would be deteriorated, if conventional means were used?
3. Select an attribute in the contradiction table corresponding to step 2b.
4. Using the contradiction table, identify the principles in the intersection of the row (attribute improved) and column (attribute deteriorated) for overcoming the technical contradiction.

Here we list the 40 principles as a reference.

1. Segmentation
   - Divide an object into independent parts.
   - Make an object easy to disassemble.
   - Increase the degree of fragmentation (or segmentation) of an object.
2. Taking out
   - Separate an interfering part (or property) from an object, or single out the only necessary part (or property) of an object.

3. Local quality
   - Change an object's structure from uniform to nonuniform; change an external environment (or external influence) from uniform to nonuniform.
   - Make each part of an object function in conditions most suitable for its operation.
   - Make each part of an object fulfill a different and useful function.
4. Asymmetry
   - Change the shape of an object from symmetrical to asymmetrical.
   - If an object is asymmetrical, increase its degree of asymmetry.
5. Merging
   - Bring closer together (or merge) identical or similar objects; assemble identical or similar parts to perform parallel operations.
   - Make operations contiguous or parallel, and bring them together in time.
6. Universality
   - Make a part or object perform multiple functions to eliminate the need for other parts.
7. Nested doll
   - Place one object inside another; place each object, in turn, inside the other.
   - Make one part pass through a cavity in the other.
8. Antiweight
   - To compensate for the weight of an object, merge it with other objects that provide lift.
   - To compensate for the weight of an object, make it interact with the environment (e.g., use aerodynamic, hydrodynamic, buoyancy, and other forces).
9. Preliminary antiaction
   - If it will be necessary to perform an action with both harmful and useful effects, this action should be replaced later with antiactions to control harmful effects.
   - Create beforehand stresses in an object that will oppose known undesirable working stresses later on.
10. Preliminary action
    - Perform, before it is needed, the required change of an object (either fully or partially).
    - Prearrange objects so that they can come into action from the most convenient place without losing time for their delivery.
11. Beforehand cushioning
    - Prepare emergency means beforehand to compensate for the relatively low reliability of an object.

12. Equipotentiality
    - In a potential field, limit position changes (e.g., change operating conditions to eliminate the need to raise or lower objects in a gravity field).
13. The other way around
    - Invert the action(s) used to solve the problem (e.g., instead of cooling an object, heat it).
    - Make movable parts (or the external environment) fixed, and fixed parts movable.
    - Turn the object (or process) upside down.
14. Spheroidality
    - Instead of using rectilinear parts, surfaces, or forms, use curvilinear ones; move from flat surfaces to spherical ones, from parts shaped as a cube (parallelepiped) to ball-shaped structures.
    - Use rollers, balls, spirals, domes.
    - Go from linear to rotary motion; use centrifugal forces.
15. Dynamics
    - Allow (or design) the characteristics of an object, external environment, or process to change to be optimal or to find an optimal operating condition.
    - Divide an object into parts capable of movement relative to each other.
    - If an object (or process) is rigid or inflexible, make it movable or adaptive.
16. Partial or excessive actions
    - If 100 percent of an effect is hard to achieve using a given solution method, then by using slightly less or slightly more of the same method, the problem may be considerably easier to solve.
17. Another dimension
    - Move an object in two- or three-dimensional space.
    - Use a multistory arrangement of objects instead of a single-story arrangement.
    - Tilt or reorient the object; lay it on its side.
    - Use another side of a given area.
18. Mechanical vibration
    - Cause an object to oscillate or vibrate.
    - Increase its frequency (even up to the ultrasonic).
    - Use an object's resonance frequency.
    - Use piezoelectric vibrators instead of mechanical ones.
    - Use combined ultrasonic and electromagnetic field oscillations.
19. Periodic action
    - Instead of continuous action, use periodic or pulsating actions.
    - If an action is already periodic, change the periodic magnitude or frequency.
    - Use pauses between impulses to perform a different action.

20. Continuity of useful action
    - Carry on work continuously; make all parts of an object work at full load, all the time.
    - Eliminate all idle or intermittent actions or work.
21. Skipping
    - Conduct a process, or certain stages of the process (e.g., destructive, harmful, or hazardous operations) at high speed.
22. Blessing in disguise
    - Use harmful factors (particularly, harmful effects of the environment or surroundings) to achieve a positive effect.
    - Eliminate the primary harmful action by adding it to another harmful action to resolve the problem.
    - Amplify a harmful factor to such a degree that it is no longer harmful.
23. Feedback
    - Introduce feedback (referring back, cross-checking) to improve a process or action.
    - If feedback is already used, change its magnitude or influence.
24. Intermediary
    - Use an intermediate carrier article or intermediary process.
    - Merge one object temporarily with another (which can be easily removed).
25. Self-service
    - Make an object serve itself by performing auxiliary helpful functions.
    - Use waste resources, energy, or substances.
26. Copying
    - Instead of an unavailable, expensive, fragile object, use simpler and inexpensive copies.
    - Replace an object or process with optical copies.
    - If visible optical copies are already used, move to infrared or ultraviolet copies.
27. Cheap short-living
    - Replace an expensive object with a multitude of inexpensive objects, compromising certain qualities (such as service life, for instance).
28. Mechanics substitution
    - Replace a mechanical means with a sensory (optical, acoustic, taste, or smell) means.
    - Use electric, magnetic, and electromagnetic fields to interact with the object.
    - Change from static to movable fields, from unstructured fields to those having structure.
    - Use fields in conjunction with field-activated (e.g., ferromagnetic) particles.

29. Pneumatics and hydraulics
    - Use gas and liquid parts of an object instead of solid parts (e.g., inflatable, filled with liquids, air cushion, hydrostatic, hydroreactive).
30. Flexible shells and thin films
    - Use flexible shells and thin films instead of three-dimensional structures.
    - Isolate the object from the external environment using flexible shells and thin films.
31. Porous materials
    - Make an object porous or add porous elements (inserts, coatings, etc.).
    - If an object is already porous, use the pores to introduce a useful substance or function.
32. Color changes
    - Change the color of an object or its external environment.
    - Change the transparency of an object or its external environment.
33. Homogeneity
    - Make objects interacting with a given object of the same material (or a material with identical properties).
34. Discarding and recovering
    - Make portions of an object that have fulfilled their function go away (discard by dissolving, evaporating, etc.) or modify these directly during operation.
    - Conversely, restore consumable parts of an object directly during operation.
35. Parameter changes
    - Change an object's physical state (e.g., to a gas, liquid, or solid).
    - Change the concentration or consistency.
    - Change the degree of flexibility.
    - Change the temperature.
36. Phase transitions
    - Use phenomena occurring during phase transitions (e.g., volume changes, loss or absorption of heat).
37. Thermal expansion
    - Use thermal expansion (or contraction) of materials.
    - If thermal expansion is being used, use multiple materials with different coefficients of thermal expansion.
38. Strong oxidants
    - Replace common air with oxygen-enriched air.
    - Replace enriched air with pure oxygen.
    - Expose air or oxygen to ionizing radiation.
    - Use ozonized oxygen.
    - Replace ozonized (or ionized) oxygen with ozone.

39. Inert atmosphere
    - Replace a normal environment with an inert one.
    - Add neutral parts or inert additives to an object.
40. Composite materials
    - Change from uniform to composite (multiple) materials.

Example 9.5: Using 40 Principles and the Contradiction Matrix to Improve Wrench Design

When we use a conventional wrench to undo an overtightened or corroded nut (as shown in the following picture), one of the problems is that the corners of the nut receive a concentrated load, so they may wear out quickly. You can reduce the clearance between the wrench and nut, but it will be difficult to fit the wrench onto the nut. Is there anything we can do to solve this problem?

Clearly we want to reduce the clearance between the wrench and nut to improve operation reliability; however, this leads to the deterioration of operations. From the TRIZ standpoint, a technical contradiction is present when a useful action simultaneously causes a harmful action.

A problem associated with a technical contradiction can be resolved either by finding a tradeoff between the contradictory demands or by overcoming the contradiction. Tradeoff or compromise solutions do not eliminate the technical contradictions, but rather soften them, thus retaining harmful (undesired) actions or shortcomings, in the system. An engineering problem becomes an invention one when it has a technical contradiction that cannot be overcome by conventional means and tradeoff solutions are not acceptable. The 40 principles and the contradiction matrix are important tools for overcoming contradictions.

Step 1. Build contradiction model: Look into the problems and find a pair of contradictions. The contradiction should be described using 2 of the 39 parameters for technical contradictions. In this problem, the contradiction is
- Things we want to improve: Reliability (parameter 27)
- Things are getting worse: Ease of operation (parameter 33)

Step 2. Check contradiction matrix: Locate the parameter to be improved in the row and the parameter to be deteriorated in the column in the contradiction matrix for inventive principles. The matrix offers the following principles 27, 17, and 40 (see the following partial matrix).

What should be improved?	25. Waste of time	26. Quantity of substance	27. Reliability	28. Measurement accuracy	29. Manufacturing precision	30. Harmful action at object	31. Harmful effect caused by the object	32. Ease of manufacture	33. Ease of operation	34. Ease of repair	35. Adaptation
25. Waste of time		35 38 18 16	10 30 4	24 34 28 32	24 26 28 18	35 18 34	35 22 18 39	35 28 34	4 28 10 34	32 1 10	35 28
26. Quantity of substance	35 38 18 16		18 3 28 40	3 2 28	33 30	35 33 29 31	3 35 40 39	29 1 35 27	35 29 10 25	2 32 10 25	15 3 29
27. **Reliability**	10 30 4	21 28 40 3		32 3 11 23	11 32 1	27 35 2 40	35 2 40 26		**27 17 40**	1 11	13 35 8 24
28. Measurement accuracy	24 34 28 32	2 6 32	5 11 1 23			28 24 22 26	3 33 39 10	6 35 25 18	1 13 17 34	1 32 13 11	13 35 2
29. Manufacturing precision	32 26 28 18	32 30	11 32 1			26 28 10 36	4 17 34 26		1 32 35 23	25 10	

Step 3. Interpret principles: Read each principle and construct analogies between the concepts of principle and your situation, and then create solutions to your problem. Principle 17 (another dimension) indicates that the wrench problem may be resolved by moving an object in a two- or three-dimensional space or using a different side of the given area. From principle 27 (cheap short-living) and principle 40 (composite material), we may replace an expensive object with a multitude of inexpensive objects and change from uniform material to composite material.

Step 4. Resolve the problem: The working surface of the wrench can be redesigned in a nonuniform shape by applying principle 17 (see the following picture). Principles 27 and 40 can be used together. The idea is to attach soft metal or plastic pads on the wrench working surfaces when tightening or undoing expensive nuts.

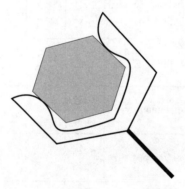

9.5 TRIZ Applications in the Service Industry

TRIZ research started with the study of patents. Most TRIZ principles and methods are based on knowledge accumulated in technical areas. However, many researchers have been studying how to extend TRIZ to nontechnical areas.

Since the 1970s, there have been continuous efforts made by Russian TRIZ researchers to extend TRIZ into nontechnical areas; these efforts were documented in detail in an excellent paper by Zlotin et al. (1999). The original 40 inventive principles were developed based on the study of a huge number of patents to find inventive solutions for technical problems. Darrell Mann and Ellen Domb (1999) and Mann (2004) studied more than 800 business case studies and developed the first generation 40 inventive business principles. In the software industry, Kevin Rae (2001) developed 40 inventive principle analogies of TRIZ in the context of software and computing. Inventive principles have also been adapted in service operation management (Zhang et al. 2003), quality management (Retseptor 2003), and education (Marsh et al. 2002).

In his book *Hands-on Systematic Innovation for Business and Management* Mann (2004) tries to extend all major TRIZ methods, such as functional modeling and analysis, trends of technological evolution, separation principles, inventive principles and trimming to the business and management area.

In summary, the extension of TRIZ into nontechnical areas is still young and in an early stage. We will expect more breakthroughs in the development of systematic innovation methods in the service industry in the next few years.

In this book, we will concentrate on introducing the 40 inventive business principles. We hope this will lay a solid foundation for readers to apply inventive principles to their own service application areas.

9.6 Business Inventive Principles

Similar to the 40 regular inventive principles, the 40 inventive business principles are aimed to resolve business contradictions. In the regular 40 inventive principles, the contradictions are expressed by a matrix of 39 by 39 engineering parameters. Darrell Mann identified 31 business parameters, and business contradictions are expressed by a matrix of 31 by 31 business parameters. These 31 business parameters are illustrated in Table 9.2.

Again, the business contradiction table and the 40 inventive business principles do not offer a direct solution to the problem; they only suggest the most promising directions for searching for a solution. The problem solver has to interpret these suggestions and find how to apply them to a particular situation.

When using the business contradiction table and 40 inventive business principles, following this simple procedure will be helpful:

1. Decide the attribute to be improved, and use one of the 31 parameters in the contradiction table to standardize or model this attribute.
2. Answer the following questions:
    a. How can this attribute be improved using conventional means?
    b. Which attribute would be deteriorated, if conventional means were used?

Table 9.2 Thirty-one Business Parameters (*Mann 2004*)

Categories	Parameters
Research and development	R&D specification/capacity/means
	R&D cost
	R&D time
	R&D risk
	R&D interface
Production	Production specification, capacity, and means
	Production cost
	Production time
	Production risk
	Production interface
Supply	Supply specification, capacity, and means
	Supply cost
	Supply time
	Supply risk
	Supply interface
Support	Support specification, capacity, and means
	Support cost
	Support time
	Support risk
	Support interface
Customer	Customer revenue, demand, and feedback
	Amount of information
	Communication Flow
	Harmful factors affecting systems

(*Continued*)

Table 9.2 Thirty-one Business Parameters (*Mann 2004*) (*Continued*)

Categories	Parameters
Customer (*Cont.*)	System-generated harmful factors
	Convenience
	Adaptability and versatility
	System complexity
	Control complexity
	Tension/stress
	Stability

3. Select an attribute in the contradiction table corresponding to step 2b.
4. Use the contradiction table, identify the principles in the intersection of the row (attribute improved) and column (attribute deteriorated) for overcoming the technical contradiction.

Here we list the 40 inventive business principles as a reference (Mann 2004).

Principle 1. Segmentation

A. Divide an Object into Independent Parts

- Divide an organization into different product centers.
- Autonomous profit centers.
- Use a work breakdown structure for a large project.
- Franchise outlets.
- Image/value/satisfaction segmentation of customer purchase-related preferences.
- Kano diagram excitement, performance, and threshold product attribute parameters.
- Marketing segmentation by demographics, sociographics, psychographics, lifestyles, etc. (creation of microniches).
- Segmentation of idea management process into fertilization, seeding, and incubation phases.
- Strength, weakness, opportunity, and threat (SWOT) analysis.

B. Make an Object Easy to Disassemble

- Flexible pensions
- Use of temporary workers on short-term projects
- Flexible manufacturing systems
- Modular furniture and offices
- Container shipment

C. Increase the Degree of Fragmentation or Segmentation

- Quality circles
- Empowerment by segmentation of decision making
- Distance learning (also see principle 2)
- Virtual office or remote working (also see principle 2)
- Creative segmentation—high-performance small car, cordless power tool

Principle 2. Taking Out

Separate an Interfering Part or Property from an Object, or Single Out the Only Necessary Part (or Property) of an Object

- Break down barriers between departments (point 9 of Deming's 14 points).
- Eliminate exhortations (point 10 of Deming's 14 points).
- Eliminate targets (point 11 of Deming's 14 points).
- Drive out fear (point 8 of Deming's 14 points).
- Separate the *people* from the *problem*.
- Lean manufacturing.
- USP advertising.
- Just-in-time inventory management.
- Activity-based costing instead of allocation cost accounting.
- Separate development and production activities—skunkworks, tiger-teams, etc.
- Smart software learns user preference and filters out nonuseful information.
- Semantic processors used to extract knowledge from text.

Principle 3. Local Quality

A. Change an Object's Structure from Uniform to Nonuniform, Change an External Environment (or External Influence) from Uniform to Nonuniform

- Move away from rigid salary structures and job grading.
- Flexible working hours.

- Franchise fast-food outlets have local dishes in addition to normal product range.
- Casual (dress-down) days.
- Introduce "Corporate Jester" (e.g., British Airways) as a method of encouraging out-of-the-box thinking.
- Red team/blue team proposal preparation structures.
- Quiet work areas.

B. Make Each Part of an Object Function in Conditions Most Suitable for Its Operation

- Empowerment of individuals.
- Have each employee's workplace customized to his or her ergonomic and psychological needs.
- Working hours phased to accommodate people working on international, shifted time-zone projects.
- Customizable software.

C. Make Each Part of an Object Fulfill a Different and Useful Function

- Organizational division by function rather than product.
- Staff specialists in centers of excellence.
- Position factory or distribution center near customers.
- Hire local people to acquire cultural knowledge of local customers.
- Kids areas in restaurants, etc.

Principle 4. Asymmetry

A. Change the Shape of an Object from Symmetrical to Asymmetrical

- Proportionately more P or more S in the Deming PDSA cycle.
- Skewed normal distributions.
- Use a different marketing approach for each class of clients. (Combine with segmentation and local quality—make each class smaller to be sure the approach is exactly tailored to it.)
- Budget for different departments individually rather than using a constant percentage increase or reduction for all departments.

B. If an Object Is Asymmetrical, Change Its Degree of Asymmetry

- 360° appraisals.
- More equitable two-way dialog between management and workers.
- Shift away from calendar-influenced sales bias [e.g., shift from annual to bi-annual car registration dates (to reduce August sales peak), greeting card companies, etc.].

- Honda's 4M—man maximum, machine minimum—product design philosophy.
- Bigger customer focus group or Internet focus group.
- On-line, web-cam shopping—"one store serves the world."
- Collaboration with complementary organizations when competing for business with other directly competitive companies.

Principle 5. Merging

A. Bring Closer Together (or Merge) Identical or Similar Objects, Assemble Identical or Similar Parts to Perform Parallel Operations

- Personal computers in a network.
- Cell-based manufacturing.
- Toyota JIT.
- Common-interest group.
- Multiscreen cinemas.
- Shopping malls.
- Merge companies with related products.
- Internet Cafe.
- The Joiner Triangle—quality, scientific approach, all-one-team.
- "Young engineers have ideas, old engineers have bad experiences"— Japanese saying.

B. Make Operations Contiguous or Parallel; Bring Them Together in Time

- Theory of constraints.
- Enlist customer help in designing the product (Boeing 777—Working Together Teams).
- Multimedia presentations.
- Call centers.

Principle 6. Universality

A. Make an Object or Structure Perform Multiple Functions; Eliminate the Need for Other Parts

- Multiskilling of work force.
- Team leader acts as recorder and timekeeper.
- One-stop shopping—supermarkets sell insurance, banking services, fuel, newspapers, etc.
- Rapid reaction forces in the military—cross-trained, equipment versatility, etc.

- Semco—members of managerial staff set their own salaries; shop floor workers set their own productivity targets; part of change agent's job is to eliminate need for his or her job.
- Industry standard.

Principle 7. Nested Doll

A. Place One Object Inside Another; Place Each Object, in Turn, Inside the Other

- Store-in-store
- Profit centers inside an organization
- Hierarchical organization structures
- Four levels of knowledge [(1) basic skills, (2) know-how, (3) process management, (4) strategic vision] contained in effective company (e.g., Sony) training schemes

B. Make One Part Pass through a Cavity in the Other

- Plug holes in organization structure.
- Use 5 Whys question sequence (Liker 2004) to break through layers of problems to get to root cause.
- Expose traditionally inward facing job-holders to external events and customers (e.g., engineers shadow marketing people during customer visits).
- Door sensors count customers going into and out of a store or office, etc. (use data for market profiling, etc.).
- Casino hotel architecture (Las Vegas style): The guest must pass through the gaming area to get to the restaurant, the hotel registration, even the lavatories!

Principle 8. Antiweight

A. To Compensate for the Weight (Downward Tendency) of an Object, Merge It with Other Objects That Provide Lift

- In a merger of two companies, one lifts the other with whatever its stronger features are (distribution system, marketing, methods, capital, etc.).
- Companies increase flagging sales by making connections with other rising products.
- Attaching the word *new* is the most powerful way of enhancing the sales of fast-moving consumer goods.

B. To Compensate for the Weight (Downward Tendency) of an Object, Make It Interact with the Environment (e.g., Use Global Lift Forces)

- A small company is lifted by a resource—use of a transportation network, etc.—to the level of the larger companies.
- Political parties boost poll ratings by attaching themselves to popular causes.
- Attach product and service marketing to customer and business driving forces (megatrends—aging population, desire for flexibility, simplicity, etc.).

Principle 9. Preliminary Antiaction

A. If It will be Necessary to Perform an Action with Both Harmful and Useful Effects, this Action Should be Replaced with Antiactions to Control Harmful Effects

- When making a public announcement, include all the information, not just the harmful parts (e.g., Johnson & Johnson's handling of the Tylenol tampering case).
- Use formal risk assessment methods to quantify risk and identify mitigation actions before (and during) a project.
- Customer trials and segmented launch of (high-risk) new products (e.g., film companies film several endings to a movie and trial them with different audiences before finalizing selection).
- Use of voluntary redundancy, pay cuts, short-time working, or job-sharing as alternatives to downsizing.

B. Create Beforehand Stresses in an Object That will Oppose Known Undesirable Working Stresses Later on

- Epson product development engineers spend time as sales and then service staff before they are allowed to work on product development activities. Prior to a layoff, prepare compensation, outplacement, and communication packages for all affected employees.
- Team-building tasks are done before the real project starts.
- Negotiate upfront stage payment in long-term contract.

Principle 10. Preliminary Action

A. Perform, Before It Is Needed, the Required Change of an Object (Either Fully or Partially)

- Project preplanning.
- Perform noncritical path tasks early (where circumstances permit).

- Dialog with employees before embarking on change management activities.
- Use story-boarding to facilitate creative problem solving (i.e., gathering the data before the creativity session).

B. Prearrange Objects Such That they can Come into Action from the Most Convenient Place and Without Losing Time for Their Delivery

- Kanban arrangements in a just-in-time factory.
- Cell-based manufacturing.
- Publish an agenda before meetings.
- "If I had 8 hours to chop down a tree, I'd spend 6 hours sharpening my axe"—Abraham Lincoln.
- Benetton "retarded differentiation"—clothing is knitted before it is dyed; color only applied when the season's popular colors emerge.
- Dealer-fit car accessories—CD player, alloy wheels, air-conditioning, etc.

Principle 11. Beforehand Cushioning

A. Prepare Emergency Means Beforehand to Compensate for the Relatively Low Reliability of an Object

- Contingency planning.
- Establish a worst-case, fall-back position prior to negotiation—"best alternative to a negotiated agreement."
- Back up computer data.
- Run antivirus software frequently (and update it frequently).
- Encourage short, effective meetings by removing the chairs.
- Put clauses in contracts requiring arbitration or mediation to avoid litigation.
- "Eighty percent of a successful production is in the casting"—Lindsay Anderson

Principle 12. Remove Tension

A. In a Potential Field, Limit Position Changes (e.g., Change Operating Conditions to Eliminate the Need to Raise or Lower Objects in a Gravity Field)

- Make horizontal career changes to broaden skills.
- Team members distribute their own merit award money (rather than often divisive management dictating distribution of money).

- Force-field analysis—group discussion of the phrase "forces push in various directions." This team-building and problem-solving technique is discussed in Reference 3.
- Beware of the Peter principle—every employee tends to rise to his or her level of incompetence.
- Single union agreement.

Principle 13. The Other Way Round

A. Invert the Action(S) Used to Solve the Problem (e.g., Instead of Cooling an Object, Heat It)

- Bring the mountain to Mohammed, instead of bringing Mohammed to the mountain.
- Expansion instead of contraction during recession.
- Benchmark against the worst instead of the best.
- Blame the process not the person.
- "I used to think that anyone doing anything weird was weird. I suddenly realized that anyone doing anything weird wasn't weird at all, and it was the people saying they were weird that were weird"—Paul McCartney.

B. Make Movable Parts (or the External Environment) Fixed, and Fixed Parts Movable

- Home shopping.
- Home banking.
- Park-and-ride schemes in busy cities.
- Do not make changes just because they are fashionable management fads.
- "If you obey all the rules, you miss all the fun"—Katherine Hepburn.

C. Turn the Object (or Process) Upside Down

- The cash-till assistant is the most important part of a retail organization.
- Computer help lines were often originally set up with relatively no-technical staff at the front end, directing calls to progressively more technically able staff the more complicated the problem is. Latest logic suggests reversing this trend, i.e., place the most qualified staff as first point of contact (e.g., IBM).
- Product-rather than function-based organization structure.
- "Ready, Fire, Aim"—Tom Peters.
- Mercedes Benz vision changed from "the best or nothing" to "the best for our customers;" i.e., shift from internally to externally focused vision statement.

- The Peter pyramid (Peter 1986).
- Corporate unlearning—acquiring the ability to forget about the past where appropriate.
- "Ours is the age that is proud of machines that think and suspicious of men who try to"—H. Mumford Jones.
- Russian government pays inventors for patent applications; the West makes the inventor pay to apply.
- Chairperson of company spends time in the complaints department answering customer complaints.
- "When you reach the top, that's when the climb begins"—Michael Caine.

Principle 14. Curvature

A. Instead of Using Rectilinear Parts, Surfaces, or Forms, Use Curvilinear Ones; Move from Flat Surfaces to Spherical Ones; from Parts Shaped as a Cube (Parallelepiped) to Ball-Shaped Structures

- Ergonomic desk and workstation designs.
- Take the shortest path to the customer—around the organization rather than point-to-point through the bureaucracy.
- "Form the wagons into a circle"—John Wayne.

B. Use Rollers, Balls, Spirals, Domes

- Mobile factory
- Mobile car service—mechanic comes to you rather than you going to garage
- Mobile library
- "Meals on wheels" and home-delivery pizza

C. Go from Linear to Rotary Motion; Use Centrifugal Forces

- Rotate leadership of a team.
- Establish a sphere of influence, and then market to that sphere.
- Quality circles.
- Circular work cells.
- Levi Strauss' Information Service (IS) department's organizational chart resembles a solar system, with the names of 20 managers appearing once on a large circle and, in many cases, also on one of four smaller circles intersecting the large one. The small circles represent action groups focusing on specific tasks, including customer service and business systems.

Principle 15. Dynamics

A. Allow (or Design) the Characteristics of an Object, External Environment, or Process to Change to be Optimal or to Find an Optimal Operating Condition

- Empowerment.
- Customer response teams.
- Continuous process improvement.
- Rapid reaction force.
- Swatch design proliferation—design for specific market niches.
- "Cafeteria" benefits—where employees pick which types of insurance and health system, etc., they want.

B. Divide an Object into Parts Capable of Movement Relative to Each Other

- Work teams are oriented to achieve the same goal, but work at different rates on different objectives.
- Geographically or functionally independent business units.
- Conglomerate structures.

C. If an Object (or Process) is Rigid or Inflexible, Make It Movable or Adaptive

- Gallery Furniture on-line shopping—customer is able to control and move cameras to point to different products in different parts of the store from his or her home computer (www.galleryfurniture.com).
- Flexible organizational structure (chaocracy).

Principle 16. Partial or Excessive Actions

If 100 Percent of an Objective Is Hard to Achieve Using a Given Solution Method, then by Using Slightly Less or Slightly More of the Same Method, the Problem may be Considerably Easier to Solve

- When going into a new market, do "saturation" advertising by all media—mail, newspapers, local magazines, local radio, local TV, billboards, etc.
- Communicate more often and with more information than you think necessary.
- "If it ain't broke, improve it anyway"—Japanese process management philosophy.
- "The most important numbers are the ones you'll never know"—W. E. Deming (i.e., is it possible to ever know what 100 percent means?).
- "Communication is and should be hellfire and sparks, as well as sweetness and light"—Aman Vivian Rakoff.

Principle 17. Another Dimension

A. To Move an Object in a Two- or Three-Dimensional Space

- Make 360° appraisals.
- Use multidimensional organizational hierarchy charts. Use 3D charts to show "hard" and "soft" relationships or 4D charts to include an element of time or movement.
- Distribute responsibility and authority. For example, the quality department advises on technical details and conducts audits, but everyone is responsible for quality. Another good example of this is the safety office.

B. Use a Multistory Arrangement of Objects Instead of a Single-Story Arrangement

- Organizational hierarchy.
- Multistack storage systems use the height of a building and save floor space.
- Employees "disappear" from customers in a theme park, descend into a tunnel, and walk to their next assignment, where they return to the surface and magically reappear.
- Standing on the shoulders of giants . . .
- "When two people meet, there are really six people present. There is each man as he sees himself, each man as he wants to be seen, and each man as he really is"—Michael De Saintamo.

C. Tilt or Reorient the Object; Lay It on Its Side

- Horizontal (peer) communication.
- Horizontally integrated manufacture.
- Switch from vertical to horizontal (lateral) thinking, and vice versa.
- Shift from line to project management dominance in a matrix organization (and vice versa depending on prevailing market conditions).
- Shift from a portrait to landscape report format.

D. Use Another Side of a Given Area

- View your organization from the outside, either directly or by using consultants, and mystery shoppers, etc.
- Look at the selling process in new ways. Instead of selling carpets to its commercial and industrial customers, Interface now offers what it calls the "Evergreen lease." Its customers no longer buy carpets or pay an

installation fee;—they just pay a monthly service fee that guarantees they will always have clean, attractive carpets.
- "The seeing of objects involves many sources of information beyond those meeting the eye when we look at an object. It generally involves knowledge of the object derived from previous experience, and this experience is not limited to vision but may include the other senses: touch, taste, smell, hearing, and perhaps also temperature or pain"—R. L. Gregory.
- "You can't teach an old dogma new tricks"—Dorothy Parker.

Principle 18. Resonance

A. Cause an Object to Oscillate or Vibrate

- Use the process of Hoshin planning to get the whole organization "vibrating."
- "A good manager doesn't try to eliminate conflict; he tries to keep it from wasting the energies of his people. If you're the boss and your people fight you openly when they think that you are wrong—that's healthy"—Robert Townsend.
- "The things we fear most in organizations—fluctuations, disturbances, imbalances— are the primary sources of creativity"—Margaret J. Wheatley.

B. Increase Its Frequency

- Communicate frequently, in multiple modes (newsletter, Intranet, staff meetings, etc.).
- "I don't think that you should ever manage anything that you don't care passionately about"—D. Coleman, VP and CFO of Apple (Leigh et al. 1993).
- "He inspired in us the belief that we were working in a medium that was powerful enough to influence the world"—Lillian Gish on D. W. Griffiths (Leigh et al. 1993).

C. Use an Object's Resonant Frequency

- Use strategic planning (policy deployment, Hoshin Kanri) to select the right frequency and get the organization resonating at that frequency to accomplish a breakthrough strategy.
- Creating extraordinary unity of purpose in a work team. A good example of this is Don Petersen's story of the Ford versus Mazda competition to win the transmission job for the FWD Taurus.
- *Kansei*—Japanese term for resonance or oneness between product and user.

D. Use Piezoelectric Vibrators Instead of Mechanical Ones

E. Use Combined Ultrasonic and Electromagnetic Field Oscillations
(Use External Elements to Create Oscillation or Vibration)

- Bring new blood into the team.
- Hire a consultant.

Principle 19. Periodic Action

A. Instead of Continuous Action, Use Periodic or Pulsating Actions

- Use batch manufacturing.
- Tidal traffic flow schemes can ease transport into and out of busy areas.
- Change team leadership periodically (e.g., countries take turns leading the European Union).
- Introduce sabbaticals to refresh people's points of view.

B. If an Action is Already Periodic, Change the Periodic Magnitude or Frequency

- Audit at irregular intervals.
- Use monthly or weekly feedback instead of annual reviews.
- Use flexible savings schemes that pay higher interest rates when fewer withdrawals are made.

C. Use Pauses Between Impulses to Perform a Different Action

- Get work done between meetings.
- Perform maintenance work during vacations.
- Provide 24-hour car service operation—evening pickup and return of serviced car by breakfast the following morning (customer perspective).

Principle 20. Continuity of Useful Action

A. Carry on Work Continuously; Make All Parts of an Object Work at Full Load All the Time

- Run the bottleneck operations in a factory continuously to reach the optimum pace (from theory of constraints).
- Institute constant improvement (point no. 5 of Deming's 14 points (Deming 1982).
- Continuous on-line monitoring of elevators by Otis (total maintenance responsibility).

- Provide 24-hour car service operation—evening pick-up and return of serviced car by breakfast the following morning (garage perspective).
- "The power of a waterfall is nothing but a lot of drips working together."(Leigh et al. 1993)

B. Eliminate All Idle or Intermittent Actions or Work

- Multiskilling to enable working in bottleneck functions to improve work flow.
- Conduct training during pauses in work.
- Institute 24-hour shift patterns.
- "Life-long learning."
- "The more I practice, the luckier I get"—Gary Player.

Principle 21. Skipping

Conduct a Process or Certain Stages (e.g., Destructive, Harmful, or Hazardous Operations) at High Speed

- "Incrementalism is innovation's worst enemy"—Nicholas Negreponte, MIT Media Lab (Peters 1997).
- "Don't be afraid to take a big step if one is indicated. You can't cross a chasm in two small jumps"—David Lloyd George.
- "Fail fast; learn fast."
- *Fast cycle—full participation*—a method of involving the whole organization simultaneously and rapidly in a major change, such as a reorganization.
- Get through painful processes quickly (e.g., firing someone).
- Use rapid prototyping.
- "If you want to succeed, double your failure rate"—J. R. Watson, IBM founder.

Principle 22. Blessing in Disguise or Turn Lemons into Lemonade

A. Use Harmful Factors (Particularly, Harmful Effects of the Environment or Surroundings) to Achieve a Positive Effect

- Recast an attack on you as an attack on the problem.
- Making a fuss over customers who have experienced a problem with your goods or services tends to reinforce their overall positive feeling about you—to a level greater than that when no problem had occurred.
- Collect information to understand the harm, and then formulate a positive action to remove it.
- Use the *provocations* method of encouraging new ideas.
- "The Extra Mile will have no traffic jams"—Unknown.

B. Eliminate the Primary Harmful Action by Adding It to Another Harmful Action to Resolve the Problem

- Eliminate fear of change by introducing fear of competition.
- Put a problem person on an assignment in another area where he or she can do well and not be a problem to the original group.
- Use the loss-leader strategy for increasing sales.
- Keep traffic out of cities by introducing cheap Park and Rides and expensive downtown parking charges.
- Make potentially polluting industries place flow intakes downstream of flow outlets on a river.

C. Amplify a Harmful Factor to Such a Degree that It Is No Longer Harmful

- The famous software solution is That's not a bug, it's a feature.
- Benevolent dictatorship (Roberts 1989).
- Reduce resourcing levels to such an extent that new ways of doing the job have to be discovered.
- Restrict the supply of goods to create scarcity value (e.g., some sports car manufacturers seek to maintain a multiple-year waiting list for their vehicles).

Principle 23. Feedback

A. Introduce Feedback (Referring Back, Cross-Checking) to Improve a Process or Action

- Use statistical process control (SPC)—measurements are used to decide when to modify a process.
- Create budgets—measurements are used to decide when to modify a process.
- Enlist customers in the design process.
- Customer surveys, customer seminars, etc.
- Active transition management—a way of controlling the product development process between the research, development, and production phases.
- Electronic bulletin boards.
- Supermarket loyalty cards—these provide customer shopping profile information.
- "What you measure is what you get"—Joe Juran.

B. If Feedback Is Already Used, Change Its Magnitude or Influence

- Change a management measure from budget variance to customer satisfaction.
- Expose designers as well as marketers to customers.

- Multicriteria decision analysis (valid apples-and-oranges comparisons).
- Toshiba's medical systems division split into R&D, engineering, and manufacturing sectors. As a product is being developed, key personnel and leadership physically move from one sector to another to actively manage transitions between product development stages.
- "Open the kimono"—everything out in the open communication.
- "Supravision" rather than supervision.
- Coevolutionary marketing; e.g., Amazon.com invites readers to write on-line book reviews. Other readers often prefer these views to professional reviewer evaluations; therefore people visit the site more often.
- Motorola has an open dissent policy where employees fill in a minority report to senior management when ideas they consider valuable are unsupported by colleagues and immediate superiors.
- Use of half-life as a measure of improvement (e.g., the time taken to cut product development time in half) to encourage large-scale thinking.

Principle 24. Intermediary

A. Use an Intermediary Carrier Article or Intermediary Process

- Use of impartial body during difficult negotiation.
- PO (provocative operator)—a place between yes and no. This construct was devised by Edward DeBono to help avoid premature discarding of ideas.
- Subcontract noncore business (e.g., cleaning and transportation services).
- Franchisee acts as intermediary between corporate vision and customer.
- Travel agent.
- UPS distribution system using core sorting center.
- KLM "feeder" airline concept—short flights from Germany and England pull passengers away from national airlines so that passengers will fly long distances with KLM using Holland as a hub.
- Video Plus—program video using simple codes to represent channels, dates, and times.

B. Merge One Object Temporarily with Another (Which Can be Easily Removed)

- Introduce specialist troubleshooting or firefighting teams.
- Hire consultant.

- Use bridge loan arrangements to help cash flow.
- Subcontract occasional services, e.g., accounts, cleaning, and transport.

Principle 25. Self-Service

A. Make an Object Serve Itself by Performing Auxiliary Helpful Functions

- Quality circles.
- Self-help groups.
- Brand image circularity. For example, Harvard Business School produces bright people; these people enhance the school's reputation, and hence lots of people apply; hence Harvard only takes on very bright people; bright people in equals bright people out; and so the circle reinforces itself.
- "Cookies" on the Internet gather data useful for future marketing activities, while performing a useful service for the web surfer.
- Bar codes in supermarkets provide instant pricing information, but the system also gathers information to assist future marketing decisions.
- Edward DeBono's suggested to Ford UK that it buy national car parks and then only let Ford cars into the parking lots. Thus motorists buying a Ford would also be buying a parking place in every city.

B. Use Waste (or lost) Resources, Energy, or Substances

- Rehire retired workers for jobs where their experience is needed.
- Loan out temporarily underutilized workers to other organizations (load-capacity balancing across companies). For example, this can be a win-win situation in football: the football player stays match-fit, and the loaner team saves wages and fills a skills shortage for another team.
- Industrial ecosystems—e.g., plan factories so that waste heat from one operation provides power for another operation; install cogeneration equipment so that waste heat can generate electricity that can be used for your own operations or sold to the electric power utility.
- Brown-field developments.
- The Body Shop recycles used containers brought back by customers which helps promote a corporate green image.
- Recycle all packaging material.
- Scan mail into data systems and recycle the paper.

Principle 26. Copying

A. Instead of Using an Unavailable, Expensive, Fragile Object, Use Simpler and Inexpensive Copies

- Experience virtual reality via computer instead of taking an expensive vacation.

- Listen to an audiotape instead of attending a seminar.
- Rapid prototyping (e.g., stereolithography).
- Scan rare historic books, and documents so that they are accessible to all and the original remains protected.
- Lascaux II—reproduction of Lascaux cave paintings which is open to visitors.

B. Replace an Object, or Process with Optical Copies

- Virtual product service manuals.
- Videoconferencing instead of physical travel.
- Use a central electronic database instead of paper records in cases where multiple users would benefit from simultaneous access to data, e.g., medical records, customer data, and engineering drawings.
- Keep your personal calendar on a website so you (and others if desired) can access it from any computer and it cannot get lost.

C. If Optical Copies are Used, Move to infrared or ultraviolet
(Use an Appropriate Out-of-the-Ordinary Illumination and Viewing Situation)

- Evaluate employee morale using multiple methods such as interviews and questionnaires (two different "wavelengths").
- Evaluate customer satisfaction using multiple techniques.
- Have your customers and suppliers benchmark you.

Principle 27. Cheap Short-Living Objects

A. Replace an Expensive Object with a Multiple of Inexpensive Objects, Compromising Certain Qualities (Such as Service Life, for Instance)

- Use disposable paper objects to avoid the cost of cleaning and storing durable objects, e.g., plastic cups in motels, disposable diapers, and many kinds of medical supplies.
- Numerical simulation–operational analysis (virtual war-gaming, virtual business development, strategic planning modeling).
- Using a flight simulator reduces pilot training costs.

Principle 28. Another Sense

A. Replace a Mechanical Means with a Sensory (Optical, Acoustic, Taste, or Smell) Means

- "Our goal is that when you turn out the lights and climb into bed, you think you are at the Hilton"—CEO of budget motel chain.
- Have retail customers enter data by means of a touch screen, instead of filling out a form that must be keyed in by employees (e.g., wedding registries use this).

- Electronic voting.
- Supermarkets pump bakery odors around the store to help advertise bread products.

B. Use Electric, Magnetic, and Electromagnetic Fields to Interact with the Object

- Mrs. Fields Cookies has a morning video-Internet conference with all franchisees, electronic communication replaces memos, etc.
- Automatic global positioning system (GPS) sensors inform a central control point where (e.g., delivery trucks or taxis) are located.
- Electronic tagging.
- Pagers.

C. Change from Static to Movable Fields, from Unstructured Fields to those Having Structure

- Mind maps.
- Tidal traffic flow schemes or high occupancy vehicle (HOV) lanes.
- Management by walking around (MBWA).

D. Use Fields in Conjunction with Field-Activated (e.g., Ferromagnetic) Particles

- Mind-mapping software tools.
- Intelligent tidal traffic flow control (e.g., using roadside sensors).
- Use a radio transponder payment system for traffic control (Highway 91 in California). The fee ranges from $0.50 to $3.50 depending on how heavy the traffic is on the free part of the highway. Radio signals deduct the payment from the users' account when the car enters the special traffic lane.

Principle 29. Fluidity

Make Solid Things into Fluid Things

- "Water logic" versus "rock logic"—fluid, flowing, gradually built-up logic versus permanent, hard-edged, rocklike alternatives.
- Flexible (fluid) organizational structure versus old fixed hierarchical structures.
- Liquidation of assets.
- Introduction of "breathing spaces" into contracts.

Principle 30. Thin and Flexible

A. Use Flexible Shells and Thin Films Instead of Three-Dimensional Structures

- The thinnest film is a single molecule thick. Likewise, the thinnest organization structure is one employee thick. Get faster customer

service by having the single customer service agent have all the necessary data easily available, so the customer only deals with the single, flexible "shell" of the organization not the whole bulky volume.
- Card transactions instead of money—e.g., vending machines in companies use employee ID card and charges are debited direct from salary.
- Cardboard police—two-dimensional policemen or police cars over freeway bridges are used as a means of slowing down traffic.
- Inflatable passenger for lone drivers out late at night.

B. Isolate the Object from the External Environment Using Flexible Shells and Thin Films

- Office workers in open areas can use flexible curtains to shut themselves off from the visual chaos of the open area when they need to concentrate rather than communicate.
- Use trade-secret methods to separate company proprietary knowledge from general knowledge.
- Umbrella organizations.
- "We like to delegate and leave people as free as possible, so we try to push management decisions down the line. We run Rolls-Royce with a very thin corporate structure"—Lord Tombs of Brailes, ex-chairman of Rolls-Royce.

Principle 31. Holes

A. Add Holes to a System or Object

- Think of the customer-facing layers of a company as a porous membrane that filters information flow both into and out of the organization.
- Improve internal communications by creating an Intranet that is accessible by all hierarchical layers, giving workers access to the CEO and vice versa.
- Trickle-down economics.
- Government leaks—used as a way of gauging public reaction to (usually) controversial issues.

B. If a System or Object Already has Holes, Use the Pores to Introduce a Useful Substance or Function

- Empower the customer-facing layer (information is the thing that fills the pores—see inventive business principle 30A).
- Use mind maps, self-patterning capabilities, etc., to improve the information and knowledge intake and filtering abilities of the brain.
- Media relations department turns spin doctor and/or marketing feedback gatherer.

Principle 32. Color Changes

A. Change the Color of an Object or Its External Environment

- Red/blue proposal preparation teams.
- Use of lighting effects to change mood in a room or office.
- Creation of corporate colors creating a strong brand image through use of bespoke colors—BP green, British Telecom red phone boxes, Ford blue, etc.
- Use colors to communicate a state of alert (green, black, amber, red, etc.).
- Highlighter pens.

B. Change the Transparency of an Object or Its External Environment

- Transparent organizations
- Transparent communications
- Importance of creating clear, concise mission statement (Martin 1993)
- Smoke screen or misinformation to disguise confidential R&D and other activities

Principle 33. Homogeneity

A. Make Objects Interact with a Given Object of the Same Material (or Material with Identical Properties)

- Colocated project teams.
- Internal customers.
- Product branding and product families.
- Boeing "Working Together Teams" bring customers and suppliers into the design loop.
- Common data transfer protocols between different organizations.
- "The best way to make a silk purse from a sow's ear is to begin with a silk sow. The same is true of money"—Augustine's Law #1 (Augustine 1983).

Principle 34. Discarding and Recovering

A. Make Portions of an Object that Have Fulfilled their Functions Go Away (Discard by Dissolving, Evaporating, etc.) or Modify them Directly During Operation

- Flexible, variable-sized project teams.
- Balance load and capacity by using contract labor.
- Consultants.
- Contract hire of specialized equipment, facilities, etc.

B. Conversely, Restore Consumable Parts of an Object Directly in Operation

- Need to periodically reenergize continuous improvement initiatives ("enthusiasm injections")
- Lifelong learning (where individuals are given responsibility for managing their own personal continuing education and ensuring skills remain up-to-date)

Principle 35. Parameter Changes

A. Change an Object's Physical State (e.g., to a Gas, Liquid, or Solid)

- Virtual prototyping
- Numerical simulation
- Virtual shopping, e.g., Amazon.com
- Telephone banking
- Electronic voting in elections

B. Change the Concentration or Consistency

- Change the team structure (e.g., football teams use substitutes).
- Stores introduce special offers and other promotions.

C. Change the Degree of Flexibility

- Introduce intelligence into on-line catalogs (e.g., first-generation catalogs were replicas of previous paper versions, but the latest generation incorporates search engines, expert systems, etc.)
- Software with options for beginner through expert usage.
- Moves away from fixed clothing size partitions, e.g., Levi's Personal Pairs—a customer at a participating store chooses which fabric he or she wants and then is measured. Those measurements are transmitted instantly to a Levi's plant in Tennessee where the data controls a laser cutter. The bar-coded pieces are stitched on the regular assembly line and mailed directly to the customer. (The custom Levi's, which customers love, run about $15 more than off-the-rack ones.)

D. Change Emotional and Other Parameters

- Get customers excited ("hot") about the product by giving them ownership of the change.
- Get employees excited about the future of the company by using full involvement strategic planning, stock options, etc.
- "A fired-up team wins games even if it's not the best team. A fired-up company can achieve the same result" (Martin 1993).

Principle 36. Paradigm Shift

Use Phenomena Occurring During Disruptive Shift in an Economy
(Awareness of Macroscale Business Phenomena)

- Awareness of the requirements of different stages—conception, birth, development, maturity, retirement—of a project (e.g., shifting labor requirements, shifting budget requirements).
- Transition from a bull to a bear market.
- Tendency to relax after receiving a quality award, innovation award, etc.
- Forming, storming, norming, and performing phases of team development; e.g., take advantage of enthusiasm dip during storming and norming phases.

Principle 37. Relative Change

A. Use the Relative Differences that Exist in an Object or System to do Something Useful

- Match personalities on work teams.
- Some organizations create creative tension by employing two independent teams to develop a new product or process and then have them compete.
- "It seems safe to say that significant discovery, really creative thinking, does not occur with regard to problems about which the thinker is lukewarm"—Mary Henle.

B. Make Different Parts of a System Act Different in Response to Changes

- Expand or contract marketing efforts depending on the product's "hotness"—rate of sales and profitability.
- Match personalities on work teams.

Principle 38. Enriched Atmosphere

A. Replace a Normal Atmosphere with an Enriched One

- Create risk and revenue sharing partnerships.
- Have guest speakers at a seminar.
- Use internal subject matter experts.
- Use simulations and games instead of lecture-style training.
- Use case studies in training.
- Injection of new blood or a new challenge into a team.

- Consider personal chemistry issues when assembling a project team; find people who will spark off interesting reactions with each other.
- Deming's four stages of learning—unconscious incompetence, conscious incompetence, conscious competence, unconscious competence.
- Focus teams on a single project only (give them an enriched environment full of success factors).
- "Leadership is a potent combination of strategy and character. But if you must be without one, be without strategy"—General H. Norman Schwartzkopff.

B. Exposing a Highly Enriched Atmosphere with One Containing Potentially Unstable Elements

- Corporate jester.
- "I like Bartok and Stravinsky. It's a discordant sound and there are discordant sounds inside a company. As president you must orchestrate the discordant sounds into a kind of harmony. But you never want too much harmony. One must cultivate a taste for finding harmony within discord or you will drift away from the forces that keep a company alive"—Takeo Fujisawa, Honda cofounder.

Principle 39. Calm Atmosphere

A. Replace a Normal Environment with an Inert One

- Move away from the (normal) disruptive performance appraisal, merit award, and reward environment to an (emotionally neutral) more fair system of working practice.
- Hare brain, tortoise mind (Claxton 1997)
- Take time-outs during negotiation.
- Have away-days and team-building days.
- Hold corporate retreats.
- Operations room, e.g., for planning organizational change, proposal submissions, and contract tendering, etc.

B. Add Neutral Parts or Inert Additives to an Object

- Use of neutral third parties during difficult negotiations (e.g., Senator George Mitchell in Northern Ireland and ACAS)
- Introduction of quiet areas into the workplace
- Rest breaks and pause-for-reflection breaks in meetings

Principle 40. Composite Structures

Change from Uniform to Composite (Multiple) Structures (Awareness and Utilization Of Combinations of Different Skills and Capabilities)

- Create multidisciplinary project teams.
- Do training with a combination of lecture, simulations, on-line learning, video, etc.
- Employ different personality types (e.g., Myers-Briggs) on a team.
- Hard person–soft person negotiating team.
- Mix of thinking skills in a project team.
- Positional players in a football team.
- Combined high risk–low risk investment strategy.

Example 9.5: Shorten the Product Development Duration for Complex Products

Many manufacturers of complex products, for example, automobiles, suffer from long product development time and cost. Quick product development time is usually desired, but when a product is too complex, a quick product development cycle is very difficult to achieve. Product complexity also makes product development cost very high.

We can try to figure out some ways to resolve these difficulties by using inventive business principles. First, we can model this problem as two pairs of contradictions as follows:

1. The contradiction of system complexity versus R&D time
2. The contradiction of system complexity with R&D cost

For contradiction 1, by referring to the business contradiction matrix (see Appendix B at the end of this chapter) we can find the following inventive business principles:

$$5, 6, 25, 10, 2, 37$$

By closely examining these principles, we find the following items highly relevant.

Principle 5: Merging
A. *Bring closer together (or merge) identical or similar objects; assemble identical or similar parts to perform parallel operations.*
B. *Make operations contiguous or parallel; bring them together in time.*

Based on principle 5, the following approaches can be proposed for the product development situation:

1. Develop similar components or subsystems in parallel.
2. Develop noninterfering/components or subsystems in parallel.

Principle 6: Universality

Make an object or structure perform multiple functions; eliminate the need for other parts.

Based on principle 6 the following approaches can be proposed for the product development situation:

3. Simplify parts and reduce the part count.

Principle 25. Self-service

B. Use waste (or lost) resources, energy, or substances.

Based on principle 25B, the following approaches can be proposed for the product development situation:

4. Rehire retired workers for jobs where their experience is needed.
5. Borrow temporarily underutilized workers from other organizations.

Principle 10. Preliminary Action

A. *Perform, before it is needed, the required change of an object (either fully or partially).*

Based on principle 10A, the following approaches can be proposed for the product development situation:

6. Prepare preliminary works on key development activities long before these activities' starting times.

Principle 2. Taking Out

Separate an interfering part or property from an object, or single out the only necessary part (or property) of an object.

Based on principle 2, the following approaches can be proposed for the product development situation:

7. Break down barriers between departments and form a fully empowered product development team.

For contradiction 2, by referring to the business contradiction matrix (see Appendix B at the end of this chapter), we can find the following inventive business principles:

$$5, 2, 35, 1, 29$$

In resolving contradiction 1, principles 5 and 2 have already been explored. Now we explore principles 35, 1, and 29. By closely examining these principles, we can find the following items highly relevant:

Principle 35. Parameter Changes
A. *Change an object's physical state (e.g., to a gas, liquid, or solid).*

Based on principle 35A, the following approaches can be proposed for the product development situation:

8. Use more virtual prototypes and computer simulation.

Principle 1. Segmentation
A. Divide an object into independent parts.
B. Make an object easy to disassemble.

Based on principle 1A and B, the following approaches can be proposed for the product development situation:

9. Reduce the dependencies among subsystems of the product.
10. Promote modular design practices.

Principle 29. Fluidity
Make solid things into fluid things.

Based on principle 29, the following approaches can be proposed for the product development situation:

11. Use a flexible organizational structure for the product development team, and adjust team structures as the product development goes through different stages.

In summary, by using the 40 inventive business principles, the following 11 suggestions are proposed to reduce product development time and cost:

1. Develop similar components and subsystems in parallel.
2. Develop noninterfering components or subsystems in parallel.
3. Simplify parts and reduce the part count.
4. Rehire retired workers for jobs where their experience is needed.
5. Borrow temporarily underutilized workers from other organizations.
6. Prepare preliminary works on key development activities long before these activities' starting times.
7. Break down barriers between departments and form a fully empowered product development team.
8. Use more virtual prototypes and computer simulation.
9. Reduce the dependencies among subsystems of the product.
10. Promote modular design practices.
11. Use a flexible organizational structure for the product development team, and adjust team and structures as the product development goes through different stages.

Appendix A: Contradiction Table of Inventive Principles

What should be Improved? \ What is deteriorated?	1. Weight of movable object	2. Weight of fixed object	3. Length of movable object	4. Length of fixed object	5. Area of movable object	6. Area of fixed object	7. Volume of movable object	8. Volume of fixed object	9. Speed	10. Force	11. Stress, pressure	12. Shape	13. Object's composition stability
1. Weight of movable object			15 8 29 34		29 17 38 34		29 2 40 28		2 8 15 38	8 10 18 37	10 36 37 40	10 14 35 40	1 35 19 39
2. Weight of fixed object				10 1 29 35		35 30 13 2		5 35 14 2		8 10 19 35	13 29 10 18	13 10 29 14	26 39 1 40
3. Length of movable object	8 15 29 34				15 17 4		7 17 4 35		13 4 8	17 10 4	1 8 35	1 8 10 29	1 8 15 34
4. Length of fixed object		35 28 40 29				17 7 10 40		35 8 2 14		28 10	1 14 35	13 14 15 7	39 37 35
5. Area of movable object	2 17 29 4		14 15 18 4				7 14 17 4		29 30 4 34	19 30 35 2	10 15 36 28	5 34 29 4	11 2 13 39
6. Area of fixed object		30 2 14 18		26 7 9 39						1 18	10 15 36 37		2 38
7. Volume of movable object	2 26 29 40		1 7 35 4		1 7 4 17				29 4 38 34	15 35 36 37	6 35 36 37	1 15 29 4	28 10 1 39
8. Volume of fixed object		35 10 19 14	19 14	35 8 2 14						2 18 37	24 35	7 2 35	34 28 35 40
9. Speed	2 28 13 38		13 14 8		29 30 34		7 29 34			13 28 15 19	6 18 38 40	35 15 18 34	28 33 1 18
10. Force	8 1 37 18	18 13 1 28	17 19 9 36	28 10	19 10 15	1 18 36 37	15 9 12 37	2 36 18 37	13 28 15 12		18 21 11	10 35 40 34	35 10 21
11. Stress, pressure	10 36 37 40	13 29 10 18	35 10 36	35 1 14 16	10 15 36 28	10 15 36 37	6 35 10	35 24	6 35 36 21	36 35 21		35 4 15 10	35 33 2 40
12. Shape	8 10 29 40	15 10 26 3	29 34 5 4	13 14 10 7	5 34 4 10		14 4 15 22	7 2 35	35 15 34 18	35 10 37 40	34 15 10 14		33 1 18 4
13. Object's composition stability	21 35 2 39	26 39 1 40	13 15 1 28	37	2 11 13	39	28 10 19 39	34 28 35 40	33 15 28 18	10 35 21 16	2 35 40	22 1 18 4	
14. Strength	1 8 40 15	40 26 27 1	1 15 8 35	15 14 28 26	3 34 40 29	9 40 28	10 15 14 7	9 14 17 15	8 13 26 14	10 18 3 14	10 3 18 40	10 30 35 40	13 17 35
15. Duration of moving object's operation	19 5 34 31		2 19 9		3 17 19		10 2 19 30		3 35 5	19 2 16	19 3 27	14 26 28 25	13 3 35
16. Duration of fixed object's operation		6 27 19 16		1 40 35				35 34 38					39 3 35 23
17. Temperature	36 22 6 38	22 35 32	15 19 9	15 19 9	3 35 39 18	35 38	34 39 40 18	35 6 4	2 28 36 30	35 10 3 21	35 39 19 2	14 22 19 32	1 35 32
18. Illumination	19 1 32		2 35 32	19 32 16	19 32 26		2 13 10		10 13 19	26 19 6		32 30	32 3 27
19. Energy expense of movable object	12 18 28 31		12 28		15 19 25		35 13 18		8 15 35	16 26 21 2	23 14 25	12 2 29	19 13 17 24
20. Energy expense of fixed object		19 9 6 27								36 37			27 4 29 18

Theory of Inventive Problem Solving (TRIZ)

What should be Improved? \ What is deteriorated?	14. Strength	15. Duration of moving object's operation	16. Duration of fixed object's operation	17. Temperature	18. Illumination	19. Energy expense of movable object	20. Energy expense of fixed object	21. Power	22. Waste of energy	23. Loss of substance	24. Loss of information	25. Waste of time	26. Quantity of substance
1. Weight of movable object	28 27 18 40	5 34 31 35		6 29 4 38	19 1 32	35 12 34 31		13 36 18 31	6 2 34 19	5 35 3 31	10 24 35	10 35 20 28	3 26 18 31
2. Weight of fixed object	28 2 10 27		2 27 19 6	28 19 32 22	19 32 35		18 19 28 1	15 19 18 22	18 19 28 15	5 8 13 30	10 15 35	10 20 35 26	19 6 18 26
3. Length of movable object	8 35 29 34	19		10 15 32 19		8 35 24		1 35	7 2 35 39	4 29 23 10	1 24	15 2 29	29 35
4. Length of fixed object	15 14 28 26		1 40 35	3 35 38 18	3 25			12 8	6 28	10 28 24 35	24 26 14	30 29 14	
5. Area of movable object	3 15 40 14	6 3		2 15 16	15 32 19 13	19 32		19 10 32 18	15 17 30 26	10 35 2 39	30 26	26 4 6 13	29 30 6 13
6. Area of fixed object	40			2 10 19 30	35 39 38			17 32	17 7 30	10 14 18 39	30 16	10 35 4 18	2 18 40 4
7. Volume of movable object	9 14 15 7	6 35 4		34 39 10 18	10 13 2	35		35 6 13 18	7 15 13 16	36 39 34 10	2 22	2 6 34 10	29 30 7
8. Volume of fixed object	9 14 17 15		35 34 38	35 6 4				30 6		10 39 35 34		35 16 32 18	35 3
9. Speed	8 3 26 14	3 19 35 5		28 30 36 2	10 13 19	8 15 35 38		19 35 38 2	14 20 19 35	10 13 28 38	13 26		10 19 29 38
10. Force	35 10 14 27	19 2		35 10 21		19 17 10	1 16 36 37	19 35 18 37	14 15	8 35 40 5		10 37 36	14 29 18 36
11. Stress, pressure	9 18 3 40	19 3 27		35 39 19 2		14 24 10 37		10 35 14	2 36 25	10 36 3 37		37 36 4	10 14 36
12. Shape	30 14 10 40	14 26 9 25		22 14 19 32	13 15 32	2 6 34 14		4 6 2 14		35 29 3 5		14 10 34 17	36 22
13. Object's composition stability	17 9 15	13 27 10 35	39 3 35 23	35 1 32	32 3 27 15	13 19	27 4 29 18	32 35 27 31	14 2 6 30	2 14 40		35 27 28 10	15 32 35 27
14. Strength		27 3 26		30 10 35 40	35 19	19 35 10	35	10 26 35 28	35 28 31 40	29 3 28 10	29 10 27	3 35 10 40	29 3 29 10 35
15. Duration of moving object's operation	27 3 10			19 35 39	2 19 4 35	28 6 35 18		19 10 35 38		28 27 3 18	10	20 10 28 18	3 35 10 40
16. Duration of fixed object's operation				19 18 36 40				16		27 16 18 38	10	28 20 10 16	3 35 31
17. Temperature	10 30 22 40	19 39	19 18 36 40		32 30 21 16	19 15 3 17		2 14 17 25	21 36 35 38	21 36 29 31		35 28 21 18	3 17 30 39
18. Illumination	35 19	2 19 6		32 35 19		32 1 19	32 35 1 15	32	19 16 1 6	13 1	1 6	19 1 26 17	1 19
19. Energy expense of movable object	5 19 9 35	28 35 6 18		19 24 3 14	2 15 19			6 19 37 18	12 22 15 24	35 24 18 5		35 38 19 18	34 23 16 18
20. Energy expense of fixed object	35					19 2 35 32			28 27 18 31				3 35 31

Appendix A: Contradiction Table of Inventive Principles (*Continued*)

What should be Improved? \ What is deteriorated?	27. Reliability	28. Measurement accuracy	29. Manufacturing precision	30. Harmful action at object	31. Harmful effect caused by the object	32. Ease of manufacture	33. Ease of operation	34. Ease of repair	35. Adaptation	36. Device complexity	37. Measurement or test complexity	38. Degree of automation	39. Productivity
1. Weight of movable object	3 11 1 27	28 27 35 26	28 35 26 18	22 21 18 27	22 35 31 39	27 28 1 36	35 3 2 24	2 27 28 11	29 5 15 8	26 30 36 34	28 29 26 32	26 35 18 19	35 3 24 37
2. Weight of fixed object	10 28 8 3	18 26 28	10 1 35 17	2 19 22 37	35 22 1 39	28 1 9	6 13 1 32	2 27 28 11	19 15 29	1 10 26 39	25 28 17 15	2 26 35	1 28 15 35
3. Length of movable object	10 14 29 40	28 32 4	10 28 29 37	1 15 17 24	17 15	1 29 17	15 29 35 4	1 28 10	14 15 1 16	1 19 26 24	35 1 26 24	17 24 26 16	14 4 28 29
4. Length of fixed object	15 29 28	32 28 3 10	2 32	1 18		15 17 27	2 25	3	1 35	1 26	26		30 14 7 26
5. Area of movable object	29 9	26 28 32 3	2 32	22 33 28 1	17 2 18 39	13 1 26 24	15 17 13 16	15 13 10 1	15 30	14 1 13	2 36 26 18	14 30 28 23	10 26 34 2
6. Area of fixed object	32 35 40 4	26 28 32 3	2 29 18 36	27 2 39 35	22 1 40	40 16	16 4	16	15 16	1 18 36	2 35 30 18	23	10 15 17 7
7. Volume of movable object	14 1 40 11	25 26 28	25 28 2 16	22 21 27 35	17 2 40 1	29 1 40	15 13 30 12	10	15 29	26 1	29 26 4	35 34 16 24	10 6 2 34
8. Volume of fixed object	2 35 16		35 10 25	34 39 19 27	30 18 35 4	35		1		1 31 26	2 17		35 37 10 2
9. Speed	11 35 27 28	28 32 1 24	10 28 32 25	1 28 35 23	2 24 35 21	35 13 8 1	32 28 13 12	34 2 28 27	15 10 26	10 28 4 34	3 34 27 16	10 18	
10. Force	3 35 13 21	35 10 23 24	28 29 37 36	1 35 40 18	13 3 36 24	15 37 18 1	1 28 3 25	15 1 11	15 17 18 20	26 35 10 18	36 37 10 19	2 35	3 28 35 37
11. Stress, pressure	10 13 19 35	6 28 25	3 35	22 2 37	2 33 27 18	1 35 16	11	2	35	19 1 35	2 36 37	35 24	10 14 35 37
12. Shape	10 40 16	28 32 1	32 30 40	22 1 2 35	35 1	1 32 17 28	32 15 26	2 13 1	1 15 29	16 29 1 28	15 13 39	15 1 32	17 26 34 10
13. Object's composition stability		13	18	35 24 18 30	35 40 27 39	35 19	32 35 30	2 15 10 16	35 30 34 2	2 35 22 26	35 22 39 23	1 8 35	23 35 40 3
14. Strength	11 3	3 27 16	3 27	18 35 37 1	15 35 22 2	11 3 10 32	32 40 28 2	27 11 3	15 3 32	2 13 28	27 3 15 40	15	29 35 10 14
15. Duration of moving object's operation	11 2 13	3	3 27 16 40	22 15 33 28	21 39 16 22	27 1 4	12 27	29 10 27	1 35 13	10 4 29 35	19 29 39 35	6 10	35 17 14 19
16. Duration of fixed object's operation	34 27 6 40	10 26 24		17 1 40 33	22	35 10	1	1	2		25 14 6 35	1	20 10 16 38
17. Temperature	19 35 3 10	32 19 24	24	22 33 35 2	22 35 2 24	26 27	26 27	4 10 16	2 18 27	2 17 16	3 27 35 31	26 2 19 16	15 28 35
18. Illumination		11 15 32	3 32	15 19	35 19 32 39	19 35 28 26	28 26 19	15 17 13 16	15 1 19	6 32 13	32 15	2 26 10	2 25 16
19. Energy expense of movable object	19 21 11 27	3 1 32		1 35 6 27	2 35 6	28 26 30	19 35	1 15 17 28	15 17 13 16	2 29 27 28	35 38	32 2	12 28 35
20. Energy expense of fixed object	10 36 23			10 2 22 37	19 22 18	1 4				19 35 16 25			1 6

Theory of Inventive Problem Solving (TRIZ)

What should be Improved? \ What is deteriorated?	1. Weight of movable object	2. Weight of fixed object	3. Length of movable object	4. Length of fixed object	5. Area of movable object	6. Area of fixed object	7. Volume of movable object	8. Volume of fixed object	9. Speed	10. Force	11. Stress, pressure	12. Shape	13. Object's composition stability
21. Power	8 36 38 31	19 26 17 27	1 10 35 37		19 38	17 32 13 38	35 6 38	30 6 25	15 35 2	26 2 36 35	22 10 35	29 14 2 40	35 32 15 31
22. Waste of energy	15 6 19 28	19 6 18 8	7 2 6 13	6 38 7	15 26 17 30	17 7 30 18	7 18 23	7	16 35 38	36 38			14 2 39 6
23. Loss of substance	35 6 23 40	35 6 22 32	14 29 10 39	10 28 24	35 2 10 31	10 18 39 31	1 29 30 36	3 39 18 31	10 13 28 38	14 15 18 40	3 36 37 10	29 35 3 5	2 14 30 40
24. Loss of information	10 24 35	10 35 5	1 26	26	30 26	30 16		2 22	26 32				
25. Waste of time	10 20 37 35	10 20 26 5	15 2 29	30 24 14 5	26 4 5 16	10 35 17 4	2 5 34 10	35 16 32 18		10 37 36 5	36 37 4	4 10 34 17	35 3 22 5
26. Quantity of substance	35 6 18 31	27 26 18 35	29 14 35 18		15 14 29	2 18 40 4	15 20 29		35 29 34 28	35 14 3	10 36 14 3	35 14 2	15 2 17 40
27. Reliability	3 8 10 40	3 10 8 28	15 9 14 4	15 29 28 11	17 10 14 16	32 35 40 4	3 10 14 24	2 35 24	21 35 11 28	8 28 10 3	10 24 35 19	35 1 16 11	
28. Measurement accuracy	32 35 26 28	28 35 25 26	28 26 5 16	32 28 3 16	26 28 32 3	26 28 32 3	32 13 6		28 13 32 24	32 2	6 28 32	6 28 32	32 35 13
29. Manufacturing precision	28 32 13 18	28 35 27 9	10 28 29 37	2 32 10	28 33 29 32	2 29 18 36	32 28 2	25 10 35	10 28 32	28 19 34 36	3 35	32 30 40	30 18
30. Harmful action at object	22 21 27 39	2 22 13 24	17 1 39 4	1 18	22 1 33 28	27 2 39 35	2 22 37 35	34 39 19 27	21 22 35 28	13 35 39 18	22 2 37	22 1 3 35	35 24 30 18
31. Harmful effect caused by the object	19 22 15 39	35 22 1 39	17 15 16 22		17 2 18 39	22 1 40	17 2 40	30 18 35 4	35 28 3 23	35 28 1 40	2 33 27 18	35 1	35 40 27 39
32. Ease of manufacture	28 29 15 16	1 27 36 13	1 29 13 17	15 17 27	13 1 26 12	16 40	13 29 1 40	35	35 13 8 1	35 12	35 19 1 37	1 28 13 27	11 13
33. Ease of operation	25 2 13 15	6 13 1 25	1 71 13 12		1 17 13 16	18 16 15 39	1 16 35 15	4 18 39 31	18 13 34	28 13 35	2 32 12	15 34 29 28	32 35 30
34. Ease of repair	2 27 35 11	2 27 35 11	1 28 10 25	3 18 31	15 13 32	16 25	25 2 35 11	1	34 9	1 11 10	13	1 13 2 4	2 35
35. Adaptation	1 6 15 8	19 15 29 16	35 1 29 2	1 35 16	35 30 29 7	15 16	15 35 29		35 10 14	15 17 20	35 16 1 8	15 37 1 8	35 30 14
36. Device complexity	26 30 34 36	2 26 35 39	1 19 26 24	26	14 1 13 16	6 36	34 26 6	1 16	34 10 28	26 16	19 1 35	29 13 28 15	2 22 17 19
37. Measurement or test complexity	27 26 28 13	6 13 28 1	16 17 26 24	26	2 13 18 17	2 39 30 16	29 1 4 16	2 18 26 31	3 4 16 35	36 28 40 19	35 36 37 32	27 13 1 39	35 30 39 30
38. Degree of automation	28 26 18 35	28 26 35 10	14 13 28 17	23	17 14 13		35 13 16		28 10	2 35	13 35	15 32 1 13	18 1
39. Productivity	35 26 24 37	28 27 15 3	18 4 28 38	30 14 26 7	10 26 34 31	10 35 17 7	2 6 34 10	35 37 10 2		28 15 10 36	10 37 14	10 10 34 40	35 3 22 39

Appendix A: Contradiction Table of Inventive Principles (*Continued*)

What should be Improved? \ What is deteriorated?	14. Strength	15. Duration of moving object's operation	16. Duration of fixed object's operation	17. Temperature	18. Illumination	19. Energy expense of movable object	20. Energy expense of fixed object	21. Power	22. Waste of energy	23. Loss of substance	24. Loss of information	25. Waste of time	26. Quantity of substance
21. Power	26 10 28	19 35 10 38	16	2 14 17 25	16 6 19	16 6 19 37			10 35 38	28 27 18 38	10 19	35 20 10 6	4 34 19
22. Waste of energy	26			19 38 7	1 13 32 15			3 38		35 27 2 37	19 10	10 18 32 7	7 18 25
23. Loss of substance	35 28 31 40	28 27 3 18	27 16 18 38	21 36 39 31	1 6 13	35 18 24 5	28 27 12 31	28 27 18 38	35 27 2 31			15 18 35 10	6 3 10 24
24. Loss of information		10	10		19			10 19	19 10			24 26 28 32	24 28 35
25. Waste of time	29 3 28 18	20 10 28 18	28 20 10 16	35 29 21 18	1 19 26 17	35 38 19 18	1	35 20 10 6	10 5 18 32	35 18 10 39	24 26 28 32		35 38 18 16
26. Quantity of substance	14 35 34 10	3 35 10 40	3 35 31	3 17 39		34 29 16 18	3 35 31	35	7 18 25	6 3 10 24	24 28 35	35 38 18 16	
27. Reliability	11 28	2 35 3 25	34 27 6 40	3 35 10	11 32 13	21 17 27 19	36 23	21 11 26 31	10 11 35	10 35 29 39	10 28	10 30 4	21 28 40 3
28. Measurement accuracy	28 6 32	28 6 32	10 26 24	6 19 28 24	6 1 32	3 6 32		3 6 32	26 32 27	10 16 31 28		24 34 28 32	2 6 32
29. Manufacturing precision	3 27 40	3 27		19 26	3 32	32 2		32 2	13 32 2	35 31 10 24		32 26 28 18	32 30
30. Harmful action at object	18 35 37 1	22 15 33 28	17 1 40 33	22 33 35 2	1 19 32 13	1 24 6 27	10 2 22 37	19 22 31 2	21 22 35 2	33 22 19 40	22 10 2	35 18 34	35 33 29 31
31. Harmful effect caused by the object	15 35 22 2	15 22 33 31	21 39 16 22	22 35 2 24	19 24 39 32	2 35 6	19 22 18	2 35 18	21 35 22 2	10 1 34	10 21 29	1 22	3 24 39 1
32. Ease of manufacture	1 3 10 32	27 1 4	35 16	27 26 18	28 24 27 1	28 26 27 1	1 4	27 1 12 24	19 35	15 34 33	32 24 18 16	35 28 34 4	35 23 1 24
33. Ease of operation	32 40 3 28	29 3 8 25	1 16 25	26 27 13	13 17 1 24	1 13 24		35 34 2 10	2 19 13	28 32 2 24	4 10 27 22	4 28 10 34	12 35
34. Ease of repair	1 11 2 9	11 29 28 27	1		4 10 15 13	15 1 28 16		15 10 32 2	15 1 32 19	2 35 34 27		32 1 10 25	2 28 10 25
35. Adaptation	35 3 32 6	13 1 35	2 16	27 2 3 35	6 22 26 1	19 35 29 13		19 1 29	18 15 1	15 10 2 13		35 28	3 35 15
36. Device complexity	2 13 28	10 4 28 15		2 17 13	24 17 13	27 2 29 28		20 19 30 34	10 35 13 2	35 10 28 29		6 29	13 3 27 10
37. Measurement or test complexity	27 3 15 28	19 29 25 39	25 34 6 35	3 27 35 16	2 24 26	35 38	19 35 16	19 1 16 10	35 3 15 19	1 18 10 24	35 33 27 22	18 28 32 9	3 27 29 18
38. Degree of automation	25 13	6 9		26 2 19	8 32 13	2 32 13		28 2 27	23 28	35 10 18 5	35 33	24 28 35 30	35 13
39. Productivity	29 28 10 18	35 10 2 18	20 10 16 38	35 21 28 10	26 17 19 1	35 10 38 19	1	35 20 10	28 10 29 35	28 10 35 23	13 15 23		35 38

Theory of Inventive Problem Solving (TRIZ)

What should be Improved? \ What is deteriorated?	27. Reliability	28. Measurement accuracy	29. Manufacturing precision	30. Harmful action at object	31. Harmful effect caused by the object	32. Ease of manufacture	33. Ease of operation	34. Ease of repair	35. Adaptation	36. Device complexity	37. Measurement or test complexity	38. Degree of automation	39. Productivity
21. Power	19 24 26 31	32 15 2	32 2	19 22 31 2	2 35 18	26 10 34	26 35 10	35 2 10 34	19 17 34	20 19 30 34	19 35 16	28 2 17	28 35 34
22. Waste of energy	11 10 35	32		21 22 35 2	21 35 2 22		35 32 1	2 19		7 23	35 3 15 23	2	28 10 29 35
23. Loss of substance	10 29 39 35	16 34 31 28	35 10 24 31	33 22 30 40	10 1 34 29	15 34 33	32 28 2 24	2 35 34 27	15 10 2	35 10 28 24	35 18 10 13	35 10 18	28 35 10 23
24. Loss of information	10 28 23			22 10 1	10 21 22	32	27 22				35 33	35	13 23 15
25. Waste of time	10 30 4	24 34 28 32	24 26 28 18	35 18 34	35 22 18 39	35 28 34 4	4 28 10 34	32 1 10	35 28	6 29	18 28 32 10	24 28 35 30	
26. Quantity of substance	18 3 28 40	3 2 28	33 30	35 33 29 31	3 35 40 39	29 1 35 27	35 29 10 25	2 32 10 25	15 3 29	3 13 27 10	3 27 29 18	8 35	13 29 3 27
27. Reliability		32 3 11 23	11 32 1	27 35 2 40	35 2 40 26		27 17 40	1 11	13 35 8 24	13 35 1	27 40 28	11 13 27	1 35 29 38
28. Measurement accuracy	5 11 1 23			28 24 22 26	3 33 39 10	6 35 25 18	1 13 17 34	1 32 13 11	13 35 2	27 35 10 34	26 24 32 28	28 2 10 34	10 34 28 32
29. Manufacturing precision	11 32 1			26 28 10 36	4 17 34 26		1 32 35 23	25 10		26 2 18		26 28 18 23	10 18 32 39
30. Harmful action at object	27 24 2 40	28 33 23 26	26 28 10 18			24 35 2	2 25 28 39	35 10 2	35 11 32 31	22 19 29 40	23 19 29 40 34	33 3	22 31 13 24
31. Harmful effect caused by the object	24 2 40 39	3 33 26	4 17 34 26							19 1 31	2 21 27 1	2	22 35 18 39
32. Ease of manufacture		1 35 12 18		24 2			2 5 13 16	35 1 11 9	2 13 15	27 26 1	6 28 11 1	8 28 1	35 1 10 28
33. Ease of operation	17 27 8 40	25 13 2 34	1 32 35 23	2 25 28 39		2 5 12		12 26 1 32	15 34 1 16	32 26 12 17		1 34 12 3	15 1 28
34. Ease of repair	11 10 1 16	10 2 13	25 10	35 10 2 16		1 35 11 10	1 12 26 15		7 1 4 16	35 1 13 11		34 35 7 13	1 32 10
35. Adaptation	35 13 8 24	35 5 1 10		35 11 32 31		1 13 31	15 34 1 16	1 16 7 4		15 29 37 28	1	27 34 35	35 28 6 37
36. Device complexity	13 35 1	2 26 10 34	26 24 32	22 19 29 40	19 1	27 26 1 13	27 9 26 24	1 13	29 15 28 37		15 10 37 28	15 1 24	12 17 28
37. Measurement or test complexity	27 40 28 8	26 24 32 28		22 19 29 28	2 21	5 28 11 29	2 5	12 26	1 15	15 10 37 28		34 21	35 18
38. Degree of automation	11 27 32	28 26 10 34	28 26 18 23	2 33	2	1 26 13	1 12 34 3	1 35 13	27 4 1 35	15 24 10	34 27 25		5 12 35 26
39. Productivity	1 35 10 38	1 10 34 28	32 1 18 10	22 35 13 24	35 22 18 39	35 28 2 24	1 28 7 19	1 32 10 25	1 35 28 37	12 17 28 24	35 18 27 2	5 12 35 26	

Appendix B: Business Contradiction Matrix (*Mann 2004*)

What should be Improved? \ What is deteriorated?	1. R&D spec/capability/means	2. R&D cost	3. R&D time	4. R&D risk	5. R&D interface	6. Production spec/capability/means	7. Production cost	8. Production time
1. R&D spec/capability/means		2 4 15 38	21 38 35 23 15	3 9 24 23 36 11	3 13 24 33 38 25	23 29 35 4 13 5	2 26 29 40	35 6 10 2 20
2. R&D cost	2 4 15 38		26 34 1 10 3	27 9 34 16 37	13 26 35 10 1	26 35 17 27 34 3	26 35 17 27 34 3	10 2 6 15
3. R&D time	21 38 35 23 15	26 34 1 10 3		1 29 10 40 11	15 25 35 1 40	5 6 20 35 2	5 29 35 2	7 26 10 15 3
4. R&D risk	3 9 24 23 36 11	27 9 34 16 37	1 29 10 40 11		6 29 15 14 17 25	24 35 10 3 13 11	5 35 40 23 1 12	5 40 20 15
5. R&D interface	3 13 24 33 38 25	13 26 35 10 1	15 25 35 1 40	6 29 15 14 17 25		5 6 17 40 33 10 26	15 23 29 5 13	15 40 23 3 24 13
6. Production spec/capability/means	23 29 35 4 13 5	5 2 27 1	5 6 20 35 2	24 35 10 3 13 11	5 6 17 40 33 10 26		15 25 3 10 5 8	1 35 21 15 4 10
7. Production cost	2 26 29 40	26 35 17 27 34 3	5 29 35 2	5 35 40 23 1 12	15 23 29 5 13	15 25 3 10 5 8		1 24 29 10 27 3 14
8. Production time	35 6 10 2 20	10 2 6 15	7 26 10 15 3	5 40 20 15	7 5 37 10	1 35 21 15 4 10	1 24 29 10 27 3 14	
9. Production risk	3 5 10 2 23 12	6 7 23 26 13	6 15 7 37 13 9	11 23 39 7 9 33	7 5 37 10	6 27 35 22 12 37	26 10 1 3 25 12	10 27 15 6 3 22 29
10. Production interface	5 7 37 1 4	15 35 10 25 24	25 23 35 29 2 13	7 3 17 23 24	28 40 6 29 13 31 30	3 25 17 35 12 13	26 1 37 25 2 28	10 15 38 20 27 6 3
11. Supply spec/capability/means	6 2 35 25 3	23 6 11 28	11 6 23 19 18 2	5 35 13 26 6	6 35 15 13 14	7 13 22 6 35 12 13	5 2 30 35 17 8 25	5 17 16 3 10
12. Supply cost	15 6 15 13	10 5 35	5 13 23 25	1 11 2 34	2 33 3 15 10	15 35 13 22	5 35 3 12 17 24	5 2 35 13 25
13. Supply time	21 35 2 39	10 19 35 22	10 25 7 2	1 2 11 38 15	5 2 35 10 12	35 5 13 22	2 35 24 10 13 5	3 10 23 40 13 4
14. Supply risk	11 39 30 31	11 1 32 16	23 7 29 2 24 37	13 22 25 9 35 26	5 35 13 40 3 9	15 16 3 2 24 6	2 13 10 26 29	13 2 35 10 24
15. Supply interface	11 26 2 5 13	10 38 13	1 17 40 38 24 2	13 22 25 9 35 26	28 40 6 15 29	10 25 3 33	12 3 35 5 10 7	23 12 3 24 13 7

What should be Improved? \ What is deteriorated?	9. Production risk	10. Production interface	11. Supply spec/capability/means	12. Supply cost	13. Supply time	14. Supply risk	15. Supply interface	16. Support spec/capability/interface
1. R&D spec/capability/means	3 5 10 2 23 12	5 7 3 7 1 4	6 2 35 25 3	15 6 1 5 13	21 35 2 39	11 39 30 31	11 26 2 5 13	36 11 2 35 27
2. R&D cost	6 7 23 26 13	15 35 10 25 24	23 6 11 28	10 5 35	10 19 35 22	11 13 2 16	10 38 13	27 6 1 10
3. R&D time	6 15 7 37 13 9	25 23 35 29 2 13	11 6 23 19 18 2	5 13 23 25	10 25 7 2	23 7 29 2 24 37	1 17 40 38 24 2	6 10 3 35 20
4. R&D risk	11 23 39 7 9 33	7 3 17 23 24	5 35 13 26 6	1 11 2 34	1 2 11 38 15	13 22 25 9 35 26	35 13 22 25 9 26	6 1 26 37 15
5. R&D interface	7 5 3 37 10	28 40 6 29 13 31 30	6 35 15 13 14	2 33 3 15 10	5 23 5 10 12	5 35 13 40 3 9	28 40 6 15 29	6 13 35 21 12
6. Production spec/capability/means	6 27 35 2 12 37	3 2 5 17 35 12 13	7 13 22 6 35	15 35 13 22	35 5 13 22	15 16 3 2 24 6	10 25 3 33	35 23 1 24
7. Production cost	26 10 1 3 25 12	26 1 37 25 2 28	5 2 30 35 17 8 25	5 35 31 2 17 24	2 35 24 10 13 5	2 13 10 26 29	12 3 35 5 10 7	1 35 10 29 27
8. Production time	10 27 15 6 3 22 29	10 15 38 20 27 6 3	5 17 16 3 1 0	5 2 35 13 25	3 10 23 40 13 4	13 2 35 10 24	23 12 3 24 13 7	1 35 10 38 29 25 13
9. Production risk		5 6 23 20 7 10 25	5 25 3 35 2 10	5 35 23 25 2	13 22 25 1 10	5 26 35 2 25	5 10 40 2 4 25	13 35 2 15 24
10. Production interface	5 6 23 20 7 10 25		6 2 3 7 40 10	5 30 10 15 2 12	5 35 6 13 17 10 24	23 33 5 26 2	33 5 2 26 10	23 11 40 2 32 29
11. Supply spec/capability/means	5 25 3 35 2 10	6 2 37 40 10		7 35 19 1 10 29	35 1 13 2 24	7 8 11 10 24 12 25	6 30 15 40 12 2	11 23 35 1 29 17
12. Supply cost	5 35 23 25 2	5 30 10 15 2 12	7 35 19 1 10 29		3 24 38 10 19	27 3 19 24 8	1 28 6 38 4	35 24 5 13 27
13. Supply time	13 22 25 1 10	5 35 6 13 17 10 24	35 1 13 2 24	3 24 38 10 19		10 29 15 13 2 3	5 19 3 15 10 18	25 10 29 19 4
14. Supply risk	5 26 35 2 25	23 33 5 26 2	7 8 11 10 24 12 25	27 3 19 24 8	10 29 15 13 2 3		5 10 25 37 2 14 38	1 35 6 24 25
15. Supply interface	5 10 40 2 4 25	33 5 2 26 10	6 30 15 40 12 2	1 28 6 38 4	5 19 3 15 10 18 5 10 25 37 2 14 38			10 31 24 35 3

Appendix B: Business Contradiction Matrix (Mann 2004) (Continued)

What should be Improved? \ What is deteriorated?	17. Support cost	18. Support time	19. Support risk	20. Support interface	21. Customer revenue/demand/feedback	22. Amount of information	23. Communication flow	24. Harmful factor affecting system
1. R&D spec/capability/means	15 35 28 25 29	5 2 6 27 25	15 27 40 12 27	11 2 5 9 26	14 13 2 27 10	37 13 25 10 39	6 25 31 29 7 23	11 25 2 26 3
2. R&D cost	6 1 25 10 27	6 1 25 10 27	10 25 2 22	6 10 1 7 20	7 25 30 21 10 9 2	37 25 28 2 32	6 18 37 13 25 22	35 27 3 28 2
3. R&D time	7 15 40 26 5	7 40 1 26 15	23 24 2 37 7	6 10 26 24 2 38	7 19 21 29 30	7 2 37 20 25	6 26 18 19 40	26 2 35 24 11
4. R&D risk	11 7 28 35	1 2 32 28 7	40 36 6 10 26 13	6 10 7 26 13	36 13 25 22 37 3	13 10 26 25 4 37	30 6 31 4 9 13 22	35 2 15 26 3
5. R&D interface	6 7 40 38 13	6 38 20 10 37	5 35 40 13	28 40 6 7 30	4 7 25 40 13 35 28	1 6 3 40 25	2 6 35 3 25 18	3 26 35 28 24
6. Production spec/capability/means	13 10 17 2 27 34	5 6 10 12 27 25	6 10 2 27 12	6 40 10 27	5 15 35 25 33	13 32 15 23 24 18 16	6 2 13 25 10	22 24 35 13 24 2
7. Production cost	3 2 35 10 27	27 3 10 25 24	10 25 27 3 35	10 35 7 24 25	7 13 1 24 25	26 27 25 34 37	6 35 37 18	2 35 5 34 15
8. Production time	3 13 25 5 35	35 25 5 4 19	35 29 13 25 2 31	13 9 26 23 7	13 1 37 17 31 29	13 15 23 25 3 37	2 37 18 19 25	22 35 3 13 24
9. Production risk	3 35 19 24	24 14 13 35 2	7 5 3 10 25	5 35 33 7 25 10	13 22 7 13 24 39	5 25 3 37 32 26 13	25 38 3 26 10 13	35 2 26 34 25
10. Production interface	23 10 3 13 22	23 13 10 1 2	10 14 2 25 29	40 33 6 10 26 2	7 5 10 40 4 2 25	2 37 4 13 37 25	2 28 3 37 32 25 10	3 26 35 28 10 24
11. Supply spec/capability/means	23 11 2 6 26	23 11 26 2 7	11 23 24 2 9 17	23 11 2 25 35 32	10 3 25 5 15	13 4 28 37 17 7	5 25 23 10 35 28	13 17 29 2 35 15
12. Supply cost	27 5 35 25 10 2	10 27 30 35 2 5	10 12 2 27 7 5	10 24 25 1 6	2 35 13 25 26 16	28 35 2 37 34 7	35 6 1 27 25 12 28	11 35 2 12 31 30
13. Supply time	25 27 10 2	27 2 13 35 10	10 25 35 6 13	24 5 35 25 7 10	35 13 25 1 22 26	28 2 37 32 35 7	6 31 25 35 37 16	35 3 29 2 10 12
14. Supply risk	19 10 5 27 2	2 27 10 5 25	24 25 10 7 1	5 35 2 13 19	25 22 2 35 10 17	5 37 15 6 32	6 16 13 35 7 2	2 13 35 31 24 12
15. Supply interface	5 10 26 1 13 25	29 30 2 25 5 32	5 25 10 9 2 35	5 6 38 40 25 10	13 25 39 24 7 17	3 6 37 28 32 35	2 3 13 4 12 25	3 35 13 14 39

292

What should be Improved? \ What is deteriorated?	25. System-generated harmful factors	26. Convenience	27. Adaptability/versatility	28. System complexity	29. Control complexity	30. Tension/stress	31. Stability
1. R&D spec/capability/means	25 29 2 37 13	15 35 25 16 28	30 25 29 1 35	17 25 1 19 35	25 15 19 35	3 2 25 35 9	25 2 15 36 29
2. R&D cost	28 26 2 22 8 35	25 2 35 1 29	35 28 19 1 8	5 2 35 1 29	25 19 35 27 2 18	1 19 35 27 2 18	11 25 27 15 2
3. R&D time	26 2 15 19 35 40	1 2 15 19 25 28	15 1 35 14 4	5 6 25 10 2 37	25 28 15 2 6 37	2 39 24 10 4 13	10 3 35 22 27
4. R&D risk	2 3 35 15 12 9	26 3 11 24 5 13 40	2 40 31 28 35 29 7	28 30 35 1 17	25 1 3 37 40 12 24	1 23 2 25 13 39	9 14 1 24
5. R&D interface	3 26 35 37 2 40	16 13 25 28 37	29 37 40 1 35 17 30	25 28 1 3 10	6 28 1 3 40 25 13 9	35 3 37 32 9 18	15 17 25 3 4 36
6. Production spec/capability/means	35 22 18 39	2 15 1 5 28 7 10 13 16 12	1 15 17 2 28 38	12 17 27 26 1 28 24 13	28 1 13 16 25 37	35 1 3 10 16	35 1 23 3 19 13 5 39 40
7. Production cost	1 35 27 10 2	1 25 2 27 29	1 30 10 38 29 35	35 5 1 2 29 25	6 3 25 10 32 37	1 35 2 25 13 17	10 1 35 27
8. Production time	35 22 18 10 24 2	19 2 35 26 13 30	10 15 30 7 2 29 25 13	25 28 2 35 10 15	25 37 3 13 28	2 20 12 25 3 13 14	10 15 29 2 19 7
9. Production risk	25 10 39 24 29	3 26 6 11 35	2 40 38 30 35 29	25 2 26 5 29 35	30 12 25 40 2 37	25 9 24 39 7 19	9 1 37 3 19
10. Production interface	3 26 35 29 24	5 19 28 32 2 10	29 1 17 40 38	10 18 28 2 35	18 28 19 15 40 2 25	3 40 19 1 24	11 25 1 3 4
11. Supply spec/capability/means	10 1 34 35 15 13	35 3 13 2 15	13 17 7 15 19	29 30 35 17 3	6 5 28 37 3 25	2 23 5 30 10 13 35	15 5 25 10 35
12. Supply cost	30 2 15 3 5 13	10 35 2 12 31 30	1 17 40 3 29	35 19 1 25 2	22 2 37 4 32 25	10 3 25 7 40	19 3 25 10 4
13. Supply time	25 10 29 13 12 21	24 35 28 1 29	15 1 10 27 7	38 24 16 15 3	28 32 25 2 37	1 10 15 25 24 2 19	35 3 5 27 20 18
14. Supply risk	2 15 19 23 40 24	5 16 10 13 25 2	15 17 40 3 29 25	2 4 15 28 35 32	2 28 15 24 37	1 19 13 10 39	9 13 1 25 14
15. Supply interface	2 30 40 22 26	5 25 3 40 20	29 28 30 3 15	28 5 3 25 37 40	25 8 22 28 32 37	5 3 17 29 13 35 2	33 15 23 17 7

Appendix B: Business Contradiction Matrix (Mann 2004) (Continued)

What should be Improved? \ What is deteriorated?	1. R&D Spec/capability/means	2. R&D cost	3. R&D time	4. R&D risk	5. R&D interface	6. Production spec/capability/means	7. Production cost	8. Production time
16. Support spec/capability/means	36 11 2 35 27	27 6 1 10	6 10 3 35 20	6 1 26 37 15	6 1 3 35 21 12	35 23 1 24	1 35 10 29 27	1 35 10 38 29 25 13
17. Support cost	15 35 28 25 29	6 1 25 10 27	7 15 40 26 5	1 1 7 28 35	6 7 40 38 13	13 10 17 27 34	3 2 35 10 27	3 13 25 5 35
18. Support time	5 2 6 27 25	6 1 25 10 27	7 40 1 26 15	1 2 32 28 7	6 38 20 10 37	5 6 10 12 27 25	27 3 10 25 24	35 25 5 4 19
19. Support risk	15 27 40 12 27	10 25 2 2	23 24 2 37 7	40 36 6 10 26 13	5 35 40 13	6 10 2 27 12	10 25 27 3 35	35 29 13 25 2 31
20. Support interface	11 2 5 9 26	6 10 1 7 20	6 10 26 24 2 38	6 10 7 26 13	28 40 6 7 30	6 40 10 27	10 35 7 24 25	13 9 26 23 7
21. Customer revenue/demand/feedback	14 13 22 7 10	7 25 30 21 10 9 2	7 19 21 29 30	36 13 25 22 37 3	4 7 25 40 13 35 28	5 15 35 25 33	7 13 1 24 25	13 1 37 17 31 29
22. Amount of information	37 13 25 10 39	37 25 28 2 32	7 2 37 20 25	1 3 10 26 25 4 37	1 6 3 40 25	13 32 15 23 24 18 16	26 27 25 34 37	13 15 23 25 3 37
23. Communication flow	6 25 31 29 7 23	6 18 37 13 25 22	6 26 18 19 40	30 6 31 4 9 13 22	26 35 3 25 18	6 2 13 25 10	6 35 37 18	2 37 18 19 25
24. Harmful factors affecting system	11 25 2 26 3	35 27 3 28 2	26 2 35 24 11	35 2 15 26 3	3 26 35 28 24	22 24 35 13 24 2	2 35 5 34 15	22 35 3 13 24
25. System-generated harmful factors	25 29 2 37 13	28 26 2 22 8 35	26 2 15 19 35 40	23 35 15 12 9	3 26 35 37 2 40	35 22 18 39	1 35 27 10 2	35 22 18 10 24 2
26. Convenience	15 35 25 16 28	25 2 35 1 29	1 2 15 19 25 28	26 3 11 24 5 13 40	16 13 25 28 37	2 15 15 28 7 10 13 16 12	1 25 2 27 29	19 2 35 26 13 30
27. Adaptability/versatility	30 25 29 1 35	35 28 19 1 8	15 1 35 14 4	2 40 31 28 35 29 7	29 37 40 1 35 17 30	1 5 17 2 28 38	1 30 10 38 29 35	10 15 30 7 2 29 25 13
28. System complexity	17 25 1 19 35	5 2 35 1 29	5 6 25 10 2 37	28 30 35 1 17	25 28 1 3 10	12 17 27 26 1 28 24 13	35 5 1 2 29 25	25 28 2 35 10 15
29. Control complexity	25 15 19 35	25 19 35 27 2 18	25 28 15 2 6 37	25 1 3 37 40 12 24	6 28 1 3 40 25 13 9	28 1 13 16 25 37	6 3 25 10 32 37	25 37 3 13 28
30. Tension/stress	3 2 25 35 9	1 1 9 35 27 2 18	2 39 24 10 4 13	1 23 2 25 13 39	35 3 37 32 9 18	35 1 3 10 16	1 35 2 25 13 17	2 2 10 12 25 13 14
31. Stability	25 2 15 36 29	11 25 27 15 2	10 3 35 22 27	9 14 1 24	15 17 25 3 4 36	35 1 23 3 19 13 5 39 40	10 1 35 27	10 15 29 2 19 7

What should be Improved? / What is deteriorated?	9. Production risk	10. Production interface	11. Supply spec/capability/means	12. Supply cost	13. Supply time	14. Supply risk	15. Supply interface	16. Support spec/capability/interface
16. Support spec/capability/means	13 35 2 15 24	23 11 40 2 32 29	11 23 35 1 29 17	35 24 5 13 27 17	25 10 29 19 4	1 35 6 24 25	10 31 24 35 3	2 25 10 35 15
17. Support cost	3 35 19 24	23 10 3 13 22	23 11 2 6 26	27 5 35 25 10 2	25 27 10 2	19 10 5 27 2	5 10 26 1 13 25	22 25 15 3 32
18. Support time	24 14 13 35 2	23 13 10 1 2	23 11 26 27	10 27 30 35 2 5	27 2 13 35 10	2 27 10 5 25	29 30 2 25 5 32	13 22 10 35 4 6
19. Support risk	7 5 3 10 25	10 14 2 25 29	11 23 24 29	10 12 2 27 7 5	10 25 35 6 13	24 25 10 7 1	5 25 10 9 2 35	28 25 5 7 2 24
20. Support interface	5 35 33 7 25 10	40 3 6 10 26 2	23 11 2 25 35 32	10 24 25 1 6	24 5 35 25 7 10	5 35 2 13 19	5 6 38 40 25 10	28 25 7 22 5 13
21. Customer revenue/demand/feedback	13 22 7 13 24 39	7 5 10 40 4 2 25	10 3 25 5 15	2 35 13 25 26 16	35 13 25 1 22 26	25 22 2 35 10 17	13 25 39 24 7 17	28 25 7 22 5 13
22. Amount of information	5 25 3 37 32 28 13	2 37 4 13 37 25	13 4 28 37 17 7	28 35 2 37 34	28 2 37 32 35 7	5 37 15 6 32	3 6 37 28 32 35	10 28 3 25 37 4
23. Communication flow	25 38 3 26 10 13	2 28 3 37 32 25 10	5 25 23 10 35 28	35 6 12 7 25 12 28	6 31 25 35 37 16	6 16 13 35 7 2	2 3 13 4 12 25	10 28 37 3 7
24. Harmful factors affecting system	35 2 26 34 25	3 26 35 28 10 24	13 17 29 2 35 15	11 35 2 12 31 30	35 3 29 2 10 12	2 13 35 31 24 12	3 35 13 14 39	27 35 34 2 40
25. System-generated harmful factors	25 10 39 24 29	3 26 35 29 24	10 1 34 35 15 13	30 2 15 3 5 13	25 10 29 13 12 21	2 15 19 23 40 24	2 30 40 22 26	2 35 40 24 26 39
26. Convenience	3 26 6 11 35	5 19 28 32 2 10	35 3 13 2 15	10 35 2 12 31 30	24 35 28 1 29	5 16 10 13 25 2	5 25 3 40 20	27 17 40 3 8
27. Adaptability/versatility	2 40 38 30 35 29	29 1 17 40 38	13 1 7 17 15 19	1 17 40 3 29	15 1 10 27 7	15 17 40 3 29 25	29 28 30 3 15	35 13 8 24 29
28. System complexity	25 2 26 5 29 35	10 18 28 2 35	29 30 35 17 3	35 19 1 25 2	38 24 16 15 3	2 4 15 28 35 32	28 5 3 25 37 40	13 35 1 2 9
29. Control complexity	30 12 25 40 2 37	18 28 19 15 40 2 25	6 5 28 37 3 25	22 2 37 4 32 25	28 32 25 2 37	2 28 15 24 37	25 8 22 28 32 37	11 13 2 35 25
30. Tension/stress	25 9 24 39 7 19	3 40 19 1 24	2 23 5 30 10 13 35	10 3 25 7 40	1 10 15 25 24 2 19	1 19 13 10 39	5 3 17 29 13 35 2	11 35 24 19 2 25
31. Stability	9 1 37 3 19	11 25 1 3 4	15 5 25 10 35	19 3 25 10 4	35 3 5 27 20 18	9 13 1 25 14	33 15 23 17 7	25 26 1 10 12

Appendix B: Business Contradiction Matrix (Mann 2004) (Continued)

What should be Improved? \ What is deteriorated?	17. Support cost	18. Support time	19. Support risk	20. Support interface	21. Customer revenue/demand/feedback	22. Amount of information	23. Communication flow	24. Harmful factor affecting system
16. Support spec/capability/means	2 25 10 35 15	22 25 15 3 2	13 22 10 35 4 6	28 25 5 7 2 24	28 25 7 22 5 13	10 28 3 25 37 4	10 28 37 37	27 35 34 2 40
17. Support cost		5 4 25 10 17 14 13	27 35 25 14 1 31	26 25 37 3 24 2	24 25 37 3 7 28 18	28 3 17 37 32 4	25 1 28 32 20 35	1 35 22 25 17
18. Support time	5 4 25 10 17 14 13		15 29 9 19 1 18 35 31	15 29 10 1 35 30	7 20 24 35 25 26	1 2 15 35 25 4 37	6 31 2 35 28 37	35 15 1 3 10
19. Support risk	27 35 15 14 1 31	15 29 9 19 1 18 35 31		5 6 40 33 7 24	20 7 4 13 35 25 24	25 3 28 35 37 10	29 31 6 2 30 15 10	25 35 11 15 19 1
20. Support interface	26 25 37 3 24 2	15 29 10 1 35 30	5 6 40 33 7 24		16 17 40 13 10 25	1 3 37 2 28 7 4	2 3 15 18 25	11 24 35 5 21 14
21. Customer revenue/demand/feedback	24 25 37 7 28 18	7 20 24 35 25 26	20 7 4 13 35 24	16 17 40 13 10 25		2 29 3 35 13 1 37 28 4	29 31 30 7 13 17 38	39 3 5 17 26 35
22. Amount of information	28 3 17 37 32 4	1 2 15 35 25 4 37	25 3 28 35 37 10	1 3 37 2 28 7 4	2 29 3 35 13 1 37 28 4		2 37 3 4 31 28 7	22 10 1 2 35
23. Communication flow	25 1 28 32 20 35	6 31 2 35 28 37	29 31 6 2 30 15 10	2 3 15 18 25	29 31 30 7 13 17 38	2 37 3 4 31 28 7		6 30 15 28 13 36 2
24. Harmful factors affecting system	1 35 22 25 17	35 15 1 3 10	25 35 11 15 19 1	11 24 35 5 21 14	39 3 5 17 26 35	22 10 1 2 35	6 30 15 28 13 36 2	
25. System-generated harmful factors	2 24 35 22 13 31 10	35 15 29 3 1 19	25 3 4 35 15 19	25 13 22 10 17	38 10 6 5 35 24	10 21 22 29 19	1 28 4 35 7 24	35 3 24 4 13 31 15
26. Convenience	25 1 12 26 10 15	5 25 13 2 10	2 3 25 10 16 5 31	7 5 6 20 26 2 30	28 27 35 40 1 30	27 25 4 10 22 13 6 19	25 1 19 29 35 18	2 25 28 39 15 10
27. Adaptability/versatility	17 35 15 13 2	3 30 40 29 17	1 30 40 17 14 15	29 30 17 14 18 1	40 17 16 14 15	15 10 2 13 29 3 4	25 6 37 40 15 19	35 11 22 32 31
28. System complexity	35 1 25 2 17	28 15 17 32 37	13 35 4 2 37	28 17 29 37 10 4 13	25 1 2 19 10 4	10 25 13 40 2	1 25 4 37 6 18	22 19 29 40 35 15 10
29. Control complexity	15 25 19 28 37	28 25 37 15 3 14	10 15 1 34 37	25 15 10 30 29	25 2 7 37 6 4 19	2 7 25 19 1 40 37	25 1 19 37 10	3 15 2 22 25 9 28 26
30. Tension/stress	35 24 10 2 25 31 19	2 24 10 40 25 8	10 11 39 1 24 35	10 8 2 24 6 21 13	2 10 12 24 25	2 28 35 10 24 31	3 4 6 7 13 36	11 25 30 2 35 28
31. Stability	1 35 2 29 10	10 15 2 30 29 12	10 35 7 9 19 1	1 11 1 40 13 22 23	10 40 29 30 28 26	11 13 25 2 24	37 1 39 40 9 31	35 24 30 18 33

What should be Improved? / What is deteriorated?	25. System-generated harmful factors	26. Convenience	27. Adaptability/versatility	28. System complexity	29. Control complexity	30. Tension/stress	31. Stability
16. Support spec/capability/means	2 35 40 24 26 39	27 17 40 3 8	35 13 8 24 29	13 35 1 2 9	11 13 2 35 25	11 35 24 19 2 25	25 26 1 10 12
17. Support cost	2 24 35 22 13 31 10	25 1 12 26 10 15	17 35 15 1 3 2	35 1 25 2 17	15 25 19 28 37	35 24 10 2 25 31 19	1 35 2 29 10
18. Support time	35 15 29 3 1 19	5 25 13 2 10	3 30 40 29 17	28 15 17 32 37	28 25 37 15 3 1 4	2 24 10 40 25 8	10 15 2 30 29 12
19. Support risk	25 3 4 35 15 19	2 3 25 10 16 5	1 30 40 17 14 15	13 35 4 2 37	10 15 1 34 37	10 11 39 1 24 35	10 35 7 9 19 1
20. Support interface	25 13 22 10 17	7 5 6 20 26 2 31	29 30 17 14 18 1	28 17 29 37 10 4 13	25 15 10 30 29	10 8 2 24 6 21 13	1 1 1 40 13 22 23
21. Customer revenue/demand/feedback	38 10 6 5 35 24	28 27 35 40 1 30	40 17 16 14 15 1	25 1 2 19 10 4	25 2 7 37 6 4 19	2 10 12 24 25	10 40 29 30 28 26
22. Amount of information	10 21 22 29 19	27 25 4 10 22 13 6 19	15 10 2 13 29 3 4	10 25 13 40 2	2 7 25 19 1 40 37	2 28 35 10 24 31	1 1 13 25 2 24
23. Communication flow	1 28 4 35 7 24	25 1 19 29 35 18	25 6 37 40 15 19	1 25 4 37 6 18	25 1 19 37 10	3 4 6 7 13 36	37 1 39 40 9 31
24. Harmful factors affecting system	35 3 24 4 13 31 15	2 25 28 39 15 10	35 11 22 32 31	22 19 29 40 35 15 10	3 15 2 22 25 9 28 26	11 25 30 2 35 28	35 24 30 18 33
25. System-generated harmful factors		1 15 13 34 31 16	3 1 29 15 10 24	25 3 15 22 10 13	25 3 15 22 10 23 13	11 25 12 8 37 35	35 40 27 39 2
26. Convenience	1 15 13 34 31 16		15 34 1 16 29 24 17	26 27 32 9 12 24 17	25 5 10 12 24 28 3	10 5 14 12 13 35	32 35 30 25 13 19 3
27. Adaptability/versatility	3 1 29 15 10 24	15 34 1 16 29 36 19		15 29 28 5 37 6 35 25	25 1 28 37 3	17 40 30 3 15 19 16	35 30 14 34 2 19 10
28. System complexity	25 3 15 22 10 23 13	26 27 32 9 12 24 17	15 29 28 5 37 6 35 25		25 19 1 28 37 26	1 10 2 24 4 19	2 2 22 35 17 19 26 24
29. Control complexity	25 3 15 22 10 23 13	25 5 10 12 24 28 3	25 15 1 28 37 3	25 19 1 28 37 3 26		11 24 35 2 40 25	11 28 32 37 25 24
30. Tension/stress	11 25 12 8 37 35	10 5 14 12 13 35	17 40 30 3 15 19 16	1 10 2 24 4 19	11 24 35 2 40 25		29 35 11 24 19 13
31. Stability	35 40 27 39 2	32 35 30 25 13 19 3	35 30 14 34 2 19 10	2 22 35 17 19 26 24	11 28 32 37 25 24	29 35 11 24 19 13	

Chapter 10

Design and Improvement of Service Processes— Process Management

10.1 Introduction

There are two key aspects in delivering service to customers: the service product and the service process. Designing and improving the main service process and other supporting processes is a key task in achieving superior service quality.

What is a process? Caulkin (1989) defines it as being a "continuous and regular action or succession of actions, taking place or carried on in a definite manner, and leading to the accomplishment of some result; a continuous operation or series of operations." Keller et al. (1999) defines the process as "a combination of inputs, actions and outputs." Anjard (1998) further defines it as being "a series of activities that takes an input, adds value to it and produces an output for a customer." This view is summarized in Fig. 10.1.

Processes involve a series of steps by which the inputs are converted into outputs, which may be goods, information, or services. The quality of outputs is entirely dependent upon the quality with which the processes are executed. In the manufacturing industry, the quality of a manufactured product depends on the quality of the process used to manufacture it. In a restaurant, the taste of a meal, the time from order to delivery, the cost, and customer satisfaction are all highly dependent on the quality of the service process. The quality of processes implies that the correct steps are used in the right order, the correct tools are used in the process, the correct technique is applied, and everything is performed at the right time.

A process that achieves maximum quality and efficiency and uses minimum cost to run is often said to have achieved process excellence. The process

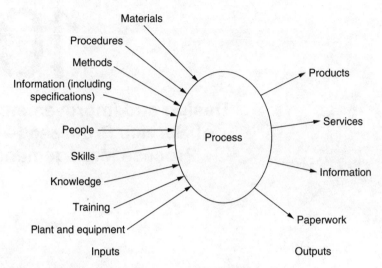

Figure 10.1 A Diagram of a Process (*Oakland 1994*)

excellence implies in part that waste is minimized. Wastes include unnecessary steps, unnecessary works, unnecessary movements, and unnecessary consumption of resources. Minimization of waste brings about more efficient application of resources, the work force, space, raw materials, and time. Efficient use of time and all other available resources can only be achieved through superior design and continuous improvement of processes. The method for designing and continuously improving processes is process management. By enhancing the efficiency and effectiveness, process management offers the potential to improve customer satisfaction and, ultimately, to offer increased profits, high growth, and long-term business stability. As a matter of fact, most organizations, both large or small, operate based on a variety of processes.

In fact, everything that we do uses some process, whether the process is documented or not, and whether it is followed precisely or not. Figure 10.2 shows a typical business operation model for any company. The common processes in a business operation often include

Core processes
- Service delivery process for delivering services to customers
- Manufacturing process for transforming raw materials into finished goods
- Product development process for designing new products and enhancing existing ones
- Marketing process for marketing products or services to customers

Design and Improvement of Service Processes—Process Management

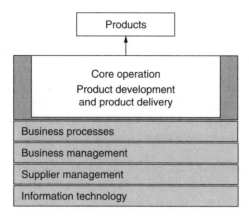

Figure 10.2 Business Operation Model

Supporting processes
- Accounting and finance process for running financial transactions
- Personnel process for hiring, firing, and promotion of employees
- Purchasing and supplier management process for acquiring supplies and services needed to run other processes

In fact, processes are everywhere in an organization. Processes found in organizations are involved in some product or service development, some production or operations to provide a service, some customer support, some marketing activities, some sales activities, and so on. Depending on the nature of the company's business, certain processes are more important than others. For service-oriented companies, service delivery processes are clearly very important. In an industry where frequent new product introductions are critical for maintaining market share; for example, the movie industry, the design and development processes may be the most important processes. The most important consideration in deciding what are important processes is who the customer is. The most important processes are those that directly affect the customer value creation.

Optimization of processes, especially the key processes, is extremely important for the bottom line of service organizations. Process management is a systematic approach to optimizing processes in terms of the following process metrics:
- Quality
- Throughput

- Efficiency
- Response time and speed
- Work-in-process (WIP) inventory
- Process cost

Other metrics such as safety may also be important. Most organizations are motivated to manage their processes through several of these metrics. In process management, the goal is usually to maximize profits, maintain a high level of customer satisfaction, and achieve long-term process stability.

Process management is a disciplined, systematic approach used to identify process issues, analyze them, design and improve the process, and maintain the improved process. Process management involves the following five phases:

1. Process mapping
2. Process diagnosis
3. Process design
4. Process implementation
5. Process maintenance

A variety of tools are used in process management. These tools include lean manufacturing and operation, process failure mode and effect analysis, computer simulation, data collection, and statistical analysis. The key for success in process management is to select the most appropriate tools to be used at each process management phase. We now briefly discuss each phase.

1. *Process mapping:* A *process map* is a schematic model for a process. "A process map is considered to be a visual aid for picturing work processes which show how inputs, outputs and tasks are linked" (Anjard 1998). Process mapping is used to develop a process map for the process under study. There are a number of different methods of process mapping; these methods include a process flowchart, IDEF0 process mapping, and value stream mapping. Once the process has been mapped and documented, the process diagnosis can then proceed.
2. *Process diagnosis:* Process diagnosis is used to try to identify the key problems in process performance and the root causes for these problems. The tools used in process diagnosis include process failure mode and effect analysis, cause-and-effect diagrams, lean manufacturing and operation principles, and value stream mapping analysis.
3. *Process design:* Process design involves the following activities:
   - Understanding the requirements of the process and translating the requirements into process design options

- Developing design alternatives for the process
- Analyzing the alternatives to establish one candidate as the best process to be implemented

Process design employs many tools including quality function deployment, lean manufacturing principles, value stream mapping analysis, computer simulation for evaluating the candidate process designs, and decision analysis for resolving complex tradeoffs.

4. *Process implementation:* Process implementation involves final validation of the process and controlled dissemination throughout the organization. This includes procuring and installing tools and equipment required for the process, as well as training activities required for the correct application of this new process.
5. *Process maintenance:* Process maintenance involves ongoing monitoring of the process and periodic improvement to ensure process performance remains high, despite changing internal and external conditions.

This chapter is devoted to covering all-important aspects of process management. Section 10.2 discusses some basic aspects of processes. Section 10.3 discusses several types of processes, their performance metrics, and process features. Section 10.4 describes several commonly used process mapping approaches. Section 10.5 discusses lean operation principles, Section 10.6 gives a detailed description of how process management works. Section 10.7 gives a process management case study.

10.2 Process Basics

10.2.1 What Is a Process?

Processes occur in a wide variety of forms and in many different areas, but the common factor is that every process consists of several steps that are used to convert some kind of input into output. Inputs may be raw materials or information, and outputs may be finished goods or services.

A process must be clearly distinguished from the tools and resources that are used in the process. The essence of a process is in the act of *doing*—the steps involved in doing something. The tools and resources are the means whereby the doing is accomplished. The reason why the act of doing has to be separated from the means for doing is that there may be a variety of tools, resources, techniques, and methods used to achieve the same objective. In as much as the process steps are distinguishable from the process tools and resources, any analysis of a process must pay adequate attention to the tools and resources because they too will impact greatly on the output or result of the process.

Also note that the product of a process is different from the process itself. The process is the act of doing; the product is the result of doing. The product is the footprint of the process that is left behind.

10.2.2 Understanding Processes

A process is similar to a project. Both processes and projects use a series of steps or operations to convert inputs into outputs. The key distinction between processes and projects is that processes are often fairly repetitive and follow the same or similar steps; on the other hand, projects may be viewed as single-pass activities or may be considered to be performed only intermittently and to follow a potentially different sequence each time. On the whole, however, projects and processes are more alike than they are dissimilar. Therefore, to a great extent both terms may be used interchangeably.

A unique and particularly interesting perspective is that processes are the integrating elements of any system. In other words, processes are the glue that holds a system together. All the tools as well as equipment and personnel in a system are inherently independent entities. The only way to get them to work together and accomplish something meaningful is through a series of process steps. This occurs as the process is performed, so flows and interactions take place between the process steps as well as between the resources that are in use.

If the processes are well designed, then there will be synergy between the independent elements of the process. This is what results in excellent process performance. If the process is poorly designed, then the result will be entropy and, therefore, poor process performance. In integration, synergy implies that the integrated whole is greater than the sum of the independent elements or parts. On the other hand, entropy implies that the whole is less than the sum of the independent parts. The final lesson, therefore, is that most system problems are due to poor integration, and any attempt to enhance integration, whether by using management techniques or computer integration, must focus on improving the processes that tie all the elements together.

Because of the sequential nature of processes, some dependency will often exist between two or more process steps. Such dependencies are known as dynamic interactions because they are not fixed; rather, their interaction patterns change constantly over time. For example, queues may form during processes, such as those in manufacturing, when one process step or operation is working faster than the next operation in the sequence. If there is any variability in the operating rates, then the length of the queue may also fluctuate.

Dynamic interactions make it so that we cannot analyze one process step in isolation of the others, because many of the occurrences at one process step may have been caused by another process step located upstream (or even downstream). By the same token, any event at a process step may have ripple effects, through dynamic interactions, both upstream and downstream. Process starving is one type of dynamic interaction where one process step cannot work even though it is in good operating condition, because an upstream operation is not providing its needed input. Process blocking is another type of dynamic interaction where an operation has to stop because it cannot send on its output due to a problem downstream. Also, the stack-up of variation on a part as it flows through a process where each operation modifies a particular attribute on the part represents another type of dynamic interaction. To obtain the variability of the overall process, we cannot just sum up the variability of each operation as if the variabilities were independent. The interaction effects must also be accounted for.

By considering the time dimension of a process, we can observe the manner by which it evolves. A process is composed of one or more steps used to transform some input into one output or a series of outputs. Each step or operation may take some time to complete; this is the duration of the process step. By observing the process in detail, a number of discrete points in time, called event times, can be identified when certain changes occur in the process. Events may, for example, include the starting and ending times for each process step. Milestones are also event times and are highlighted in order to focus on the occurrences at that event time. For example, the midpoint of a project may be considered to be the milestone point at which project progress is reviewed and certain adjustments can be made.

10.2.3 Process Resources and Constraints

A central issue in the design of a process is determining what tools and other resources are needed for the process to be at its best. Resources are generally defined to include key personnel with specified skills, equipment with specified capabilities, and ancillaries necessary for the process to operate. Such ancillaries may include space facilities or certain consumable materials that are needed.

The *personnel* needed by a process may vary according to the types of skills that are required, but practically all processes will call for some degree of human participation. Personnel may be operators of the process or equipment used in the process, they may be managers that manage the process, or they may be repair crews fixing equipment when failures occur. The attitude, skill, and experience of these personnel will almost always play a major role in determining what the process does, how it accomplishes the job, and the degree of excellence in the process. Selecting the right personnel for the right task is perhaps the most important step of process design.

The *equipment* for a process includes all machinery, computer systems, or other technological infrastructure that is used by the process. The equipment resources that are selected for a process must be well matched to the requirements of the process. Process designers must strive to avoid the "technology for technology's sake" mentality that may lead to the purchase of certain equipment that will not really serve the needs of the process. Selection of equipment is important, and justification for the selection must not be limited to the traditional return on investment (ROI) and internal rate of return (IRR) analysis. Rather, the impact of each type of equipment under consideration on the entire process must be considered.

The technique or method used with each type of equipment is also an important aspect of the role of equipment. Having correct equipment but using it incorrectly may not be any better than not having it at all. This emphasizes the importance of training. Equipment users should be trained and retrained in the correct methods or techniques. In addition to regular training, any lessons that are learned while procuring, installing, operating, and maintaining the equipment should be documented and made available to all involved.

Ultimately, the personnel and equipment required by a process can only be made available with an adequate amount of financial resources. Providing these resources is the responsibility of management in the organization. If adequate finances are not available, this will act as a major constraint on the ability to achieve the objective of the process.

Other constraints may also exist; for example, governmental regulations that stipulate what can and cannot be done in a process. It is important for process designers to distinguish clearly between constraints that are desirable and those that are not desirable. Desirable constraints may also be known as controls of the process. Of those controls that are undesirable and act to hamper the process, it is important to know which ones can be influenced or minimized and which are beyond the ability of process designers to change or modify. Scarce resources must not be spent on attempting to relieve process constraints that are unchangeable.

Process controls include all rules and regulations as well as guidelines that are provided for the process, By carefully designing and implementing process controls, it may be possible to improve the performance of a process. However, process designers must proceed cautiously. The field of process design is filled with examples of process controls that turn out to have unforeseen side effects, and those side effects may be worse (much worse in some cases) than the original problem itself.

10.2.4 Process Documentation

Processes are central to everything that we do, and it is important that they be well documented. Process documentation is important because processes are abstract, yet it is essential to communicate about them to others. A process is very different from a concrete object that can be held, viewed, and passed around.

Good process documentation is a prerequisite that is necessary for any process analysis. In addition, training of personnel in the execution of a process also requires that the process be well documented. Any method used to document processes is known as a process description language (PDL). PDLs range from plain English recipes, to various process maps, such as flowcharts, IDEF0 charts, value stream maps, and even computer simulation models. Each type of PDL has its strengths and weaknesses. For example, plain English recipes are easy to read, but they may also be quite easily misunderstood. Process charts are highly graphical, providing a pictorial representation of processes; however, it is hard to use them to describe the full details of a process. All PDLs, both structured and unstructured methods, document, at the very minimum, what the process does and how it is done.

Additional information, provided by PDLs, may include

- Areas of emphasis in the process
- Inputs to the process
- Outputs from the process
- Roles and responsibilities for the process
- Constraints on the process

- Tools and equipment in the process
- Personnel participating in the process
- General resource goals of the process
- General resources available for the process
- Rules and regulations in the process
- Guidelines and techniques for operating the process
- Any exceptions to process guidelines, techniques, rules, or regulations

Process design uses PDLs as tools for documenting processes, and the final design of a process should be documented using a PDL. The PDL acts as the blueprint for the process in much the same way a drawing is typically used as the blueprint for a new product design.

10.2.5 Process Performance Metrics

Any process design or improvement effort must ultimately confront the questions, How good is good? and What does good mean? *Good* for a process may mean, for example, low cost, high quality, or highly flexible. Therefore, in designing and improving a process, it is essential that the process management team establish exactly what *good* means to the organization. These will be the metrics for measuring the quality of the process.

Commonly used metrics include

- Process quality
- Process efficiency
- Process throughput
- Process flexibility and agility
- Process stability and robustness
- Human factors, ergonomics, and morale
- Process cost

Most organizations are interested in a mix of two or more of these metrics. The typical organization will want its processes to provide excellent quality, while yielding low cost and a satisfactory level of employee morale. When two or more metrics are considered, especially if an improvement in one metric may cause degradation in another, then relative-importance weights must be supplied for making the tradeoffs between the different metrics.

Before a process management project begins, there must be clear agreement regarding which metrics will be considered and how interactions between different metrics will affect the overall system performance.

10.2.6 Ingredients of Process Excellence

For processes to be truly excellent, they must be designed with

1. The best steps
2. The best tools
3. The best techniques for applying the tools
4. The best timing (that is, do everything at exactly the right time)

There are however, five key ingredients that are needed in order to develop and establish processes that conform to the four requirements for process excellence. The ingredients are dependent on one another and must exist in combination not in isolation. Having a few of the factors but not the others will not yield the full benefits of process excellence.

The five ingredients of process excellence are

1. Discipline
2. Creativity
3. Skill and knowledge
4. Decision making
5. Communication

The ingredients are not listed in any particular order. They are all very important. Discipline requires a focus on the key processes with a detail-oriented attitude and a commitment to do what it takes to ensure that the processes continually become better. With discipline, an increased awareness and deep relationship with processes can be established to understand what makes the processes tick. Discipline also provides the ability to trade off short-term goals for long-term goals.

Once the processes have been well established, it will take discipline and commitment to stick to the established processes and continue to apply systematic methods for managing the entire organization. The needed discipline must be embedded into the culture of the organization so that it becomes a way of life. The foundations for such a cultural atmosphere is created by all in the organization, but the direction is greatly influenced by management's leadership. Remember, process discipline may not be fun and games, but the rewards will be great over the long term, and the effort always pays for itself many times over.

Creativity incorporates the spark from the human mind into the whole effort for the designing, improving and operating processes. Creativity comes from the unique domain of the *people asset*. Everybody is an expert at something, even the "lowly" machine operator. Considering any individual to be irrelevant to process management is to potentially overlook a major source of innovative ideas for process improvement. In general, those with the most proximity to the processes of interest will know more about its ups and downs and, if given the opportunity, might contribute many excellent ideas for making the processes better. Remember, the manager may not have all the answers, at least not always.

Creativity flourishes in an environment of freedom and autonomy, and organizations that are interested in boosting the levels of their process performance must provide such a creativity-facilitating environment so that good ideas can bubble to the top. It is difficult to generate creativity on demand, but it is certain that less-restrictive cultures, cultures that focus on longer-term rather than shorter-term value, and organizations that respect the human individual, will find ways to improve continuously. 3M is well known as an organization with a creativity-facilitating culture. The impacts of this on the bottom line are clear and have been well documented.

With discipline and creativity, an organization must also possess the right skill and knowledge. Like creativity, skill and knowledge lie exclusively within the human domain. Individuals with the required skill sets must be sought out and placed where they can influence processes in the most positive way. Knowledge includes everything from process know-how and techniques to guidelines, do's and don'ts, and procedures.

No organization knows everything that it needs to know, and there are frequent opportunities to learn from mistakes as well as successes. The importance of such knowledge must be recognized, and a lessons-learned process must be established to capture, document, and disseminate such knowledge as it is discovered; this is what it takes to become a learning organization. Furthermore, never-ending education is what it takes to have a skill base in the organization that is able to stay current with changes in technology and in society or lead in the introduction of such changes.

Decision making is another core ingredient for process excellence. Organizations with discipline, creativity, and knowledge, but poor decision-making capabilities will be severely limited in their ability to develop excellent processes. Decision making involves all negotiation and trading-off of the different perspectives represented in the organization. It involves a trading-off of global versus local issues and of remote versus imminent issues. No one perspective is universal; therefore, to make the best decisions and to find the best configuration for its processes, an organization must involve all stakeholders and balance their viewpoints in order to find out what is best.

Decision-making skills require many attributes such as good listening skills, clear communication and expression, logical thinking, as well as relevant knowledge and creativity. Many techniques such as brainstorming, the nominal group technique, and decision analysis are very useful for involving several individuals or teams in decision making to improve the organization's processes.

The last, but certainly not the least important, ingredient for process excellence is organization-wide communication. Communication is necessary to obtain the necessary buy-in and an adequate level of understanding from all the stakeholders of the processes. Stakeholders include the organization's customers, those who will implement the process, those who are impacted upon by the process, the organization as a whole, and the management team of the organization.

Through communication, the goals of the processes must be understood by all stakeholders. These goals should also be clear in the context of the organization's mission statement. In addition, communication is necessary for disseminating processes organization-wide, for access to process guidelines and procedures, for decision making, and for tracking processes to determine when corrective or preventive maintenance may be needed. Other issues that need to be communicated include

1. Pressing problems that need urgent action to be solved
2. Alternatives to be considered in finding a solution
3. The problem and decision-making context
4. Factors influencing decision making
5. The selected alternative and the explanation of why it was chosen
6. The final specification of the process

New technologies such as the World Wide Web, videoconferencing, teleconferencing, and computer-aided group decision-making systems show great potential in providing mechanisms for enhanced, more convenient, and more frequent communication between individuals and teams, even when they are separated by time and great distances.

10.3 Process Types and Process Performance Metrics

In any business enterprise, processes exist everywhere, and they are really among the most important elements. There are many kinds of processes, and they work differently. The frequently encountered processes can be classified into the following categories:

- Product design and development
- Manufacturing or production
- Office or transaction
- Service factory
- Pure service shop
- Retail service store
- Professional service
- Telephone service
- Project shop
- Logistics and distribution
- Transportation service
- Purchasing and supply chain

Each category has distinct features, and we discuss each in detail.

10.3.1 Product Design and Development Process

For the manufacturing industry, product design and development is a key process, and it is also usually the most technically sophisticated, costly, and time-consuming process. For the service industry, the product design and development process does exist, but its importance and features vary from industry to industry. For the movie industry, the products are films. Here the product development process is clearly very important and creativity is very important; however, mass production is simply the duplication of films. The software industry is also very product oriented; the products are clearly the software, and the product development process is extremely important. For the health-care industry, the service product can be defined as diagnosis-treatment-care; this diagnosis-treatment-care planning process can be treated as a product development process. The plan will be different for each patient, and it will be developed quickly after the patient is admitted and might be changed due to changing circumstances.

Processes in product design and development include the conceptual and detailed design of products by designers. Design focuses on issues such as functionality of the product as well as aesthetics and ease of manufacture. Other tasks in product development may include building and testing of prototypes using various test equipment and analysis of candidate designs. Common goals in product design and development processes include shortening the required time to design new products, more frequent introduction of new designs, producing designs that are more innovative while meeting customers' needs and reducing the cost required to design new products.

In general, product development involves activities that are used to

1. Determine that a new product is required to serve some needs
2. Conceive of a concept for the product based on the wants and needs of customers
3. Develop all technical specifications for the product
4. Devise a production process
5. Validate both the design and the production process

A simple process flowchart for a product development process is shown in Fig. 10.3.

For many industries, product development is particularly important because it is the activity that has the greatest impact on the cost, quality, market acceptability, reliability, manufacturability, and disposability for the product.

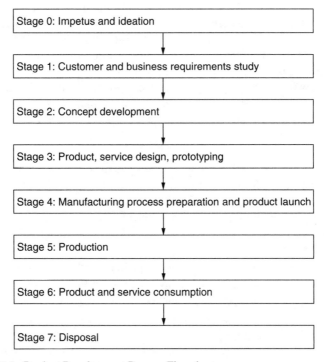

Figure 10.3 Product Development Process Flowchart

Decisions made in the product development phase of a product's life cycle will lock in most of the subsequent attributes of the product for many years to come.

As competitiveness hinges more and more on a company's ability to rapidly and efficiently bring new products to market, product development for most companies will become a paramount area for improvement. Considerable evidence exists that much of the success of Japanese automobile manufacturers has hinged on the speed, efficiency, and effectiveness of their product development processes.

Examples of Product Development Processes

Product development, movie film production

Product Development Performance Metrics

Product Development Lead Time

Most companies consider their product development lead time to be exceptionally important for determining the performance of their product development activities. Product development lead time is particularly important because this metric determines the speed with which new products can be introduced into the marketplace. Companies that have high speed in product development can introduce new products more often and adapt more quickly to changes in customer tastes. This ultimately translates into a larger market share for the company. Lead time is usually measured in months and can range from fractions of a month to tens of months, depending on the complexity and skill of a company's product development.

Efficiency

In attempting to reduce product development lead times, however, few companies can afford to ignore the efficiency of their product development. In product development, efficiency is the cost of the work force and other resources required for product development.

Robustness

In addition, the robustness (including quality, reliability, flexibility—how well the product does what it was meant to do) of the design is particularly important for evaluating any product development process.

Life Cycle Cost

Life cycle costs, including development costs; production costs; sales and distribution costs; service, support, and warranty costs; and disposal costs may be included in computing the life cycle cost for a product. Some companies even include the costs due to pollution during the production and use of the product as part of the holistic analysis of the life cycle cost. Product development has a particular vested interest in keeping the life cycle cost for any product as low as possible.

On a longer time scale, product development lead time, efficiency, robustness, and life cycle costs will contribute a great deal to the level of customer satisfaction, market share, and revenues that the company will have. These will in turn translate into profitability and influence the organization's long-term business viability.

Product Development Process Analysis

Typical activities in product development start from the initial step of market research, where the needs and wants of the market are determined, to production launch, where the product manufacturing can start in earnest. Product development activities include

- Market research
- Conceptual design
- Detailed design
- Prototype development
- Prototype testing and validation
- Process and tooling design
- Production launch

The common problems in product development process include

1. Excessive time to market
2. Excessive resources used to develop products
3. Chaos and confusion in the product development process
4. Market rejection of new product
5. Excessive product cost
6. Poor product quality and reliability

These common problems are often caused by

- Poor tradeoffs between product features, functionality, and cost
- Excessive complexity of product
- Poor understanding of the market
- Poor product and technology knowledge
- Poor coordination actions and decision making
- Excessive information flow delays
- Poor decision-making process

- Frequent information and data loss and incompatible data transmission formats
- Poor or inadequate communications
- Poor project and process management
- Market complexity (number of product types, product complexity, supply chain complexity)
- Market uncertainty or turbulence

10.3.2 Manufacturing or Production Process

Manufacturing (also known as production) involves all the activities that are used to transform raw materials and add value to them, in a series of steps, to bring them to a state where they will be purchased by end-customers. Each type of operation requires certain types of equipment and may be highly automated or mostly manual. The processes may operate on a few product types in large volume or have a high variety with small quantities of each product type. Scheduling of product flows through the production processes is important in order to produce items on a timely basis for customers. In addition, the equipment needed in manufacturing and regular maintenance activities are required to keep them in good operating condition and minimize downtime due to unanticipated failures. Manufacturing holds a special place in society as the primary means whereby prosperity is generated. As the primary engine for prosperity, it has also been recognized by economists that manufacturing productivity or efficiency is a key ingredient for long-term societal well being.

There are several common types of manufacturing processes:

- Job shop
- Cellular manufacturing
- Batch flow shop
- Line flow shop

We discuss these in detail.

Job Shop

A job shop consists of an arrangement of workstations, usually by function or by the type of process they perform. Job shops generally produce a wide variety of parts in low volume. Based on the operations required, each part type has a routing that may not be defined until the actual production time. This dynamic routing allows machines to be selected based on availability, since some operations can often be performed on more than one machine. Job shops are still the most common type of manufacturing: It is estimated

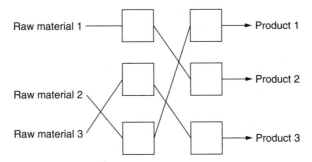

Figure 10.4 A Typical Job Shop

that 75 percent of all manufacturing is done in production batch sizes of 50 items or less. Job shops may be either a "one-of" type in which only one of something is produced (tooling or prototype shop) or a repeating type in which quantities are usually greater than one and similar jobs are produced again in the future. In the case of job shops, the general range of processes anticipated are considered in the selection of equipment. General-purpose equipment is used that is capable of providing processes for a broad range of products. The flowchart for a typical job shop is illustrated in Fig. 10.4. The boxes in the figure are machines or workstations.

Job shops tend to be very inefficient with long lead times and high work-in-process inventories. These are some of the reasons for the inefficiencies of job shops:

- Manual material movement
- Manual operations
- Long setup times
- Low equipment utilization

Examples of job shops include

- Metalworking
- Fabrication or machining operations
- Maintenance facilities for the aerospace industry

During operation, the general job shop problem is to schedule the production of N jobs on M machines. For each job, the sequence of machines is known as well as the processing time on each machine. Due dates may also be known. In scheduling, four principal goals or objectives are to be achieved:

1. Minimize job lateness or tardiness
2. Minimize the flow time or time jobs spend in production

3. Maximize resource utilization
4. Minimize production costs

The performance measures of job shops and the decision variables that can be changed in controlling a job shop include the following:

Performance measures
- Time to complete a set of jobs (rnakespan)
- Number of completed jobs that are tardy
- The average lateness of jobs that are completed
- Utilization of equipment

Decision variables
- The job selection rule
- The sequence in which jobs are processed
- The routing for a particular job if alternative routings are possible
- The resources assigned to particular jobs
- The transfer batch size
- Use of overlapped versus nonoverlapped production
- Overtime and shift policies
- Assignment of resources to workstations

Cellular Manufacturing

An alternative to a job shop is cellular manufacturing in which machines are grouped into cells according to common processes. A manufacturing cell is a group of machine tools and associated materials handling equipment that is managed by a supervisory computer. Manufacturing cells are often called group technology (GT) cells since group technology is the basis for designing the cell. *Group technology* is an approach to design based on the premise that similar things should be done similarly. Parts having similar configurations or similar processes should be produced by the same cell of machines. A cell is an independent group of machines but may be connected with other cells to form a flexible manufacturing system.

The flow of parts within the cell resembles the streamlined flow achieved in line flow manufacturing. This results in greater efficiencies by consolidating groups or families of products together and treating them, from a work flow standpoint, as a single product. A cell is an excellent way to achieve the "factory within a factory" concept and is becoming a widely adopted approach to low-volume, high-mix manufacturing. A flowchart for a typical cellular manufacturing process is illustrated in Fig. 10.5.

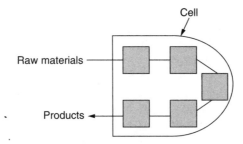

Figure 10.5 A Typical Flow Chart for Cellular Manufacturing

There are three types of job flow patterns in cellular manufacturing:
- Serial flow cells
- Random flow cells
- Virtual cells

In a serial flow cell, all parts flow through the same sequence of machines and, hence, a miniature production line is established. In random flow cells, different parts have different routings within the cell, with the effect being somewhat similar to a job shop. Machine utilization tends to be less in a random flow cell than in a serial flow cell. The concept of a virtual cell was first proposed by the National Bureau of Standards (NBS) [now the National Institute of Standards and Technology (NIST)]. This concept uses a process layout of equipment just like a job shop, rather than a cellular layout. Machines are treated logically as a group even though they are physically separated. A virtual cell functions as a cell based on the needs at the time. Individual workstations are allocated to a virtual cell on a dedicated or time-sharing basis with other virtual cells. The concept of virtual cells developed from the philosophy that changing production requirements alter the part family makeup for a given production period. When the requirements alter, the allocation of individual workstations will change.

The benefits of cellular manufacturing include the following:
- Better lead times provide fast response and more reliable delivery.
- Work in process and finished stock levels are reduced.
- Output is increased because of improved resource utilization.
- Less material handling is needed.
- Better space utilization is achieved.
- Better production planning and control is possible.
- Quality is improved, and scrap is reduced.
- Estimating, accounting, and work measurement are simplified.

Performance measures
- Machine utilization
- Production rate
- Utilization of the operator
- Utilization of the bottleneck station

Decision variables
- The number and types of machines in the work cells
- The batch size of a particular part type
- Sequencing of part types within the cell
- Material handling priorities within the cell

Batch Flow Shop

A batch flow shop, or flow shop, utilizes a product layout in which a sequence of workstations is visited in the same sequence by different product batches. Batch flow shops are similar to job shops in that many different types of discrete parts are produced in batches. In a batch flow shop, however, all flow is basically unidirectional following the same route through the manufacturing facility in a production-line fashion.

A batch flow shop is composed of one or more production lines that support a batch flow of parts. Each part may not require processing at each station and may even bypass some stations, but the same general flow is followed. All operations on parts of the same type are usually performed in the same order. Batch flow shops are most commonly found in the textile industry in which batches of different styles and sizes are processed through the same sequence of operations. In flow manufacturing, the emphasis tends to be on efficiency and streamlining flow. A typical batch flow is illustrated by Fig. 10.6.

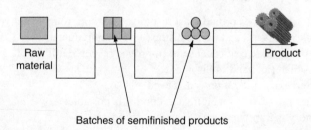

Figure 10.6 Batch Flow Shop

Performance measures
- Resource utilization
- Throughput capacity
- Work in process
- Cost

Decision variables
- Queuing between stations
- The production batch size of a particular part type
- The move batch size
- The sequence of products

Line Flow Shop (Production or Assembly Lines)

Line flow manufacturing consists of production or assembly lines and transfer lines in which products move and are processed individually rather than in a batch. Line flow systems are characteristic of many production operations in which workstations are set up by product in a serial arrangement and dedicated to manufacturing or assembling a single product. The idea is to achieve a streamlined, continuous flow of material that leads to maximum productivity. Labor and machines are highly utilized with little idle time. Since transfer lines are sufficiently different from production or assembly lines, they are treated separately.

Production and assembly operations that are of a line flow type are comprised of a serial combination of two or more production, assembly, and packaging stations typically connected by a continuous material-handling system such as a conveyor. Nonsynchronous conveyors have become the most popular and efficient material-handling system because they permit parts to maintain a continuous flow while still allowing them to queue up when necessary. Operations are often performed by hand and therefore present a special challenge to keeping the flow as continuous as possible. This is achieved by balancing the workload among stations, keeping each station busy, and reducing the variability of each operation. Usually, the more stations, the lower the cycle time, and hence the higher the throughput. An alternative to stretching out a line into more stations to increase throughput is to add parallel lines. At one extreme, a single line consisting of n serial workstations may be used. The job is broken down into as many small subtasks as possible without overproducing. At the other extreme would be n parallel lines consisting of a single station each. The entire job is performed on each single station line with as many lines as are needed to meet demand. Many lines lie somewhere in between these two extremes and consist of a mix of serial and parallel stations.

The placement and size of buffers has an impact on inventory costs and system throughput. If the entire line stopped every time a part was unavailable or a station failed, the line would be going down. Buffers allow workstations to operate independently thus cushioning the effects of scrap, part shortages, unequal production rates, workstation failures, or operator delays. However, lean manufacturing advocates strongly disagree about the use of buffers. They think the in-process inventories tie up the capital, hide the operation problems, and reduce the quality.

Many products are not produced in sufficient quantities to justify a dedicated line. Frequently a production or assembly line is used to produce a family of similar products. Products are produced in batch runs in which the line is temporarily shut down for product changeovers while machine adjustments are made for the next product. The lean manufacturing process developed quick setup procedures so that the changeover time could be reduced to a minimum.

Production and assembly lines may be either *paced*, in which movement occurs at a fixed rate and the operator must keep pace with the line, or they may be *unpaced*, in which the rate of flow is determined only by the speed of the worker. Figure 10.7 gives a typical flowchart for a line flow shop.

Examples of line flow shops include

- Appliance assembly lines
- Consumer product assembly lines
- Medical instrument assembly lines

  Performance measures
  - Average and variation in throughput capacity
  - Average and variation in work in process
  - Cost
  - Balance delay (sum of the idle time for all stations/sum of the scheduled time for all stations)
  - System efficiency (actual throughput of the system/theoretical throughput capacity of the slowest station)

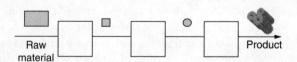

Figure 10.7 Line Flow Flowchart

- Percentage of time stations are blocked
- Percentage of time stations are starved

Decision variables
- The number and types of machines in the work cells
- The production batch size of a particular part type
- Type of material handling
- The sequence of products
- Number of stations
- Placement and size of buffers

In the service industry, there are many processes that are similar to the manufacturing process. In the restaurant business, the process of producing meals in the kitchen is very similar to the manufacturing process. All office processes, such as processing paperwork, insurance claims, and mortgage applications, involve a sequence of steps on incoming materials (paperwork), adding value to them (finishing a part of necessary procedures) in each step, until the product is finished (paperwork done). Therefore, many well-established manufacturing process management methods can be easily adapted to these service processes.

Common goals in manufacturing processes are to improve throughput, improve flexibility, increase quality, and reduce cost. Other issues such as safety may also be important. These goals and the tradeoffs that they require can be best attained by process management of the production processes.

Manufacturing Performance Metrics

Quality

Practically all manufacturing operations consider quality to be important. The importance of quality stems from customers' demands for the best value that their money can buy. Quality, quite simply, can be assessed as the degree to which customer requirements incorporated into the design for a product have been met. Any manufacturing organization that does not pay close attention to quality will slowly lose market share and disappear into oblivion.

Production Lead Time

In addition to quality, another performance metric that influences the level of customer satisfaction is the production lead time. Lead time is the time it takes one piece of raw input to move all the way through the manufacturing process, from start to finish. In general, the shorter the lead time, the better.

Work-in-Process Inventory

The lead time for a process is very closely related to the level of the work-in-process (WIP) inventory. The WIP inventory is the amount of semifinished units in between the process steps. For a given set of operations in a process, the higher the WIP, the longer the lead time is likely to be. In other words, WIP slows down the speed of a unit of product as it travels from the entry point of the process until it becomes a finished product.

Throughput

Throughput is the rate at which a process produces its output. In a production environment where customers will buy all units of product that are manufactured (especially if there are demand backlogs), then it is highly desirable to maximize throughput. The bottleneck in a production facility acts as the primary constraint on its throughput; therefore attempts to improve throughput must focus on the bottleneck operation.

Cost

In addition, the cost of products is particularly important in a manufacturing environment. Since the price for which a product can be sold is usually dependent upon the laws of demand and supply, organizations have a built-in incentive to keep costs as low as possible in order to maximize its profit. Cost typically has at least three main components: raw material, processing, and overhead costs. Each of these cost components has to be carefully managed in order to minimize the overall cost.

Flexibility

In an age of increasingly discriminating customers, where individuals are now seeking unique customized products, flexibility is becoming ever more important. Flexibility depends on the ability of a manufacturing system to handle a wide variety of product types in an efficient way. Change and market turbulence are some of the biggest challenges that a manufacturer has to face, and flexibility is the main tool that we have to cope with change, turbulence, and uncertainty.

Other manufacturing performance metrics include safety, process ergonomics, human factors, and employee morale. These are all important metrics that cannot be ignored. The bottom-line impact of these metrics is that over the long and short term they determine, directly or indirectly, the market share and revenues, as well as profitability. All of these ultimately tie together into determining what the organization's long-term business viability will be.

Design and Improvement of Service Processes—Process Management

Manufacturing Process Analysis

Typical activities in manufacturing and manufacturing support include

- Production operations
- Job setup
- Tool fabrication and repairs
- Shipping and receiving
- Shop floor control and project management
- Quality control and quality assurance
- Equipment repairs and preventive maintenance
- Scheduling
- Production planning
- Software programming
- Facilities planning

The resources needed for manufacturing range from processing equipment, like machine tools, to material-handling devices required to move material. Over and over again, the evidence has been upheld that good statistical control of production processes, training for operators and support staff, and preventive maintenance of key equipment are the primary determinants of manufacturing success.

The common problems in the manufacturing process include

1. Low throughput and efficiency
2. Excessive production lead time
3. Excessive work in process
4. Late orders
5. Excessive rework
6. Poor quality
7. Chaos, confusion, lost parts
8. Excessive production cost
9. Lost orders
10. Poor safety record

These common problems are often caused by

- High process variability
- Poor incoming raw material
- Poor operator training
- Unbalanced processes
- Excessive setup times
- Poor process design and validation
- Poor workstation and plant layout

- Lot size too small or too large
- Poor equipment maintenance
- Poor inventory management and shop floor control
- Low equipment reliability
- Poor operator discipline and motivation
- Improper equipment
- Low flexibility
- Inadequate process capacity

A number of generic strategies are available for solving the range of manufacturing problems; however, each strategy must be matched to the appropriate diagnosis. The precise manner in which a strategy is to be applied must also be determined after careful study. Strategies to be considered include

1. Comprehensive operator training programs
2. Buffer management
3. Comprehensive preventive and predictive maintenance programs
4. Quick batch changeover
5. Process simplification
6. Plant layout reorganization
7. Statistical quality control
8. Operator wage and incentive programs
9. Production scheduling system
10. Comprehensive production planning
11. Continuous improvement programs
12. Flexible manufacturing
13. Cellular manufacturing

It is necessary, at this point, to emphasize that many of these strategies have become buzzwords and have, as a result, been sometimes applied inappropriately. To avoid these expensive mistakes, each strategy should only be chosen after a thorough process management.

10.3.3 Office or Transaction Process

In the office environment there is a wide variety of transaction processes for handling the administrative activities of an organization. These include computer data entry, copying documents, filing and retrieving documents, attending meetings, performing analyses, decision making, and documenting reports. Each type of activity requires certain types of equipment and may be performed individually or in teams. The processes typically focus on information, requiring information as input and value-added information as output. The information that is processed may be stored on paper documents or be computer-based. Increasingly, attempts are being made to automate and integrate more and more of these processes in search of the so-called paperless office. Performance in office processes is often judged in terms of the process cost and efficiency. In addition, many organizations are driving toward a reduction in errors generated by the process. Some of these errors can be very expensive and may cause serious problems for the organization.

The office process has a lot of similarities with the manufacturing process. They are both step-by-step sequential processes, and each step contributes some value to the completion of the job. However, office processes are usually more dependent on people and less dependent on hardware compared with the manufacturing process.

Examples of Office Processes

Insurance, mortgage and loan

Office Process Performance Metrics

Number of Errors or Defects

The average number of errors or defects in each transaction is an important indicator of office process quality. Clearly, errors and defects will cost both customers and service providers, and they should be reduced to a minimum.

Lead Time

Lead time refers to the time from the beginning to the completion of a transaction. It is similar to the production lead time in the manufacturing process. Again, in general, the shorter the lead time the better because all customers want quick and error-free transactions. In addition, a shorter lead time means a smaller work force, so it saves service providers money.

Work-in-Process Inventory

Similar to the manufacturing process, the WIP inventory refers to semifinished paperwork and transactions between process steps. The lead time for a process is very closely related to the level of WIP. This semifinished paperwork may sit in drawers, bins, interdepartmental mail, or even become lost in the paper trail. Excessive amounts of WIP will certainly slow down the office process and will likely create errors.

Throughput

Similar to the manufacturing process, throughput is the rate at which a process produces its output. Higher throughput in the office process usually means more transactions are completed in a given time period. Clearly higher throughput means higher productivity in the office process. The bottleneck in the office process is usually the primary constraint on its throughput; therefore attempts to improve throughput must focus on the bottleneck operation.

Transaction Cost

The average cost per transaction is a good measure of the cost of operating the office process. The transaction cost depends on many factors, such as process efficiency and labor cost. It is very important to reduce transaction cost to a minimum to ensure operation profitability.

Flexibility

In the transaction process, it is natural that each transaction might be different. It is a must that the office process handle mixed transactions well, and change over from one transaction type to the next without slowing down the process.

Customer Interaction Quality

One key difference between the office process and manufacturing process is that during the execution of the office process, the process operator may directly interact with customers; during the execution of the manufacturing process, the operators usually do not interact with customers. The total experience and feeling of the customers during operator-customer interaction is a very important part of process quality and customer satisfaction.

Office Process Analysis

Typical activities in the office process include

- Data entry
- Paperwork review
- Retrieving rules and regulations
- Database updating
- Document preparation
- Decision making
- Report generation
- Application approval

The resources needed for the office process include computers, software, office equipment, shipping and handling, and most importantly, the work force. Good design of the office process, good computers and software maintenance, and training for operators and support staff are the primary determinants of office process success.

The common problems in the office process include

1. Excessive errors
2. Excessive lead time
3. Low throughput or efficiency
4. Excessive work in process
5. Overdue transactions
6. Excessive rework
7. Poor customer interaction quality
8. Chaos, confusion, lost paperwork
9. Excessive transaction costs
10. Lost paperwork or transaction

These common problems are often caused by

- Poor office process design
- Poor transaction work flow
- Poor operator training
- Unbalanced processes or bottlenecks
- Poor computer or software maintenance
- Poor operator discipline or motivation

10.3.4 Service Factory Processes

Service factories are systems in which customers are provided services using equipment and facilities requiring low labor involvement. Consequently, labor costs are low while equipment and facility costs are high. Service factories usually have both front room and backroom activities with total service being provided in a matter of minutes. Customization is done by selecting from a menu of options previously defined by the provider.

Service factory processes are similar to office processes, except that usually the customer being served will almost immediately experience the level of excellence, good or bad. For the service factory process, the quality of customer-server interaction processes may be the most critical factor for success, since poor performance in customer-server interaction will lead very quickly to desertion by customers and loss of market share. Training of personnel is often a key ingredient for achieving excellent customer-server interaction. Waiting time and service time are also two primary factors for customer satisfaction. Convenience of location is another important consideration. In the service factory process, usually the customer commitment to the provider is low because there are usually alternative providers just as conveniently located.

Examples of Service Factory Processes

Banks (branch operations); restaurants; copy centers; hair stylists; and check-in counters of airlines, hotels, and car rental agencies.

Service Factory Process Performance Metrics

Number of Errors or Defects

The number of errors or defects in each customer transaction is an important indicator of service factory process quality.

Waiting Time

In the service factory process, the average waiting time is an important process performance metric. Since waiting time does not provide any value for customers and service companies, the longer the waiting time, the poorer the performance is. The bottleneck in the service process is one of the leading causes of excessive waiting time. *Average queue length* in the service factory process is an alternative measure for waiting time. *The abandonment rate* is the portion of customers who go away due to excessive waiting time; it is also often used as a performance measure.

Service Time

The service time is the time duration used to provide the needed service. The service time might be "the smaller, the better" performance characteristic for some service factory processes; for example, in fast-food restaurants. However, for some other service factory processes, the service time for each customer depends on his or her specific needs. So the required service time will be different for different customer needs. Overall, average customer service time might be an appropriate performance measure for a service factory process.

Customer Service Quality

The customer service quality in this case has two components: one is the quality of the service, and the other is the quality of customer–service provider interaction. The quality of service is the quality of the service product, that is, the service provided. In the restaurant industry, the quality of the service includes the taste of the meal and the cleanliness of the dinnerware. The quality of customer–service provider interaction is mostly based on how happy the customer feels about the service. It includes the politeness of and the facial expressions (e.g., a smiling face) of the service provider, and even the tone of the conversation. The total experience and feeling of the customers during operator-customer interaction is a very important part of process quality and customer satisfaction.

Throughput

Similar to the manufacturing and office processes, throughput is the rate at which a process produces its output. Higher throughput in the service factory process usually means more customers are served in a given time period. Clearly higher throughput means higher productivity for the process. However, depending on the nature of the service process, there might be a tradeoff between customer service quality and throughput; the throughput may not be "the more, the better."

Service Cost

Service cost is certainly a measure of process performance. Service cost depends on many factors, such as process efficiency and labor cost. A poorly designed service process often creates a lot of waste, waiting time, errors, and bottlenecks, and it is often one of the most important contributors to excessive service cost.

Service Factory Process Analysis

Typical activities in the service factory process include

- Customer arriving
- Greeting customers
- Taking customer orders and payments
- Customer waiting
- Providing needed services
- Customer departing

The resources needed for the service factory process vary greatly depending on the nature of the service. For the restaurant industry, the resources include dining spaces, dining tables and chairs, the kitchen, and most importantly, well-qualified cooks and waiters and waitresses. Good design of the service process, equipment maintenance, facility design and layout, and training for service providers and support staff are the primary determinants of service factory process success.

The answers to the following questions may be very helpful in designing and improving a good service factory process.

- How many of each type of equipment are required to meet customer demand?
- Which layout provides the most efficient customer flow and minimizes delays?
- Which resources can be shared to assist in peak times to minimize waiting time?

- How many shifts and service providers are needed to minimize costs?
- What procedures can be used (self-service, advance ordering) to minimize service time?

The common decision variables for a service factory process include

- Number of servers during each period
- Quantities of equipment
- Size of facilities (waiting area, parking)
- Hours of operation
- Hours of cleaning and maintenance

The following statistical data are critical for designing and operating a good service factory process:

- Arrival rate of customers over the service cycle
- Length of line before balking occurs
- Length of wait before reneging occurs

The common problems in the customer service process include

1. Excessive waiting time
2. Poor customer service quality
3. Poor customer-provider interaction quality
4. Excessive errors
5. Excessive rework
6. Excessive service cost

These common problems are often caused by

- Poor service process design
- Poor service provider training
- Unbalanced processes or bottlenecks
- Poor operation management
- Poor work flow management
- Poor equipment maintenance
- Poor service provider discipline or motivation

10.3.5 Pure Service Shop Processes

In a pure service shop, service times are longer than for a service factory. Service customization is also greater. Customer needs must be identified before service can be provided. Customers may leave the location and return for pickup, to check on an order, make a payment, or for additional service at a later time. Price is often determined after the service is provided. Although front room activity times may be short, backroom activity times may be long, typically measured in hours or days. The primary consideration is quality of service. Delivery time and price are of secondary importance.

The customer's ability to describe the symptoms and possible service requirements are helpful in minimizing service and waiting times. When customers arrive, they usually all go through some type of check-in activity. At this time, a record (paperwork or computer file) is generated for the customer, and a sequence of service or care is prescribed. The duration of the service or the type of resources required may change during the process of providing the service because of a change in the status of the entity. After the service is provided, tests can be performed to ensure that the service is acceptable before releasing the entity from the facility. If the results are acceptable, the customer and the record are matched and the customer leaves the system.

Examples of Pure Service Shop Processes

Hospitals, repair shops (automobiles), equipment rental shops, banking (loan processing), Department of Motor Vehicles, Social Security offices, courtrooms, and prisons.

Pure Service Shop Process Performance Metrics

Number of Errors or Defects

The number of errors or defects in each customer transaction is an important indicator of pure service shop process quality. For example, in the health-care industry, accuracy of diagnosis and the appropriateness of the treatment are among the main health-care performance metrics. In customer help centers, wrong advice or information provided to customers will quickly cut down on the credibility of the company.

Waiting Time

In the pure service shop process, the average waiting time is an important process performance metric. Again since waiting time does not provide any value for customers and service companies, the longer the waiting time, the poorer the performance is. For example, in the health-care industry, the total length of time that a patient stays in the hospital may not be "the shorter, the better," because first of all, the patient's disease should be cured or at least reduced. Diagnosis, treatment, and care must be provided for as long as needed. However, excessive patient and doctor waiting time, excessive testing and report turnaround time, and excessive time required for administrative activity are definitely unwanted because excessive waiting time makes customers unhappy. It slows down the process, so it increases the cost; it may also reduce the throughput, so it affects the cash flow negatively. In the health-care industry, excessive waiting time may even cause complications in a patient's treatment. The bottleneck in the service process is one of the leading causes of excessive waiting time; for example, in the health-care industry, hospital bed availability is often the bottleneck for the whole hospital. *Average queue length* in the service factory process is an alternative measure for waiting time. *Abandonment rate* is the portion of customers who go away due to excessive waiting time; it is also often used as a performance measure.

Customer Service Quality

The customer service quality in this case has two components: one is the quality of the service, and the other is the quality of customer–service provider interaction. The quality of service is the quality of the service product, that is, the service provided. In the health-care industry, the quality of service includes the quality of the treatment, quality of the diagnosis, and quality of the recovery. The quality of customer–service provider interaction is mostly based on how happy the customer feels about the service. It includes the politeness of and the facial expressions (e.g., a smiling face) of the service provider, and even the tone of conversation. The total experience and feeling of the customers during operator–customer interaction is a very important part of process quality and customer satisfaction.

Service Time

The total time that a customer spends in the service facility is called the lead time. The lead time is the summation of service time and waiting time. Waiting time is definitely a waste for both customers and service-providing companies. Because in the pure service shop process the services provided to customers are highly customized, the required service times vary greatly. Achieving sufficient service quality has a higher priority than reducing service time duration. However, the non-value-added waiting times should be reduced to a minimum.

Throughput

Throughput is the rate at which a process produces its output. Higher throughput in the pure service process usually means more customers are served in a given time period. However, in the pure service shop process, the complexity and required service time of each task varies greatly; the throughput may not be "the more, the better."

Service Cost

Service cost is certainly a measure of process performance. Service cost depends on many factors, such as process efficiency and labor cost. A poorly designed service process often creates a lot of waste, waiting time, errors, and bottlenecks, and it is often one of the most important contributors to excessive service cost.

Resource Utilization

Resource utilization can be measured by the total time that a particular piece of resource is used divided by the total elapsed time. For example, the percentage of a medical doctor's time in doing value-added work, such as taking care of patients, divided by his or her total time spent in the hospital, is a measure of resource utilization of this medical doctor. It is desirable that all important resources are used at 100 percent. However, uneven resource utilization is a very common problem.

Pure Service Shop Process Analysis

Typical activities in pure service shop process include

- Customer arriving
- Greeting customers
- Taking customer orders
- Checking customer order and designing customer service
- Customer waiting
- Providing needed services
- Customer departing

The resources needed for a service factory process vary greatly depending on the nature of the service. For the health-care industry, expensive labs and diagnostic equipment; hospital infrastructure; qualified doctors, nurses, and staff; and computer information systems are all resources. Good design of the service process, equipment maintenance, facility design and layout, and training for service providers and support staff are the primary determinants of service factory process success.

The answers to the following questions may be very helpful in designing and improving a pure service shop process:

- Which layout provides the most convenient customer flow and minimizes delays?
- What is the peak capacity of the system?
- Which resources can be shared in peak times to minimize waiting time?

- How many shifts and service providers do we need to minimize costs?
- What procedures can be used (self-diagnosis, advance check-in) to minimize service time?

The common decision variables for a pure service shop process include

- Number of service providers
- Size and location of facilities (waiting areas, service areas, parking, restrooms, etc.)
- Number and type of service and transportation equipment
- Capacity of the facility
- Sequencing of customers waiting for service
- Number of staff and shift schedules
- Hours of operation
- Maintenance schedules

The common problems in pure service shop processes include

1. Excessive waiting time
2. Poor customer service quality
3. Poor customer-provider interaction quality
4. Excessive errors
5. Excessive rework
6. Excessive service cost

These common problems are often caused by

- Poor service process design
- Poor service provider training
- Unbalanced processes or bottlenecks
- Poor operation management
- Poor work flow management
- Poor equipment maintenance
- Poor service provider discipline or motivation

10.3.6 Retail Service Store Process

In retail services, the size of the facility is large in order to accommodate many customers at the same time. Customers are provided with a large number of product options from which to choose. Retail services require a high degree of labor intensity but a low degree of customization or interaction with the customer. Customers are influenced by price more so than service quality or delivery time. Customers are also interested in convenient locations, assistance with finding the products in the store, and quick checkout. Total service time is usually measured in minutes.

When customers arrive in a retail shop, they often get a cart and use that cart as a carrier throughout the purchasing process. Customers may need assistance from customer service representatives during the shopping process. Once the customer has obtained the merchandise, then he or she must get in line for the checkout process. For large items such as furniture or appliances, the customer may have to order and pay for the merchandise first. The delivery of the product may take place later.

Examples of Retail Service Store Process

Department stores, grocery stores, hardware stores, and convenience stores

Retail Service Store Process Performance Metrics

Time Waiting for Assistance

This time refers to the average time that a customer might wait to get assistance in helping to locate merchandise, have questions on product features answered, and so on.

Average Queue Length in Checkout Lines

A long queue length could be caused by fluctuations in customer arrival rates and work force management. Queue length is proportional to the *average waiting time for checkout*.

Number of Checkout Errors

Checkout errors are quite common in the retailing business, and customers do care about these errors. Checkout errors can be caused by labeling errors or data entry errors.

Customer Service Quality

The customer service quality in this case has two components: one is the quality of the service, and the other is the quality of customer–service provider interaction. In retailing service stores, the quality of service includes the promptness and the correctness of the merchandise information provided. The quality of customer–service provider interaction is mostly based on how happy the customer feels about the service. It includes the politeness of and the facial expressions (e.g., a smiling face) of the service provider, and even the tone of conversation. The total experience and feeling of the customers during operator-customer interaction is a very important part of process quality and customer satisfaction.

Average Number of Shopping Carts Available Over Time

The availability of shopping carts provides convenience for customers.

Retail Service Store Process Analysis

Typical activities in a retail service store process include

- Customer arriving
- Customer looking and collecting goods
- Customer waiting for payment
- Customer making payment
- Customer departing

The resources needed for a pure service shop process include shopping spaces, a warehouse, an information system, and checkout counters. Good design of the service process, equipment maintenance, facility design and layout, and training for service providers and support staff are the primary determinants of process success.

The answers to the following questions may be very helpful in designing and improving a retail service store process:

- Which layout provides the most convenient customer flow and minimizes delays?
- Which resources can be shared to assist in peak times to minimize waiting time?
- How many shifts and service providers do we need to minimize costs?

The common decision variables for a retail service store process include

- Number of servers
- Number of checkout stands
- Number of dock doors for delivery and pickup and number of carts
- Size of facility
- Location of merchandise, carts, and service desks
- Shifts for cashiers and customer service representatives
- Replenishment frequency and quantity of inventory
- Hours of operation
- Maintenance schedules

The common problems in pure service shop processes include

1. Excessive waiting time
2. Poor customer service quality
3. Poor customer-provider interaction quality
4. Excessive checkout errors

5. Poor in-store assistance
6. Excessive service cost

These common problems are often caused by

- Poor service process design
- Poor service provider training
- Poor store layout and labeling
- Poor scheduling
- Poor operation management
- Poor equipment maintenance
- Poor service provider discipline or motivation

10.3.7 Professional Service Process

Professional services are usually provided by a single person or a small group of experts in a particular field. The service is highly customized and provided by expensive resources. The duration of the service is long, extremely variable, and difficult to predict because the customer involvement during the process is highly variable. Processing may be performed by a single resource or multiple resources. When the customer arrives, the first process is of a diagnostic nature. Usually, an expert resource evaluates the service needed by the customer and determines the type of service, estimated service time, and cost. This diagnosis then dictates the resources that will be used to process the order. The duration of the service or the type of resources required may change during the process of providing service. This is usually a result of the customer's review of the work. After the service is provided, a final review with the customer may be done to make sure that the service is acceptable. If the results are acceptable, the customer and the record are matched and the customer leaves the system.

Examples of Professional Service Processes

Auditing services, tax preparation, legal services, architectural services, construction services, and tailoring services

Professional Performance Metrics

Number of Errors

The average number of errors or defects in each service is an important indicator of process quality. Clearly, errors and defects will cost both customers and service providers, and they should be reduced to a minimum.

Customer Service Quality

The customer service quality in this case has two components: one is the quality of the service, and the other is the quality of customer–service provider interaction. In professional service, the quality of service includes the effectiveness of the service. For example, in a tailoring service, the service quality includes the fit and style of the tailored suit. The quality of customer-service provider interaction is mostly based on how happy the customer feels about the service. It includes the politeness of and the facial expressions (e.g., smiling face) of the service provider, and even the tone of conversation. The total experience and feeling of the customers during operator-customer interaction is a very important part of process quality and customer satisfaction.

Average Service Time

The average service time for each case is an indicator of process efficiency. However, professional service is a highly customized service; the actual service times vary from case to case.

Resource Utilization

Resource utilization can be measured by the total time that a particular piece of key resource is used divided by the total elapsed time. In professional service, the most important key resources are often the key professionals, such as consultants, experts, and attorneys.

Time Spent Doing Rework

Rework time is certainly a waste for both customers and service providers. The time spent in rework should be reduced to a minimum for good process quality.

Professional Service Process Analysis

Typical activities in the retail shop service process include

- Customer arriving
- Greeting customers
- Reviewing customer orders
- Designing service
- Delivering service
- Reviewing and checking service
- Receiving payment
- Customer departing

The most important resources for professional services are well-qualified professionals. Other resources include office spaces, an information system, and specialized tools and equipment. Good design of the service process, excellent project management and discipline, continuous training of professionals and supporting staff, facility design and layout, and equipment maintenance are the primary determinants of process success.

The answers to the following questions may be very helpful in designing and improving a professional service process:

- How many of each type of equipment are required to meet a project deadline?
- Which resources can be shared to assist in making up lost time?
- How many shifts and service providers are needed to minimize costs?
- How can the available resources be scheduled to meet the deadline?

The common decision variables for a professional service process include

- Staff schedules and shifts
- Number and type of service providers
- Project review times
- Hours of operation (overtime)

The common problems in pure service shop processes include

1. Excessively long service time
2. Poor customer service quality
3. Poor customer-provider interaction quality
4. Excessive number of errors
5. Excessive service cost

These common problems are often caused by

- Poor service process design
- Poor service provider training
- Poor scheduling
- Poor project management
- Poor equipment maintenance
- Poor service provider discipline and motivation

10.3.8 Telephone Service Process

Telephone services are provided over the telephone. They are unique from other services in that the service is provided without face-to-face contact with the customer. The service may be making reservations, catalog ordering, or providing a customer support service. In a telephone service system, issues to address include the following:

- *Overflow calls.* The caller receives a busy signal.
- *Reneges.* The customer gets in but hangs up after a certain amount of time if no assistance is received.
- *Redials.* A customer who hangs up or fails to get through calls again.

The most important criteria for measuring effectiveness is service time. The customer is simply interested in getting the service or ordering the product as quickly as possible. The customer's ability to communicate the need is critical to the service time.

Calls usually arrive in the incoming call queue and are serviced based on the first in, first out (FIFO), or first in, first serve, rule. Some advanced telephone systems allow routing of calls into multiple queues for quicker service. Processing of a call is done by a single resource. Duration of the service depends on the nature of the service. If the service is an ordering process, then the service time is short. If the service is a technical support process, then the service time may be long or the call may require a callback after some research.

Examples of Telephone Services

Technical support services (hotlines) for software or hardware, mail-order services, and airline and hotel reservations

Telephone Service Performance Metrics

Service Time
Usually the average service time per call can be used as a measure of efficiency for telephone services.

Waiting Time
Average waiting time per customer is an important measure for telephone service efficiency and is closely related to customer satisfaction.

Abandonment Rate
Abandonment rate is the proportion of customers who give up waiting and abandon the service. It is usually highly correlated with waiting time and is an important measure of customer satisfaction.

Customer Service Quality

The customer service quality in this case has two components: one is the quality of the service, and the other is the quality of customer–service provider interaction. The quality of service is the quality of the service product, that is, the service provided. In the call center, the quality of the service includes the correctness of the information provided. The quality of customer–service provider interaction is mostly based on how happy the customer feels about the service. It includes politeness of the service provider and the tone of conversation. The total experience and feeling of the customers during operator-customer interaction is a very important part of process quality and customer satisfaction.

Number of Errors

The number of errors for telephone services usually means the average number of wrong pieces of information that the operators give to customers per call. It should be reduced to zero or near zero to stay in business. Operator training is the key to reduce the number of errors. Sometimes the number of errors could mean the number of telephone system errors, such as failing to redial or switching to the wrong operator.

Telephone Service Process Analysis

Typical activities in a telephone service process include

- Customer call arriving
- Customer waiting for service
- Operator talking to customers and providing services
- Customer departing

The resources for telephone services include phone systems, computer information systems, and well-trained operators. Good design of the call routine process, capacity management, and training of operators are the primary determinants of process success.

The answers to the following questions may be very helpful in designing and improving a telephone service process:

- How many shifts and service providers are needed to minimize costs?
- Which resources can be shared to assist in peak times to minimize waiting times?
- What automation technologies can be used to minimize service times?
- How can calls be routed to minimize waiting times?

The common decision variables for a telephone service process include

- Number of operators
- Capacity of the phone system
- Staff schedule and shifts
- Call routing
- Hours of operation

The common problems in a telephone service processes include

1. Excessive long waiting time
2. Poor customer service quality
3. Poor customer-provider interaction quality
4. Excessive number of errors
5. Excessive abandonment rate

These common problems are often caused by

- Poor call routine design
- Poor capacity management
- Inadequate automation
- Poor service provider training
- Poor scheduling
- Poor equipment maintenance
- Poor service provider discipline or motivation

10.3.9 Project Shop Process

A project shop process could be applicable to both manufacturing and service. In a project shop, the part or batch of parts is stationary while the resources are brought to the product for processing. The product is often quite large, such as a building or ship, a software development project, or a research and development (R & D) project; and it requires considerable time to complete. Production quantities are usually low with products often being produced from start to finish one at a time. Projects typically involve many resources where the quality of the end product and the time to finish the product are dependent on the capability of the resources as well as the planning and scheduling of those resources. Often, multiple overlapping projects must be coordinated, all requiring the use of common resources. A project process is not the same as that of a factory process, because a project is a progression through time rather than the progression of entities through space.

Because of the time and cost of the project, the quality of the end product is of utmost importance. Delivery time is also a key consideration, and penalties are sometimes assessed if due dates are not met. The project cost, although important, is less of a concern. Customer feedback is frequent and can have an impact on the priority and scheduling of remaining activities. Customer commitment is very high because it is difficult to find another producer or to start from scratch once the project is under way.

Projects are typically modeled using the critical path method (CPM), program evaluation and review technique (PERT), or Gantt charting. Many project management software packages have become available to help organize and track the progress of projects.

Examples of Project Shop Processes

Software development, R&D projects, book writing, building construction, shipbuilding, aircraft manufacturing, satellite construction

Project Shop Performance Measures

Quality of the Project

Quality of the project usually refers to the functionality and reliability of the finished entity. For example, the quality of the software includes functionality, satisfaction of customer needs, and reliability.

Time to Complete the Project

Because the time to complete the project usually means the time to market or time to users, a shorter completion time, given the project quality is not compromised, is very important.

Cost to Complete the Project

An excessive project cost could be caused by excessive project delays, poor resource allocation, waste, and rework.

Utilization of Resources

Highly utilized or overutilized resources could be indicators for bottleneck resources. Underutilized resources could be sources of excessive project costs.

Errors and Rework

Errors and rework in critical path activities will usually cause both project delays and cost overrun. Errors and rework in noncritical paths will also cause cost overrun.

Project Shop Process Analysis

A project usually consists of many activities. Some of the activities can be worked on simultaneously, and some depend on the completion of other activities. The subdivision of tasks affects the project progress because it changes the interdependence structure. Project network, CPM, and PERT can be used to model and control project execution.

The answers to the following questions may be very helpful in designing and improving a project shop process:

- What is the optimum project plan that minimizes total costs?
- What is the length and/or cost of a project given a defined set of activities?
- What is the best use of resources to minimize the delay of a project?
- How many resources are needed to meet a particular deadline?
- What is the best coordination of multiple projects to minimize delays?

The common decision variables for a project shop process include

- The subdivision of the project into individual tasks
- The time required to complete each task
- The resources required to perform each task
- The priorities with which activities access resources
- The order in which multiple projects are performed

The common problems in project shop processes include

1. Excessive long project completion time
2. Poor project quality
3. Cost overrun
4. Excessive number of errors and reworks

These common problems are often caused by

- Poor subdivision of the project into tasks
- Poor project management
- Poor coordination
- Excessive changes of plans
- Poor cost estimation
- Excessive complexity of the planned product
- Poor resource allocation

10.3.10 Logistics and Distribution Processes

Logistics and distribution processes are used for packing, warehousing, shipping, and setting up materials. Materials include raw materials, finished goods, work-in-process, as well as equipment, tools, and even personnel. The materials involved may be large or small, they may be compact or clumsy, or indeed, they may be delicate or rugged. The distances involved may be short, such as a few feet, or extremely long, as in moving items from one continent to another. The goals of logistics processes typically are to minimize damage and loss of items, to reduce costs involved, and to increase the response time for moving items around.

Typically there are different modes for the transportation aspects of logistics; certain items may be moved by air, sea, or over land by truck or rail. Depending on the items being transported, these methods each have different attributes regarding cost, speed, and damage potential. Process management can be very useful in improving the performance of these processes.

For example, in France, the Minitel system, which resulted from a process management study, has reduced logistics and shipping costs through better coordination between the different trucking companies. The Minitel system helps pair up shipments and trucks so that trucks returning to their original point of origin, which may have returned empty, are now being subcontracted to carry other items, thereby avoiding the need for extra round-trips.

Examples of Logistics and Distribution Processes

Mail and package delivery, food delivery, flower delivery, and moving services

Logistics and Distribution Processes Performance Metrics

Shipment Damage or Loss

Loss or damage of shipped items is clearly a key failure for a logistics and distribution process. Loss of items may be caused by poor handling, failure in a shipment tracking system, or failure in a sorting process or of equipment. Damaged items are mostly due to poor handling or packaging. Clearly, it is desirable to reduce shipment damage and loss to a minimum.

Delivery Delays

Delivery delay means that the items arrive in a customer's hands later than promised. It clearly makes customers unhappy and damages the shipper's credibility. Delivery delay may be caused by poor handling, poor shipment tracking, inadequate shipping capacity, bottleneck resources, and poor scheduling.

Logistic Cost

Logistic and distribution costs depend on many factors. The important factors include scheduling and routing, location and capacity of distribution centers and transshipment centers, effectiveness of the computer information system and shipment tracking system, and effectiveness of the paperwork process.

Vehicle Usage and Carrying Capacity Utilization

Capacity utilization levels of key resources, such as vehicles and handling equipment, are often indicators of bottlenecks or wastes. Overutilized or fully utilized resources often indicate that they are bottlenecks. Underutilized resources often indicate that they create waste.

Customer—Service Provider Interaction

In the logistics and distribution process, the amount of customer–service provider interaction is less than that of the customer service process. The interaction mostly happens in the shipping and receiving.

Logistics and Distribution Process Analysis

Typical activities in the logistics and distribution process include

- Taking customer shipment orders
- Verifying and registering shipment orders into the shipment tracking system
- Sorting shipment goods
- Scheduling and routing of shipments
- Loading and unloading of shipments
- Different modes of transportation (land vehicles, sea, air)
- Processing paperwork

The resources needed for logistics and distribution service processes typically include vehicles, ships, airplanes, air hubs, sorting centers, a high-capacity computer information system, a sophisticated scheduling and routing system, shipment collection and delivery centers, and well-trained employees.

The answers to the following questions may be very helpful in designing and improving a logistics and distribution process:

- How many of each vehicle are required to meet customer demand?
- Which delivery routes will maximize productivity?
- How many loads per vehicle maximizes the productivity?
- How many shifts and service providers are needed?
- What procedures can be used to minimize service or product selection time?

The common decision variables for a logistics and distribution process include

- Number of vehicles and drivers
- Facility size and layout
- Routing sequence
- Frequency of deliveries and delivery size
- Pickup loads from vendors, etc., on the backhaul or return to the warehouse where possible to avoid deadheading (i.e., driving empty).

The common problems in the logistics and distribution process include

1. Lost items
2. Damaged items
3. Shipment delays
4. Shipment errors
5. Excessive shipping cost
6. Poor customer service quality

These common problems are often caused by

- Poor logistics and distribution process design
- Poor scheduling and routing
- Unbalanced processes or bottlenecks
- Poor operation management
- Poor work flow management
- Poor equipment maintenance
- Poor service provider discipline or motivation

10.3.11 Transportation Processes

Transportation services involve the movement of people from one place to another. A fundamental difference between transportation and logistics and distribution systems is that people are being transported rather than goods. Another important difference is that the routes in transportation services tend to be fixed whereas the routes in delivery services are somewhat flexible. Customers are interested in convenient and fast transportation. Cost of transportation plays a significant role in the selection of the service. Because set schedules and routes are used in transportation, customers expect reliable service. Two types of pickup and drop-off points are used in transportation: multiple pickup and drop-off points and single pickup and drop-off points. In multiple pickup and drop-off point systems, customers enter and leave the transportation vehicle independently. In single pickup and drop-off transportation, customers all enter at one place and are dropped off at the same destination.

Examples of Transportation Processes

Airlines, railroads, cruise lines, mass transit systems, and limo services

Transportation Processes Performance Metrics

Boarding Time

This is the average total time to board a shipment of customers.

Exit Time

This is the average total time to disembark a shipment of customers.

Transportation Time

This is the average time to transport between two given locations. It is desirable to avoid longer transportation times than promised and to reduce variations of transportation times.

On-Time Arrivals

This can be measured by the percentage of on-time arrivals.

Customer Service Quality

Customer service quality includes the quality of reservations, purchasing tickets, onboard service, and other assistances.

Utilization of Transportation Vehicles

Again, overutilization of vehicles usually indicates bottlenecks; underutilization often indicates wasted resources.

Travel Cost

Excessive travel costs may be caused by wasting of resources and poor scheduling.

Transportation Process Analysis

Typical activities in a transportation process include

- Customer making reservation and/or purchasing tickets
- Customer arriving
- Customer boarding

- Scheduling and routing
- Onboard service
- Transporting
- Customers leaving

The resources needed for a transportation process typically include vehicles, ships, airplanes, air hubs, sorting centers, a high-capacity computer information system, a sophisticated scheduling and routing system, and well-trained employees.

The answers to the following questions may be very helpful in designing and improving a transportation process:

- How many of each vehicle are required to meet passenger demand?
- Which transportation routes will maximize productivity?
- How many customers can there be per vehicle?
- How can the departures be scheduled for maximizing customer convenience?
- How many shifts and service providers do we need?
- What procedures can be used (for example, preassigning seats) to minimize service time?
- How will reliability procedures (baggage checking, etc.) affect overall service time?

The common decision variables for a transportation process include

- Size and location of loading and unloading areas
- Number, size, and speed of transportation vehicles
- Scheduling of departure and arrival times
- Scheduling of vehicles and operators
- Maintenance scheduling

The common problems in a transportation process include

1. Excessive delays
2. Poor customer services
3. Excessive travel cost

These common problems are often caused by

- Poor transportation process design
- Poor scheduling and routing
- Unbalanced processes or bottlenecks
- Poor operation management
- Poor equipment maintenance
- Poor service provider discipline or motivation

10.3.12 Purchasing and Supply Process

As more and more companies move away from high levels of vertical integration to focus primarily on their core competency and purchase everything else from suppliers, the role of purchasing and supply will continue to increase. As globalization continues on its relentless path of growth, your suppliers are now just as likely to be one block away as they are to be halfway around the globe. Thus, purchasing and supply is fast evolving into global purchasing and supply. Purchasing and supply involves the activities that are used to

1. Identify the best suppliers for goods and services
2. Specify exactly what is needed and find prospective suppliers to bid on the organization's purchase requirements
3. Negotiate the best purchase terms for the organization
4. Ensure that adequate legal contracts are signed
5. Ensure that goods and services are delivered as required and in a timely manner
6. Ensure that certification for supplier payment is processed
7. Ensure that suppliers are well integrated (as required) into the organization's operations
8. Ensure that supplier morale is high and stays high

Purchasing and supply is particularly important because this is the activity that is responsible for integration of the firm's activities with those of its most important suppliers. In the U.S. automotive industry, for example, there are tier-1 suppliers who supply to the major automobile companies, tier-2 companies who supply to the tier-1 companies, and so on. Obviously, the customer does not really care about who built what components; errors made by a tier-2 supplier will give a company just as big a black eye as if the company had committed the error itself. This all means that the job of integrating supplier activities, the supply chain, in order to ensure customer satisfaction, is extremely important to every company's long-term success.

Purchasing Performance Metrics

Frequency of Material Shortage and Percent on-Time Receipts
Production activities are the areas in an organization that are most dependent on the performance of purchasing and supply. Performance metrics such as frequency of material shortages and percent on-time receipts indicate the degree to which purchasing and supply facilitates the smooth flow of production.

Expenditure on Premium Freight
Expenditure on premium freight also measures the extent to which rush orders have to be placed because of poor planning or unreliability of suppliers.

Order Processing and Information Flow Lead Time

Information flow lead time means the time needed for a piece of information to move from one point of the process to another point of the process. For example, how long it will take from order request to purchasing order issuing.

Order Processing Accuracy

To ensure high accuracy for purchasing transactions, order processing accuracy is usually expressed as number of errors per thousand transactions.

Inventory Turnover and Obsolete Material Inventory

Purchasing and supply also influences levels of inventory to a very great extent. Inventory turnover is one measure of inventory performance; obsolete material inventory is another.

These metrics interact with other factors in other areas such as manufacturing to determine the operations or production effectiveness. In addition, the material cost will depend on purchasing and supply's ability to identify the best suppliers and negotiate the most beneficial terms for the orders. Over the long term, purchasing and supply's efficiency and effectiveness will contribute to every organization's profitability, and hence to its business viability.

Purchasing and Supply Process Analysis

Typical activities in purchasing and supply include everything that is required to work with suppliers. Purchasing and supply activities include

- Vendor identification
- Vendor evaluation and selection
- Contracting
- Request for quote and request for purchase development
- Ordering and purchase order development
- Supply verification
- Payment authorization
- Vendor relationship management and supplier auditing

These are some of the standard purchasing and supply activities; however, each organization may need to customize exactly how these processes are implemented. There are factors relating to the nature of the product, the market that is served, and the type of suppliers, which will require an application of process management to improve the process.

Information is the primary commodity that the purchasing and supply process works with. The primary tool that is required for purchasing and supply is a purchasing information system. This system should be integrated with the enterprise resource planning (ERP) system for the organization, based on the need to customize the purchasing and supply information system to match the unique processes of the organization.

The common problems in the purchasing and supply process include

1. Wrong material or wrong specification delivered
2. Poor material quality
3. Material shortage
4. Material not delivered on time
5. Difficulty in finding needed material
6. Waste through spoilage, material obsolescence, and wrong item rdered
7. Wrong specification ordered
8. Excessive delay in order processing
9. Excessive delay in payment processing
10. Financial loss due to fraud, lawsuit, or inappropriate supplier payment
11. Excessive inventory on hand
12. Escalating material purchase cost

These common problems are often caused by

- Problem in coordinating action and decision making with suppliers
- Flow synchronization of long-lead-time items
- Mismatch between material flow methods (for example, lean and mass production systems)
- Poor or inadequate communication with suppliers
- Mistrust between suppliers and customers
- Cultural differences
- Market complexity (number of suppliers, number of items to order)
- Market uncertainty or turbulence
- Poor decision-making processes
- Poor product or technology knowledge
- Poor project or process management

Numerous strategies have been applied, in various situations, to enhance the efficiency and effectiveness of purchasing and supply activities. Each strategy has some unique area of application relative to the different causes of process performance problems. The following list shows a number of these strategies.

- Production planning and scheduling
- Inventory management
- Enterprise resource planning (ERP) systems
- Information sharing with suppliers (e.g., end-customer data, sales trends, and forecasts)

- Collaborative scheduling with suppliers
- Long-term contracts, alliances, partnerships, supplier partnerships, virtual organizations
- Paperwork process streamlining
- Application of purchasing cards
- Electronic commerce and electronic data interchange (EDI)
- Internet commerce and Internet-based purchasing and supply
- High-speed distribution systems
- Colocation with suppliers
- Stock pooling
- Vendor-managed inventory
- Just-in-time (JIT) or pull material flow control method
- Vendor consolidation and single sourcing
- Supply chain integration and management
- Purchasing and supply automation
- Purchasing and supply information systems
- Strategic sourcing
- Global sourcing
- Supplier development programs (e.g., technical assistance, process improvement assistance, or financing assistance)
- Purchasing and supply process management
- Statistical process control of purchasing and supply activities

Some of these strategies simply represent best practice and require little or no capital expenditure to implement. Others are quite comprehensive and require extensive changes to processes and procedures. After the necessary process management analysis has been used to identify the best strategy, adequate effort must also be put into implementation and maintenance planning for the system.

10.4 Process Mapping

Many business processes, especially service processes, are poorly defined or totally lacking in description. Many procedures are simply described by word of mouth or may reside in documents that are obsolete. In process management, it is often that by simply trying to define and map the process, we provide a means for both understanding and communicating operational details to those who are involved in the process. We also provide a baseline, or standard, for evaluating the improvement. In many cases, merely defining and charting the process as it is can reveal many deficiencies such as redundant and needless steps and other non-value-added activities. Process mapping is a visual descriptive model for a process. "A process map is considered to be a visual aid for picturing work processes which show how inputs, outputs and tasks are linked" (Anjard 1998). Process mapping is used to develop a process map for the process under study. There are a number of different methods of process mapping; these methods include a process flowchart, IDEF0 process mapping, and value stream mapping. In this section, we are going to discuss these three methods in detail.

10.4.1 Process Flowchart

The flowchart is a graphic way to describe a group of activities in a process. The basic purpose of the flowchart is to provide a graphical representation of all the activities performed in the sequence in which they are actually conducted. One advantage of the graphical flowchart is the ease with which the activities in the process and their relationships can be visualized and understood. However, a flowchart alone usually does not provide enough level of detail for each activity. Therefore, supplementary writing procedures for all activities are also provided with the flowchart to provide the details.

Flowcharts use symbols connected by arrows to describe processes. The commonly used flowchart symbols are illustrated in Table 10.1.

Example 10.1
This example illustrates a flowchart for typing a document as follows:

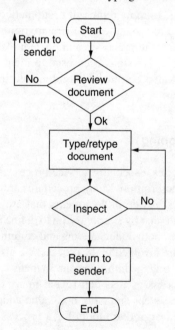

Table 10.1 Commonly Used Flowchart Symbols

Symbol	Name	Activity Represented
(rounded rectangle)	Boundary (start or end)	Identifies the beginning or end of a process. *Start* or *End* can be written inside.
(rectangle)	Activity or task	Identifies an activity or task in the process that changes input. Usually, the name of the activity is written inside.
(arrow shape)	Movement or transportation	Indicates movements of outputs between locations.
(circle)	Inspection	Identifies that the flow has stopped in order to evaluate the quality of the output or to obtain an approval to proceed.
(D-shape)	Delay	Identifies when something must wait or is placed in temporary storage.
(inverted triangle)	Storage	Identifies when an output is in storage waiting for a customer.
(diamond)	Decision	Identifies a decision or branch point in the process.
(document shape)	Document	Identifies when the output of an activity is recorded.
(cylinder)	Database	Identifies when the output of an activity is electronically stored.
(connected circles)	Connector	Indicates that an output from this flowchart will be input to another flowchart.
(arrow)	Arrow	Indicates the sequence and direction of flow within the process and usually transfers an output of one activity to the next activity.

10.4.2 IDEF0 Process Mapping

IDEF0 (International DEFinition) process mapping is a method designed to model the decisions, actions, and activities of an organization or system. It was developed by the U.S. Department of Defense, mainly for the use of the U.S. Air Force during the 1970s. Although it was developed over 30 years ago, the Computer Systems Laboratory of the National Institute of Standards and Technology (NIST) released IDEF0 as a standard for function modeling in FIPS (Federal Information Processing Standards) Publication 183, December 1993. Computer packages (for example, AI0 WIN) have been developed to aid software development by automatically translating relational diagrams into code.

An IDEF0 diagram consists of boxes and arrows. It shows the function as a box and the interfaces to or from the function as arrows entering or leaving the box. Functions are expressed by boxes operating simultaneously with other boxes, with the interface arrows constraining when and how operations are triggered and controlled. The basic syntax for an IDEF0 model is shown in Fig. 10.8.

Mapping using this standard generally involves multiple levels. The first level, the high-level map, identifies the major processes by which the company operates (Peppard and Rowland 1995). The second-level map breaks each of these processes into increasingly fine sub-processes until the appropriate level of detail is reached.

For example, Fig. 10.9 shows the first-level IDEF0 process map of a printed-circuit board (PCB) manufacturing process. Figure 10.10 shows the second-level IDEF0 process map of the PCB manufacturing process, and you can go further down with every subprocess.

There are a number of strengths and weaknesses associated with IDEF0. The main strength is that it is a hierarchical approach, so users can choose the mapping at their desired level of detail.

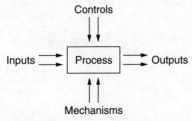

Figure 10.8 A Basic IDEF0 Process Map Template

Design and Improvement of Service Processes—Process Management

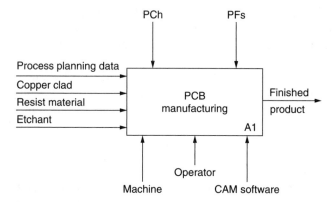

Figure 10.9 Level-1 IDEF0 Process Mapping for Printed-Circuit Board (PCB) Manufacturing

10.4.3 Value Stream Mapping

Based on process mapping, a value stream mapping can be developed to analyze how well a process works. Once a process map is established at an appropriate level of detail, the flows of products, programs, and services,

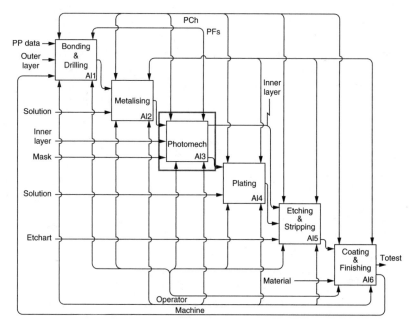

Figure 10.10 Level-2 IDEF0 Process Mapping for Printed-Circuit Board (PCB) Manufacturing Process

material; information; money; and time can be mapped. Figure 10.11 shows an example of a value stream map that maps not only material flows but also the information flows that signal and control the material flows.

After a value stream map is developed, value-adding steps are identified for each kind of flow, such as material and information flow. Non-value-adding steps (waste), value inhibitors, costs of flow, and risks to flow are also exposed, and their implications to overall process performance are analyzed.

After the problems in the existing process are identified by value stream mapping, process revision or redesign can be initiated to eliminate the deficiencies. In manufacturing processes, the revision can be made by elimination of non-value-adding steps and redesign of the layout and sequence of subprocesses, thus reducing cost and cycle time. In office processes, the revision can be made by redesigning the organizational structure, reporting mechanisms, building layout, and functional responsibilities of various departments in order to reduce non-value-added steps and paperwork travel time and mistakes, thus reducing waste and improving efficiency.

Based on the analysis, an ideal value stream map is created, in which all waste and value inhibitors are removed, the cost and risk for flow are similarly reduced to a minimum level, and we call it the ideal state. The full implementation of the ideal state may not be feasible, but it often leads to a much-improved process.

The following is a case example of a value stream mapping project (Bremer 2002) which involves a manufacturing-oriented company. The companywide information flow is illustrated by two value stream maps. The first map, Fig. 10.12, shows how management thinks the information flows in this business;

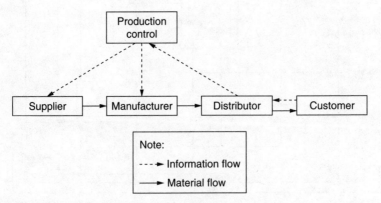

Figure 10.11 An Example of a Value Stream Map

Design and Improvement of Service Processes—Process Management

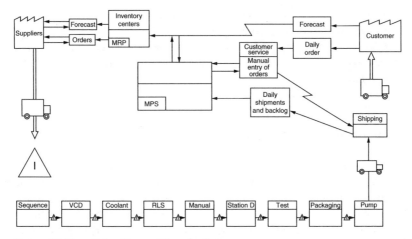

Figure 10.12 Value Stream Map of a Company

the management thinks that the flows are simple and straightforward. The second map, Fig. 10.13, shows how information really flows. It's a lot more complicated. Many process steps add no values, and they actually impede the production process. Also there are huge amounts of hidden information transactions in the process that add no value to a business from a customer's perspective.

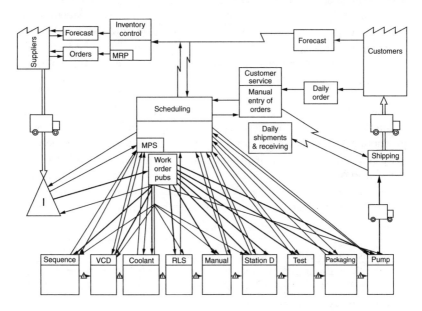

Figure 10.13 Actual Value Stream Map of the Company

Because of these problems, the company usually takes over 34 days to go from raw material to delivery to customers. After identifying these problems, the company redesigned the business process. The improved process reduced most of the hidden transactions, and the company is now able to move from raw material to delivery in five days.

10.5 Lean Operation Principles

Lean manufacturing is a very effective manufacturing strategy first developed by Toyota. During a benchmarking study for the automobile industry in the late 1990s (Womack, Jones, and Roos 1990), it was found that Toyota clearly stood above its competitors around the world with the ability it developed to efficiently design, manufacture, market, and service the automobiles it produced. This ability made a significant contribution to both the company's profitability and growth as consumers found the products to simultaneously exhibit both quality and value. The researchers found that the focus on recognizing and eliminating wasteful actions and utilizing a greater proportion of the company's resources to add value for the ultimate customer was the key to the operating philosophy. "Lean production" is first mentioned in this study and is used to describe the efficient, less wasteful production system developed by Toyota, called the Toyota production system. Lean production in comparison to mass production was shown to require one-half the time to develop new products, one-half the engineering hours to design, one-half the factory hours to produce, and one-half the investment in tools, facilities, and factory space (Monden 1993, Ohno 1990, Shingo 1989).

Although the lean manufacturing approach was originally developed in the traditional manufacturing industry, lean manufacturing mostly deals with production systems from a process viewpoint, not a hardware viewpoint. It has been found that most lean manufacturing principles can be readily adopted in other types of processes, such as product development, office, and service factory processes. When these lean manufacturing principles are applied to nonmanufacturing processes, we call them lean operation principles.

In process management practice, lean operation principles can be used to analyze the current process and discover any problems in process design and operation. Lean operation principles can also be used to guide the process redesign and improvement.

The key objective of lean operation is to eliminate all process wastes and maximize process efficiency. The key elements of lean operation include the following items:

- Waste elimination in process
- Pull-based production system
- One piece flow
- Value stream mapping
- Setup time reduction
- Work cells

Now we discuss these key elements in detail.

10.5.1 Waste Elimination in Process

In observing the mass production, Tachii Ohno (Ohno 1990, Liker 2004), an engineering genius of Toyota and the pioneer of the Toyota production system, identified the following seven wastes in production systems:

1. *Overproduction:* Producing too much, too early
2. *Waiting:* Workers waiting for machines or parts
3. *Unnecessary transport:* Unnecessary transporting of moving parts
4. *Overprocessing:* Unnecessary processing steps
5. *Excessive inventory:* Semifinished parts between operations and excessive inventory of finished products
6. *Unnecessary movement:* Unnecessary worker movements
7. *Defects:* Parts need rework or are scrap

These seven wastes are called *muda*, a Japanese term for missed opportunities or slack. These items are considered waste because in the eyes of customers, these activities do not add desired values to the products.

In lean operation principles, the seven wastes can be identified mostly by the value stream mapping method. The waste caused by overproduction can be reduced or eliminated by a pull-based production system. The waste caused by excessive inventory, waiting, and unnecessary transport can be greatly reduced by one-piece flow and work cells (cellular manufacturing). It is often necessary to use a setup time reduction technique to make one-piece flow possible. One-piece flow and work cells also make defect detection easier. Besides lean operation principles, other techniques, such as ergonomics, poke yoke (foolproof), and statistical process control should also be applied to reduce the waste caused by defects and unnecessary movements.

We will first discuss the value stream mapping method. The value stream mapping method has two stages. The first stage is to draw a current state map, which is a map of the current process. The second stage is to draw a future state map, which is a value stream map of the proposed new process. Since the drawing of the future state value stream map requires other knowledge of lean operation, such as one-piece flow and work cells, Sec. 10.5.2 discusses current state value stream mapping, Sec. 10.5.3 discusses one-piece flow and work cells; Sec. 10.5.4 discusses waste reduction and future state value stream mapping, and Sec. 10.5.5 discusses other issues in lean operation.

10.5.2 Current State Value Stream Mapping

Value stream mapping is a good method to use to chart the process and identify and quantify the waste in a process. Value stream mapping was developed to map and analyze the production process, especially the batch flow shop and flow shop processes. A value stream is all the activities (both value-added and non-value-added) required to bring a product through the main flows.

Value stream mapping is a pencil-and-paper exercise that helps you to see and understand the flow of material and information as the product makes its way through the value stream. When you want to draw a value stream map, do not use a computer; just bring a writing board, a good piece of paper, and a stopwatch. The best way is to work backward, that is, from the last step of the process to the first step of the process. In a production process, the last step is usually the shipping dock; in a restaurant kitchen process, the last step is at the point where the meal is done and the waiter is taking it away from the customer. In a production process, the first step is usually the receiving deck for the incoming materials; in a restaurant kitchen process, the first step is usually the point where the customer's order is brought into the kitchen.

Figure 10.14 is a simplified value stream map for a production process. Clearly, based on the definition of seven wastes, the staging, transportation, setup, and inspection are value-added steps; casting, machining, and assembly are value-added steps. In the figure the horizontal length of each step is proportional to the time required to do the step. The total time duration from the beginning of the process to the end of the process is often called the process lead time. Clearly, in this example, the value-added time is a small portion of the total lead time.

Design and Improvement of Service Processes—Process Management

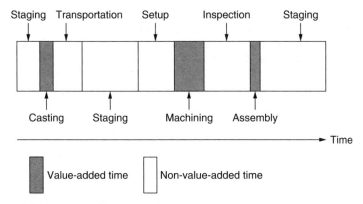

Figure 10.14 A Simplified Value Stream Map for a Production Process

We can see that this simple value stream map identified and quantified waste in the process and provided clues for process improvement. Clearly, the process can be improved if we can shorten the non-value-added time.

This kind of simple value stream mapping can also be used to analyze a service process. Figure 10.15 shows a simplified value stream map for a

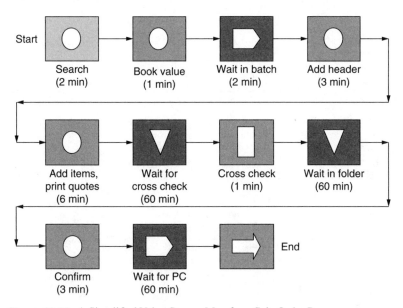

Figure 10.15 A Simplified Value Stream Map for a Sale Order Process

sale order process. The lighter shaded boxes are value-added steps; the darker shaded boxes are non-value-added steps. The first box, Search, does not really add value for customers, but it is an essential step for now, so a lighter shaded box is used.

In many service processes, the simplified value stream map as shown by Fig. 10.15 is sufficient to map the process and identify and quantify the wastes in the process. In some cases, the simplified value stream map is not sufficient to describe the process; in this case, a more formal type of value stream map illustrated in Sec. 10.4.3 can be used to map the process.

The formal value stream map uses arrows and icons to illustrate the process. There are two types of flows that are of major concern. One is the material flow, the other is the information flow. Figure 10.16 shows the commonly used icons for the material flow in value stream maps. Figure 10.17 shows the commonly used icons for the material flow in value stream maps.

In material flow, the process boxes should be identified one by one. A data box should be established for each process box. In each data box, the following data should be measured (by stopwatch) and recorded:

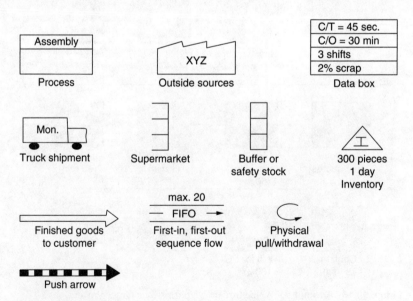

Figure 10.16 Icons Used in Material Flows in Value Stream Maps

Design and Improvement of Service Processes—Process Management   367

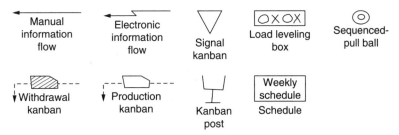

Figure 10.17 Icons Used in Information Flows in Value Stream Maps

1. *Cycle time (C/T):* Time required to produce one piece of product by a machine, station, and/or operator or the time required to repeat a given sequence of operations or events.
2. *Change over time (C/O):* Time required to switch from one product type to another product type, for example, the time it takes for a pizza maker to switch from making one type of pizza to another type of pizza.
3. *Uptime:* Proportion of time a process step is operational.
4. Production batch size (EPE).
5. Number of operators.
6. Number of product variations.
7. Scrap rate.

Figure 10.18 shows a complete value stream map for a manufacturing process. In the figure, we can see that below each process box, there is a data box. For example, in the leftmost process box, Stamping, the cycle time is 1 second, the changeover time is 1 hour, uptime = 85 percent, and the production batch size (EPE) is 2 weeks of supply, that is, the stamping press produces a big batch of parts (enough to supply for 2 weeks) in one shot. In Fig. 10.18, between the first process box, Stamping, and the second process box, S. Weld 1, there is an in-process inventory of semifinished parts. The average inventory holding time is 7.6 days. In the first process box, the value-added time is 1 second, which is equal to the stamping cycle time. From the lean operation point of view, the in-process inventory holding is a non-value-added activity. If we add all value-added time for the whole process, it is equal to 184 seconds, which is recorded at the lower-right corner of Fig. 10.18. The production lead time for the whole process is 23.5 days. Clearly, in the whole production lead time, only a tiny proportion is value-added time. The top portion of the value stream map shows the information flow pattern.

Cycle time, value creation time, and lead time are among the most important measures in lean operation management. Figure 10.19 gives good definitions and illustrations for these measures.

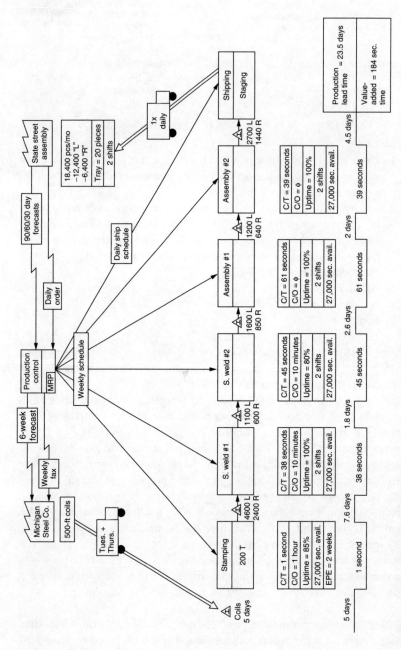

Figure 10.18 A Complete Value Stream Map for a Manufacturing Process

Design and Improvement of Service Processes—Process Management

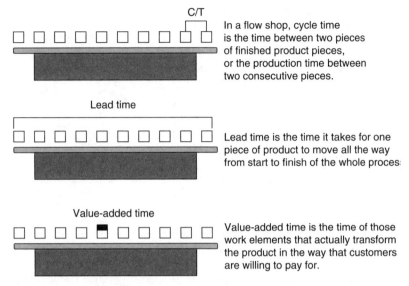

Figure 10.19 Some Important Process Metrics Used in Value Stream Mapping

10.5.3 Lean Operation Techniques

In many production systems, there are huge amounts of muda (the seven wastes) in the process. From the examples given in Sec. 10.5.2, we can see that out of the whole production lead time, the value-added time is usually only a small fraction. The ratio of value-added time over production lead time can be used as a measure of process efficiency. Specifically this can be stated as

$$\text{Process efficiency} = \frac{\text{value-added time}}{\text{total lead time}} \qquad (10.1)$$

The major goal of lean operation is to increase process efficiency. A process that has a high efficiency will have much less waste, a shorter lead time, and lower cost. As a rule of thumb, a process is considered to be lean if the process efficiency is more than 25 percent. Based on the research by Michael George (2003), the typical process efficiency and world-class efficiency for many types of processes are summarized in Table 10.2.

Clearly, the process efficiencies of typical processes are very low. A big proportion of process lead time is not used to do value-added work, but to do non-value-added work, that is, muda (the seven wastes). Lean operation attempts to redesign the process flow and layout so that the portion of process time spent doing non-value-added work is greatly reduced.

Table 10.2 Process Efficiency for Various Processes

Process Type	Typical Process Efficiency (%)	World-Class Process Efficiency (%)
Machining	1	20
Fabrication	10	25
Assembly	15	35
Continuous manufacturing	30	80
Transactional business processes	10	50
Cognitive business processes	5	25

The most frequently used techniques in lean operation include the following:

- One-piece flow
- Work cells (cellular manufacturing)
- Pull-based production
- Quick setup time reduction

We now discuss these in detail.

One-Piece Flow

As we discussed in Sec. 10.3.2, there are several types of manufacturing processes, such as the job shop, batch flow shop, and line flow processes. The job shop process is also called the "machine village," which means that similar machines are grouped together. The job flow patterns of such production systems can be quite erratic and messy, as illustrated in Fig. 10.20. The job shop process is featured by low utilization, long delays, high work-in-process inventory, and a long lead time. The advantage of the job shop process is that it can take a large variety of tasks.

Many service processes are also job shop processes. For example, in most organizations, the departments are functionally grouped together, such as the personnel, accounting, and benefit departments. If a new employee wants to finish all his or her paperwork, he or she must go through all these departments. In many organizations, paperwork has to be approved by many departments, so each piece of paperwork will first go to one department, then

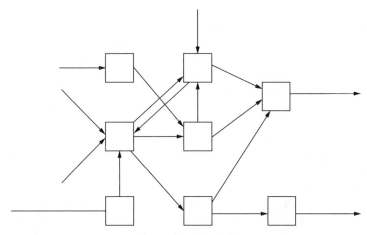

Figure 10.20 A Typical Flow Pattern of A Job Shop

through interdepartmental mail, and then to the next department; if a mistake occurred in the previous department, the paperwork could be sent back for correction. It is also quite usual that documents become lost or buried in the paper trail and eventually get lost or take a very long time to be completed. A batch flow process has a better flow pattern, as illustrated by Fig. 10.21.

However, there are still a lot of work-in-process inventories. The value stream map illustrated by Fig. 10.18 is a batch flow process. We can see clearly that most of the lead time is spent on inventory waiting in the stock. This is better than that of the job shop in terms of flow pattern, but it is still inefficient.

One-piece flow, or single-piece flow, is the solution proposed by the lean operation principle. One-piece flow is actually the line flow shop illustrated

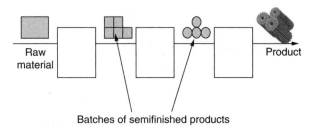

Figure 10.21 Flow Patterns of a Batch Flow Shop

by Fig. 10.7. The objective of process-oriented line flow is to convert functional layouts of machines in the factory into a series of processes, based upon the production of families, or commodities. Process-oriented flows are superior to traditional functional layouts since they reduce travel distance, required floor space, and total throughput times. A single-piece flow means that the workpiece is worked on one piece at a time, not one batch at a time. This will eliminate the work-in-process inventory completely.

On the other hand, in a single-piece line flow process, any error or defect in any process step will cause the whole line to stop. In traditional Western operation management, the work-in-process inventory, or buffer inventory, is used to temporarily feed the downstream process steps so the line will not stop. However, the Toyota production system believes that the buffer inventory has more disadvantages than benefits; buffer inventory ties up money and hides hidden problems. In the Toyota production system, zero buffer inventory is used to expose all the hidden problems in the production process; it forces you to debug all hidden problems so eventually, you will have a zero-defect production process.

Work Cell (Cellular Manufacturing)

The ideal production process setup for lean operation is a one-piece flow work cell, as illustrated by Fig. 10.22. A work cell is a U-shaped layout of several different kinds of machines that form a one-piece flow line. The U-shape is used because it saves floor space and shortens travel distance for operators.

If a production facility has to make many kinds of products, these products can be grouped into several categories such that the products are similar within each category. Each category of products will be produced by one

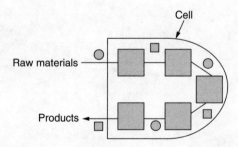

Figure 10.22 One-Piece Flow Work Cell

particular work cell; this work cell is equipped with machines that are fit to produce this category of products. This multiple-cell setup is illustrated by Fig. 10.23. This group work cell setup can handle as large a number of product varieties as that of the job shop illustrated by Fig. 10.20. However, the flows of the work cell group will be much smoother and faster than that of the job shop.

In an office process, such as insurance claim processing, there are many types of claims, and each type should be processed in a different way. Many office processes use a similar process to that of the job shop; each claim goes to a different path, and flows are really messy and erratic. A lot of errors and delays can be caused by this job shop setup. If we use the work cell group concept, we can establish several separate flow line departments, each one handling one category of claims. Within each category the claims type and paperwork procedures are similar, and each department has several operators, each one handling one step of the paperwork. All the paperwork needed for one piece of the claim will be finished after one complete flow through a work cell, as illustrated by Fig. 10.23. This departmentalization-type work flow is usually more efficient than the job shop or batch flow shop processes.

Pull-Based Production

A pull-based production system is a demand-driven production system. It is modeled after the supermarket shelf replenishment operation. On supermarket shelves, there are lots of goods, such as milk, eggs, and orange juices, that are ready for customers to pick up. The customers *pull* the goods from the shelves, and then depending on how many items are taken away,

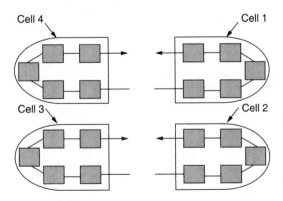

Figure 10.23 A Group of Work Cells

the inventory person in the supermarket will refill the same amount of items by pulling them from the warehouse; then the warehouse person will order roughly the same amount of items that were pulled from the warehouse.

Restaurant operation is a perfect example of pull-based production. The customer places the order, and then the kitchen produces exactly what the customer ordered. In general, the key feature for pull-based production is that the information flow direction is opposite to that of the material flow. The information flow means the production control order. In the restaurant case, the production control is the order for the kitchen to cook. This order's direction is from customer to kitchen; on the other hand, the direction of material flow is the flow of food in the restaurant case, the direction from the kitchen to the customer. Clearly, the information flow direction and material flow direction in the restaurant kitchen are opposite to each other.

The opposite of pull-based production is push-based production. The key feature for the push-based production is that the direction of information flow is the same as that of the material flow. In push-based production, each work stop sends the work downstream of the operation, that is, pushes the work downstream, without considering whether the downstream areas can make use of it. Typically, activities are planned centrally but do not reflect actual conditions in terms of idle time, inventory, and queues.

Agricultural production is a typical push-based production. Because the production cycle is very long, there is no way that farmers can produce only the amount of food based on real-time demand. The production plan is purely based on market forecasts and sometimes just based on last year's production. The production command will flow in the same direction as the work flow. It is well known that agricultural production often suffers from oversupply and market fluctuations. Clearly, pull-based production, whenever possible, will create much less overproduction, so the waste caused by overproduction can be reduced.

In value stream mapping, the symbols illustrated in Fig. 10.24 are used to describe the pull production system.

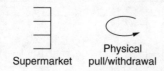

Figure 10.24 Pull Symbols in Value Stream Mapping

Quick Setup Time Reduction

When one-piece flow and a cellular manufacturing system are used, it is very important that the setup time needed between producing one type of product and another type of product be greatly reduced. Otherwise, the production system will be overwhelmed by frequent, long changeover times from one type to the next.

The Toyota production system developed many quick setup time reduction techniques. However, the key idea is to divide the setup time into two categories of elements: internal elements and external elements. The internal elements are the actions needed in the setup where the regular production has to stop. The external elements are the actions needed in the setup where the regular production does not have to stop. The key strategy in the quick setup time reduction technique is to redesign the work elements in setup so that overwhelming amounts of setup work are done externally, that is, without production stoppage.

There must be at least a thousand years of history using the quick setup time reduction technique in the restaurant industry. One of the keys for success in the restaurant business is to reduce the production lead time, that is, the time from customer order to serving the food. Nobody wants to wait in a restaurant for hours without food. The kitchen has to be able to switch over from one item to another without much delay, and the setup time for different dishes must be very fast. It is impossible to batch-produce the same dishes and save those as inventory, so one-piece flow should be strictly enforced. People in the restaurant kitchen found numerous ways to do the quick changeover. The main trick is to do a lot of preparations off-line, that is, when there is no customer order or in parallel with the cooking process. This is the same idea as that of the Toyota production system.

10.5.4 Future State Value Stream Map

As we discussed in Sec. 10.5.3, lean operation techniques can be used to generate a new process design in order to reduce the product lead time and to increase process efficiency. The value stream map for the new design is called the future state value stream map. Figure 10.25 is the future value stream map for the production system illustrated by Fig. 10.18.

In this future state value stream map, the batch size for the stamping operation is reduced from 2 weeks of supply to one shift of supply. The batch line process of welds and assemblies in the old value stream map is changed to a U-shaped work cell. Several supermarket shelf symbols in

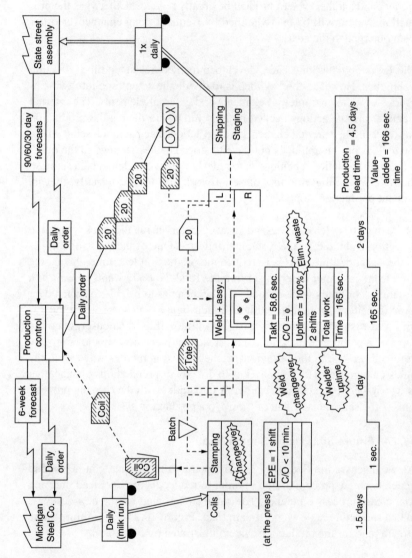

Figure 10.25 The Future State Value Stream Map

Fig. 10.25 indicate that the push-based production is changed into a pull-based production. As a result of this process redesign, the lead time is reduced to 4.5 days from the 23.5 days of Fig. 10.18.

10.6 Process Management Procedures

10.6.1 Process Management Overview

Process management is performed by following a series of steps in the correct order and applying the appropriate tools for each step. It is very important that this methodology be customized to match the requirements of different types of processes and particular process situations. For example, the process management procedures for a project shop case will be very different than the procedures for the supply chain case.

Process management involves the following five phases:

1. Process mapping
2. Process diagnosis
3. Process design
4. Process implementation
5. Process maintenance

10.6.2 Process Mapping

In the process mapping stage, the following tasks must be performed:

1. *Identify the process under study.* In any DFSS project, we are going to work on one process at a time. We need to identify which process we want to work on and define the scope of the project and the boundary of the process.
2. *Outline the process.* The process types and performance metrics discussed in Sec. 10.3 can be very helpful in identifying what type of process that we are working on, and what are the possible process features and performance metrics.
3. *Document process objectives and performance metrics.* Based on the knowledge provided by Sec. 10.3 and detailed analysis of the process under study, we can finalize what are our process objectives and performance metrics.
4. *List the process steps.*
5. *Determine the layout process sequence.*
6. *Identify the process resources for each process step.*

7. *Document the process.* We can use any or all of the process map techniques:
   a. Process flowchart
   b. IDEF0 chart
   c. Current state value stream map

The goal of the process mapping is to derive the detailed visual definition of the process and capture the strengths and weaknesses of the process that will drive the process design and improvement activities.

10.6.3 Process Diagnosis

The goal of process diagnosis is to identify the key weaknesses of the process and provide the guidelines for process redesign and improvements. The following approaches are often used in process diagnosis:

1. *Value stream map analysis:* Value stream mapping can expose non-value-added activities and process efficiency problems. By using lean operation principles to analyze the current state value map, possible improvement ideas can be generated.
2. *Process map analysis:* A real detailed process map may expose "hidden factories," that is, unnecessary loops and steps. This process map analysis may help to generate process improvement ideas.
3. *Process analysis based on process types:* The knowledge outlined in Sec. 10.3 can also be used to analyze the possible weaknesses of the process. For example, if we find that our process is an office process but we use a job shop type of layout, then we can immediately know that this process type is inefficient for the office process and probably should change to using several lean work cells.
4. *Cause and effect diagram* or fishbone diagram analysis.
5. *Data collection:* Collecting such data as the waiting time, process time, or equipment and operator utilization for each process step may help to identify the weak links and bottlenecks of the process.
6. *Process simulation:* For many service processes, discrete event simulation can be a very useful tool to evaluate the current process and identify weak links and bottlenecks.

10.6.4 Process Design (Redesign)

After the process diagnosis step, the weaknesses and bottlenecks of the process should be known. Now is the time to propose the process change and generate new designs. New designs can be generated based on

1. *Applying lean operation principles:* For example, the future state value stream map can be derived by applying lean operation principles to the current state value map.
2. *Brainstorming:* The DFSS team can use brainstorming to generate new designs.
3. *Process knowledge:* The process knowledge described in Sec. 10.3 can be used to generate design ideas. For example, if we identify our process as a project shop–type process, then the redesign solution should be based on project management techniques, such as redividing the work breakdown structure, generating a different project network, or redistributing the resource allocation.

Discrete event simulation experiments can be used as a valuable tool to try out each design alternative. The evaluation of the simulation results will help us to select the best design alternative.

10.6.5 Process Implementation

Process implementation involves final validation of the process and controlled dissemination throughout an organization. This includes procuring and installing tools and equipment required for the process, as well as training activities required for correct application of the new process.

Even after the main features of the new or improved process have been determined, the team must not rush to implement. It is likely that additional refinements in the process will be required before it is ready for "prime time." A convenient method for validating the process is to develop small pilot implementations, evaluate the performance of the pilot processes, validate them in detail, and carry out refinements in detail. Refinements may be found in different areas, from the sequence of steps in the process, to the configuration of tools and resources selected, and even the documentation of the process manual. The piloting, validation, and refinement might take anywhere from a few days to several months. The validation plan and all the necessary follow-through should be overseen by the DFSS team in order to determine that all the process needs will be met.

Once validation is complete, final documentation of the process manual can be developed. The process manual should serve as the document that will be used as a reference for operating the process as well as all training and future maintenance of the process. In documenting, reference should be made to all the process documentation information that has been developed in the mapping and design phases. The process manual should be documented in an attractive and easily understood manner. Optionally, new technologies such as the World Wide Web and Intranets may be used as the medium for access to the process manual.

Many processes will require the hiring of new employees with certain specific skills and the purchase of specialized tools or equipment. Following the specifications of the process design, such employees can be hired and the equipment purchased. Procuring long-lead-time equipment or hiring employees with hard-to-find skills should start early so that everything can be in place for the process to commence operation. Installation of equipment should follow recommended procedures, and new employees should be trained quickly so that they can be at their most effective as soon as possible.

10.6.6 Process Maintenance

Process maintenance involves ongoing monitoring of the process, as well as periodic improvements to ensure that process performance remains high despite changing internal and external conditions. After the process has been implemented, it should not just be abandoned to its fate. Rather, the process should be treated like one would treat any other valuable asset of the organization; track its performance over time and schedule periodic reexamination to determine if it needs any major or minor overhauls. This maintenance will ensure that if conditions were to change or tools and equipment were to wear out, an opportunity would be available to update the process and cope with the new requirements. For tracking and process monitoring, it is recommended that, at the minimum, the most important of the performance metrics considered in the process design should be tracked. In addition, other data providing early warnings regarding the state of the process should be collected.

For example, for critical equipment, failure times and downtime durations should be logged. Data collection systems or data tracking forms need to be designed for acquiring the data of interest. All statistical process control activities for the process are an integral part of the process maintenance. Maintenance schedules for important equipment should be based on the frequency recommended by the manufacturers of the equipment.

10.7 A Process Management Case Study

10.7.1 Background

This case study is from Mejabi (2003). USAA is an insurance company that started losing market share due to price competition from competitors and customer service complaints about excessively long waiting times. To reverse this trend the company needed to improve service levels and reduce cost by providing a fast, cost-efficient, high-quality, and customer-focused claims process. A process management approach was adopted, and a process management project was initiated. The objectives of this project were the following:

- Seek ways to redesign the process to yield simpler and more cost-effective procedures for claims adjudication.
- Reduce process lead times to have fast and effective customer service.
- Improve customer-focused process characteristics in the company.

The process involved in this project was an office process. The following process performance metrics were used in this study:

- *Service lead time:* Average response time for processing an insurance claim
- *Throughput:* Claims processed per week
- *Efficiency:* Percent net activity value to activity cost ratio

10.7.2 Process Mapping

Following a standard process management procedure, the first step was to draw a process map. The process map is illustrated in Fig. 10.26. A simplified process map is illustrated in Fig. 10.27. The drawing of process maps helps us understand how the process worked. It was found that there were two types of insurance claims: simple claims and complex claims. The labor needs and work procedures for dealing with these two kinds of claims were quite different, as illustrated in Fig. 10.27.

The drawing and study of these process maps also helps us understand the following process performance metrics:

- *Service cost:* Average cost to process a claim
- *Service lead time:* Average response time for processing a claim
- *Throughput:* Claims processed per week

10.7.3 Process Diagnosis

Several methods were used for process diagnosis. First, the claim case types were analyzed based on historical data. Fifty-two percent of claims were of the complex claims type, and 48 percent were of the simple claims type. Roughly 90 percent of claims ended up with insurance coverage, and 10 percent of claims ended up with no coverage. This case type distribution

382 Chapter Ten

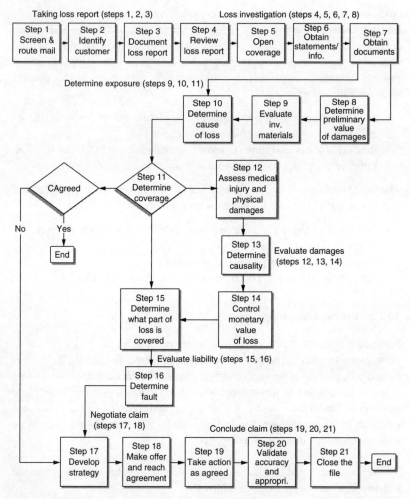

Figure 10.26 USAA Insurance Claim Processing Process Map

study helps us to understand the workload distributions for each of the process steps. It is illustrated in Fig. 10.28.

Then, cause-and-effect diagrams were used to analyze the root causes of current process problems, the low level of customer satisfaction, excessive claim processing lead time, and low throughput. Figure 10.29 is the cause-and-effect diagram for customer satisfaction. It was found that customer satisfaction was dependent on three factors: customers' perception of the fairness of the settlement, speed of payment and overall claim service quality,

Design and Improvement of Service Processes—Process Management 383

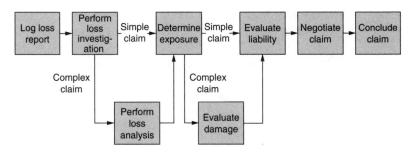

Figure 10.27 Simplified USAA Insurance Claim Processing Process Map

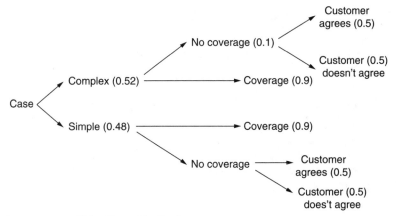

Figure 10.28 Claim Types Distribution

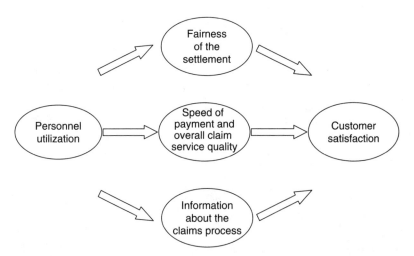

Figure 10.29 Cause-and-Effect Analysis of Customer Satisfaction

and information about the claim process. All these three factors depended on how much time relevant service providers worked with the customers. If there was an uneven workload among the service providers, some of the providers would be too busy and have very little time to explain to customers.

Figure 10.30 is the cause-and-effect diagram for the process throughput and processing lead time. Long queue times, a high work-in-process level (unfinished paperwork), and long activity processing times were the causes of the low throughput and long lead time. Inefficient workload allocation and workload balancing were the causes for the long queue times and high work-in-process level.

Another cause-and-effect analysis was conducted on cost efficiency. Based on lean operation principles, all the activities were classified as value-added or non-value-added. The following value-added activities were identified:

Customer value-added
- Identify customer
- Take action
- Access damage
- Document loss report
- Determine preliminary value of the damage

Regulatory value-added
- Open coverages
- Obtain documentation
- Determine coverages
- Make offer and reach agreement

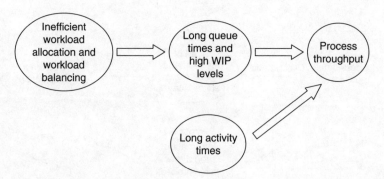

Figure 10.30 Cause-and-Effect Diagram for Throughput and Lead Time

The customer value-added activities are the activities that add value to customers; the regulatory value-added activities are the activities that are required by law or company regulations.

The non-value-added activities included

Screen and route mail
Review loss report
Obtain status and information
Evaluate investigation materials
Determine the cause of loss
Determine loss coverage
Control monetary value
Validate accuracy
Close file

Figure 10.31 is the cause-and-effect diagram for the cost efficiency. Clearly, the inefficient process flow compounded the problem of a low ratio of value-added time versus non-value-added time. Figure 10.32 illustrates the itemized cost figures for conducting value-added works versus non-value-added works.

In addition to the cause-and-effect analysis, the project team decided that a simulation study was very necessary to further analyze the process and

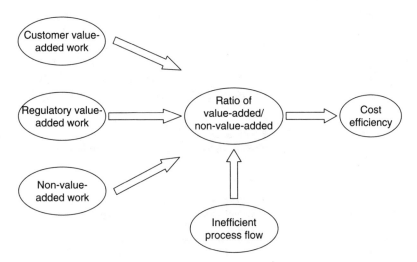

Figure 10.31 Cause-and-Effect Diagram for Cost Efficiency

386 Chapter Ten

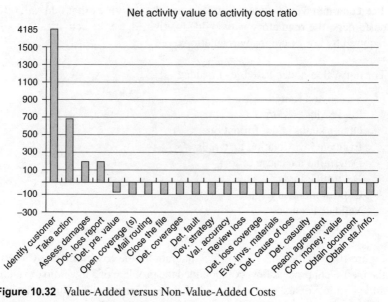

Figure 10.32 Value-Added versus Non-Value-Added Costs

determine the root cause of the current problems and to evaluate any suggested process change to improve the system performance. Specifically, the simulation analysis could

- Help develop a dynamic template for value stream analysis (resource cost: capital cost, labor cost, overhead, and activity cycle times)
- Show the effect of process changes on cost per claim, claim response time (lead time), and throughput (claims processed per week)
- Show the effect of activity processing time on cost per claim

Before the simulation study could be started, data had to be collected in order to set up the parameters for the simulation study. The following data were collected:

- WIP levels at the bottleneck activity
- Difference between highest and lowest personnel utilization of overall activities
- Throughput members served per week
- Average cost for one insurance claim
- Average cycle time to complete a claim

Table 10.3 shows a portion of simulation output on cycle times and process costs for several key activities.

Table 10.3 Simulation Results on Cycle Time and Process Costs

Activity	Cycle Time (min)	Process Cost ($/min)
Taking loss report	28.58	0.37
Perform loss investigation	52.73	0.61
Perform loss analysis	12.33	0.40
Determine exposure	12.2	0.63
Evaluate damage	18.2	1.36
Evaluate liability	14.15	0.38
Negotiate claim	12.03	0.67
Conclude claim	15.6	0.46

10.7.4 Process Design

Based on the process diagnosis conducted, the following design alternatives were developed.

    Alternative 1: As-is current USAA insurance process
    Alternative 2: Process simplification of USAA insurance process
    Alternative 3: Case type departmentalization of USAA insurance process

Alternative 1 was the current design. Further simulation analysis on this current design yielded the following results:

- Average cycle time to complete a claim was 21.4 hours (1283 minutes).
- Average number of claims in process (waiting) at the bottleneck activity was 239.
- Difference between highest and lowest personnel utilization was 93 percent.
- Throughput was 25 claims per week.

Figure 10.33 shows the service providers' utilization rate. Obviously, there were tremendous workload inbalances among different service providers.

Alternative 2 was a design simplification of the current process; it had the following features:

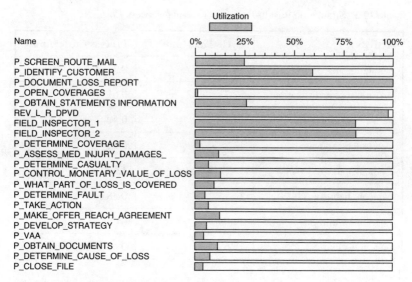

Figure 10.33 Personnel Utilization of Design Alternative 1

- Combine the "reviewing loss report and open coverage(s)," "determine cause of loss and casualty," and "develop strategy and reach agreement" activities.
- Reduce total USAA work force from 21 to 14.
- Increase the work force for the "document loss report activity" (the bottleneck) from 1 to 3.
- Eliminate denied claim call costs by moving the "determining coverage(s)" activity upward in the USAA process. Thus, an early checkpoint would be installed in the system to keep these calls from proceeding forward and adding unnecessary costs to the total claim process.

Further simulation analysis on this current design simplification yielded the following results:

- Average cycle time to complete a claim was 20.6 hours (1233 minutes).
- Average number of claims in process (waiting) at the bottleneck activity was 176.
- Difference between highest and lowest personnel utilization was 75 percent.
- Throughput was 62 claims per week.

Figure 10.34 shows the service providers' utilization rate based on simulation. Clearly, design alternative 2 had a much more balanced personnel

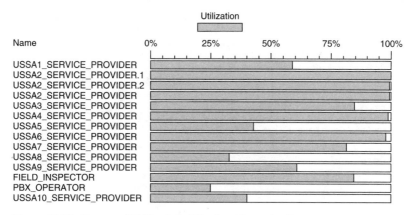

Figure 10.34 Personnel Utilization of Design Alternative 2

utilization rate; its throughput, cycle time, and work-in-process (WIP) were also improved.

Design alternative 3 was the case type departmentalization; specifically it had the following features:

- Use two types of process flow regarding two different cases. In this case, simple and complex insurance claims use two different process procedures to complete insurance claim activities.
- Increase work force assignment for "documenting loss report activity" (bottleneck of the as-is process) from 1 to 3.
- Combine "reviewing loss report" and "open coverage(s)," "determine cause of loss and casualty," and "develop strategy and reach agreement" activities all together.
- Separate the "determine preliminary value of damage" activity into two different activities:
  Field inspection
  Determining preliminary value of damage
- Reduce total USAA work force from 20 to 13.
- Move "determine coverage" activity up-front (close to the beginning of the process) in order to reduce *denied claim* costs in the process.

Further simulation analysis on this current design yielded the following results:

- Average cycle time to complete a claim was 19.2 hours (1154 minutes).
- Average number of claims in process (waiting) at the bottleneck activity was 178.

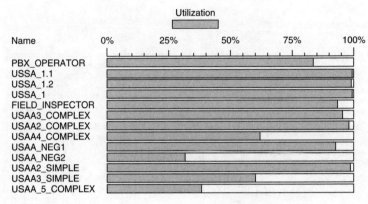

Figure 10.35 Personnel Utilization of Design Alternative 3

- Difference between highest and lowest personnel utilization was 67 percent.
- Throughput was 90 claims per week.

Figure 10.35 shows the service providers' utilization rate based on the simulation.

Design alternative 3 had an even better personnel utilization rate; its throughput, cycle time, and work-in-process (WIP) were also improved compared with design alternative 2.

Table 10.4 Comparison of Three Design Alternatives

Alternative	Evaluation Metrics		
	Throughput (Claims/Week)	Average Cost per Claim ($)	Average Lead Time (min.)
Alternative 1: as-is process	25	116.10	1283
Alternative 2: process simplification	62	105.47	1233
Alternative 3: case type departmentalization	90	102.7	1154
Importance weight	0.15	0.25	0.60

Table 10.4 lists key performance metrics for these three designs. Clearly, design alternative 3 was the winning design. Case type departmentalization could increase the throughput from the current 25 claims per week to 90 claims per week; the average cost per claim was reduced to $102.7, compared with the current level of $116.10; and the average lead time was reduced to 1154 minutes from the current 1283-minute level.

Chapter 11

Statistical Basics and Six Sigma Metrics

11.1 Introduction

Six Sigma is a *data-driven* management system with near-perfect performance objectives (Pande et al. 2000). By data-driven we mean that in Six Sigma, the real data collected in the process under study is the only source to measure the current performance, analyze the root causes for the problem, and derive improvement strategies. Near-perfect performance objectives means that in Six Sigma, we will improve the process until it achieves a very low level of defects and a very high level of performance. Clearly, it also needs the real data from the process to verify if the desired performance requirements are met.

Data analysis is a very important part of Six Sigma. In the real business and engineering process, many data collected are random variables; that is, their value will vary with some degree of uncertainty. Let us look at Example 11.1.

Example 11.1

In a semiconductor manufacturing process, we have a step where an oxide film is grown on a silicon wafer by using a furnace. In this step, a cassette of wafers is placed in a quartz "boat" and the boats are placed in the furnace. A gas flow is created in the furnace, and it is brought up to temperature and held there for a specified period of time. In this process, it is required that the most desirable oxide film thickness be 560 angstroms (Å); the specification of the oxide thickness is 560 ± 100 Å. That is, an oxidized wafer is out of specification if its thickness is either lower than 460 Å or higher than 660 Å. We collected the following film thickness data in the process:

547	563	578	571	572	575	584	549	546	584	593	567
548	606	607	539	554	533	535	522	521	547	550	610
592	587	587	572	612	566	563	569	609	558	555	577
579	552	558	595	583	599	602	598	616	580	575	

Does this process satisfy our quality requirement?

In Example 11.1, clearly the film thickness varies from wafer to wafer, so it is a random variable. A random variable is either discrete or continuous. If the set of all possible values is finite or countably infinite, then the random variable is discrete; if the set of all possible values of the random variable is an interval, then the random variable is continuous. Clearly, the film thickness variable is continuous.

The theoretical basis for modern data analysis is statistics. There are different methods in statistics that can be used to analyze data; some of them are very simple, such as descriptive statistics. Descriptive statistics can provide intuitive display and analysis of the data. Some methods are more sophisticated, such as the probability distribution models and statistical inferences; these analyses are more powerful, can provide more insights, and are able to provide credible inference and prediction about the process based on data. All popular Six Sigma performance metrics are based on the theory of statistics; therefore, familiarity with basic statistics is very essential in understanding Six Sigma metrics.

In Sec. 11.2 we review several descriptive statistical methods. In Sec. 11.3 we review several commonly used probability distribution models. In Sec. 11.4 we review some basic aspects of statistical estimation. Finally, in Sec. 12.6, we discuss Six Sigma metrics.

11.2 Descriptive Statistics

11.2.1 Graphical Descriptive Statistics

Descriptive statistics are a set of simple graphical and numerical methods that can quickly show some intuitive properties displayed in the data. The commonly used graphical descriptive statistical methods include the dot plot, histogram, and box plot.

Dot Plot

The *dot plot*, as illustrated by Fig. 11.1, is a simple yet effective diagram; each dot represents a piece of data. The dot plot can display the distribution pattern and the spread of data points.

Histogram

A *histogram* is a diagram displaying the frequency distribution. The horizontal axis is partitioned into many small segments. The number of data points (or the percentage of points) that fall in each segment is called the

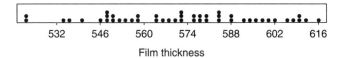

Figure 11.1 Dot Plot of Film Thickness Data

frequency and is displayed as the height of the bar corresponding to that segment. For example, the histogram for the film thickness data is displayed in the histogram illustrated in Fig. 11.2. The leftmost segment is the bracket (515, 525). In the data set, there are two data points (521 and 522) in this range, so the height of the bar is 2. We also can see that a large portion of data fall in-between 545 to 585.

Box Plot

A *box plot* is also a very useful way of displaying data. A box plot displays the minimum (lowermost point), maximum (uppermost point), median (centerline), 25th percentile (lower bar of the box), and 75th percentile (upper bar of the box). Figure 11.3 shows the box plot of the data in Example 11.1; the centerline of the box corresponds to 572, which is the median of the data. The lower bar of the box corresponds to 552, which is the 25th percentile of the data; the upper bar corresponds to 592, which is the 75th percentile of the data; the uppermost point corresponds to 616, which is the maximum of the data; and the lowermost point corresponds to 521, which is the minimum of the data.

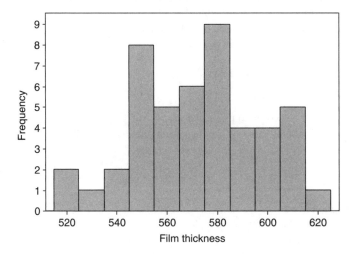

Figure 11.2 Histogram of Film Thickness Data

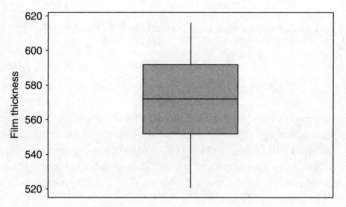

Figure 11.3 Box Plot of Film Thickness Data

11.2.2 Numerical Descriptive Statistics

Numerical descriptive statistics are numbers calculated from a data set in order to help us to create a mental image of the distribution pattern of the data. There are three types of numerical descriptive statistics:

1. The numerical measure that describes the central tendency of the data, that is, where the center of the data set is. Frequently used measures of central tendency include the mean and median.
2. The numerical measures that describe the spread of data, are also called the measures of variation. Frequently used measures of variation include variance, standard deviation, range, maximum, and minimum.
3. The numerical measures that describe the relative position of the data, are often called the measures of relative standing. The frequently used measure here is percentile points.

Measures of Central Tendency

Mean

The mean $\bar{y}$ is also called the arithmetic mean. Of a set of n measurements, $y_1, y_2, \ldots, y_n$ is the average of the measurements, specifically

$$\bar{y} = \frac{1}{n}\sum_{i=1}^{n} y_i \qquad (11.1)$$

Example 11.2
The mean for the data of Example 11.1 is

$$\bar{y} = \frac{1}{47}(547 + 563 + \cdots + 575) = 572.02$$

Clearly, the mean is simply a numerical average, which gives a good sense of where the center is for a data set. It is the most commonly used measure of central tendency. However, in some cases, it is not a preferred measure. For example, assume that in a subdivision, there are 20 families. Most of families have an annual income around $40,000; however, there is one family with an annual income of $1,000,000. If we use the arithmetic mean as the measure of central tendency, then $\bar{y}$ will be in the neighborhood of $90,000, which is by no means a family's typical midincome in this circumstance. The median would be a better measure in this case.

Median

The median of a set of measurements $y_1, y_2, \ldots, y_n$ is the middle number when the measurements are arranged in ascending (or descending) order. Specifically, let $y_{(i)}$ denote the ith value of the data set when $y_1, y_2, \ldots, y_n$ are arranged in ascending order. Then the median m is the following:

$$m = \begin{cases} y_{[(n+1)/2]} & \text{if } n \text{ is odd} \\ \dfrac{y_{(n/2)} + y_{(n/2+1)}}{2} & \text{if } n \text{ is even} \end{cases} \qquad (11.2)$$

Example 11.3
The median of the data set in Example 11.1 can be calculated as follows. First, we arrange the data in ascending order:

$$y_{(1)}, y_{(2)}, \ldots, y_{(n-1)}, y_{(n)} = 521, 522, 533, \ldots, 610, 612, 616$$

In this data set, $n = 47$, and it is an odd number, $n + 1/2 = 48/2 = 24$. It can be found that $y_{24} = 572$. Therefore, $m = 572$.

Measures of Variation

The most commonly used measures of variation are the range, the variance, and the standard deviation.

Range

The range is equal to the difference between the largest (maximum) and the smallest (minimum) measurements in a data set, specifically

$$\text{Range} = \text{maximum} - \text{minimum} \qquad (11.3)$$

Example 11.4
For the data of Example 11.1, the maximum $= y_{(n)} = y_{(47)} = 616$ and the minimum $= y_{(1)} = 521$; therefore, the range $= 616 - 521 = 95$.

The range is very easy to compute, but it only gives the distance of the two most extreme observations. It is not a good measure of variation for the whole data set. The variance and standard deviation are better measures in this aspect.

Variance

The variance of a sample of n measurements $y_1, y_2, \ldots, y_n$ is defined as

$$s^2 = \frac{1}{n-1} \sum_{i=1}^{n} (y_i - \bar{y})^2 \qquad (11.4)$$

Example 11.5
For the data of Example 11.1, the variance can be computed as

$$s^2 = \frac{1}{47-1}[(547-572.02)^2 + (563-572.02)^2 + \cdots + (575-572.02)^2] = 601.72$$

Sample variance s^2 is obviously an average of the sum of squared deviations from the mean of all observations. Squared deviation makes sense because no matter if an observation is smaller or larger than the mean, the squared deviation will always be positive. The average of the squared deviation is a measure of variation for the whole data set. However, the numerical scale and measurement unit of variance is the square of the original data. For example, if the original data is length in inches, the variance will be in the unit of squared inch, which cannot compare well with the original data.

Standard Deviation

The standard deviation is the square root of variance, specifically, the standard deviation of a sample of n measurements $y_1, y_2, \ldots, y_n$ and is defined as

$$s = \sqrt{s^2} = \sqrt{\frac{1}{n-1} \sum_{i=1}^{n} (y_i - \bar{y})^2} \qquad (11.5)$$

Example 11.6
For the data of Example 11.1, the standard deviation can be computed as

$$s = \sqrt{s^2} = \sqrt{601.72} = 24.53$$

Measure of Relative Standing

The measure of relative standing provides a numerical value or score that describes a predefined location relative to other observations in a data set.

A very commonly used measure of relative standing is the 100pth percentile, or simply called percentile points.

The *100pth percentile* of a data set is a value y located so that $100p$ percent of the data is smaller than y, and $100(1 - p)\%$ of the data is larger than y, where $0 \leq p \leq 1$.

Example 11.7
The median is the 50th percentile, because 50 percent of data points are smaller than the median and 50 percent of data are larger than the median. The 25 percent percentile is often called the lower quartile and is denoted by Q_L or Q_1; 25 percent of data will be smaller than Q_L and 75 percent of data will be larger than Q_L in a given data set. The 75th percentile is often called the upper quartile and is denoted by Q_U or Q_3; 75 percent of data will be smaller than Q_U and 25 percent of data will be larger than Q_U.

MINITAB can compute all types of descriptive statistics conveniently. The following MINITAB output is the printout of the descriptive statistics for the data set of Example 11.1.

Descriptive Statistics: Film Thickness

```
Variable          N   N*    Mean  SE Mean  StDev  Minimum      Q1
Film Thickness   47    0  572.02     3.58  24.53   521.00  552.00
                                                   Median      Q3
                                                   572.00  592.00


Variable       Maximum
Film Thickness  616.00
```

11.3 Random Variables and Probability Distributions

The data set collected in a process, such as the data set described in Example 11.1, is called a *sample* of data, because it only reflects the reality of a snapshot of the process. For example, the data set in Example 11.1 is only a small portion of the production data. If we are able to collect all the film thickness data for all wafers in the whole life cycle of the oxidation furnace, then we collected a whole *population* of data. In real-world business decision making, the population is of more interest for the decision makers. We are definitely more interested in the overall quality level for the population. Random variables and probability distributions are the mathematical tools used to describe the behavior of populations.

A random variable can be defined as a variable that takes different values following some specific probability distribution. A random variable is a discrete random variable if it can take only a countable number of values. A simple example of a discrete random variable is the number of points that a fair six-sided die will show on a toss, y. It can only be 1, 2, 3, 4, 5, or 6, with equal probabilities. A random variable can also be a continuous random variable if it can take all real numbers in a given interval. For example, a random person's height (that is, a person you randomly met on a street) is a continuous random variable.

The probability structure of a random variable, say y, is described by its probability distribution. If y is a discrete random variable, then its probability distribution is described by the probability function, often denoted by $p(y)$. If y is a continuous random variable, its probability distribution is described by its probability density function, often denoted by $f(y)$.

The properties of the probability function $p(y)$ and probability density function $f(y)$ are summarized as follows:

Discrete random variable y

$0 \leq p(y_i) \leq 1$ y_i's are the possible values that y can take

$P(y = y_i) = P(y_i)$

$\sum_{\text{All}-y_i} p(y_i) = 1$

Continuous random variable y

$$f(y) \geq 0$$
$$P(a \leq y \leq b) = \int_a^b f(y)\,dy$$
$$\int_{-\infty}^{+\infty} f(y)\,dy = 1$$

11.3.1 Expected Value, Variance, and Standard Deviation

The random variable and probability distribution deal with the population. The mean of the population (or population mean μ) is the most frequently used measure of central tendency for the population; μ is also called the expected value of the random variable y.

The expected value $E(y) = \mu$ is defined as follows:

$$\mu = E(y) = \begin{cases} \sum_{\text{All}-y} y p(y) & \text{if } y \text{ is discrete} \\ \int_{-\infty}^{+\infty} y f(y)\,dy & \text{if } y \text{ is continuous} \end{cases}$$

The population variance is often denoted by σ^2, and it is often simply called the variance. The variance of a random variable y is defined as follows:

$$\sigma^2 = \begin{cases} \sum_{\text{All-}y} (y-\mu)^2 p(y) & \text{if } y \text{ is discrete} \\ \int_{-\infty}^{+\infty} (y-\mu)^2 f(y) \, dy & \text{if } y \text{ is continuous} \end{cases}$$

Variance is often denoted by $V(y)$. From the definition of expected value, it is clear that

$$\sigma^2 = \text{Var}(y) = E[(y-\mu)^2]$$

The population standard deviation is often called the standard deviation σ, where σ is simply the square root of variance σ^2, that is,

$$\sigma = \sqrt{\sigma^2}$$

In Six Sigma–related applications, there are several probability distribution models that are often used as the basis for data-driven decision making. They are the normal distribution, exponential distribution, binomial distribution and Poisson distribution. We give an overview for each of these distributions.

Normal Distribution

The normal distribution (or Gaussian distribution) was first proposed by Gauss (1777–1855). It is often used to model the probability distribution of continuous variables that have the following properties:

1. There are many random factors that can affect the value of the random variable.
2. Each of these random factors has relatively small influence on the random variable; there is no dominate factor.

The normal distribution is the most popular distribution in quality engineering and Six Sigma. It is often used to model the follow random variables:

1. Quality characteristic of parts from suppliers
2. Students' test scores or employee performance scores

The probability density function of the normal distribution is as follows:

$$f(y) = \frac{1}{\sqrt{2\pi}\sigma} e^{-\frac{1}{2}\left(\frac{y-\mu}{\sigma}\right)^2} \quad -\infty < y < +\infty$$

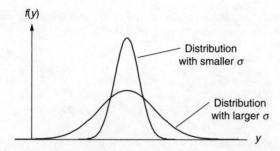

Figure 11.4 Normal Probability Density Curve

For the normal distribution,

$$E(y) = \mu \quad \text{and} \quad \text{Var}(y) = \sigma^2$$

A normal random variable y with $E(y) = \mu$ and $\text{Var}(y) = \sigma^2$ is denoted by $N(\mu, \sigma^2)$. The probability density function $f(y)$ displays a bell-shaped curve as illustrated by Fig. 11.4. The distribution is centered at μ, and the smaller σ results in a tighter curve and vice versa.

An important special case of the normal distribution is the standard normal distribution. In the standard normal distribution, $\mu = 0$ and $\sigma^2 = 1$. The standard normal random variable is often denoted by $z \sim N(0, 1)$. The standard normal distribution table is mainly used to calculate probabilities for all kinds of normal distributions.

Figure 11.5 shows that if $y \sim N(\mu, \sigma^2)$, then $P(\mu - \sigma \leq y \leq \mu + \sigma) = P(-1 \leq z \leq 1) = 0.6826 = 68.27\%$; that is, 68.27 percent of observations from a

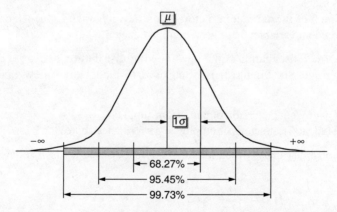

Figure 11.5 Percentage Distribution Properties of Normal Random Variable

normal population will locate within a distance of one standard deviation from the mean. Similarly, $P(\mu - 2\sigma \leq y \leq \mu + 2\sigma) = P(-2 \leq z \leq 2) = 0.9545 = 95.45\%$, that is; 95.45 percent of observations from a normal population will locate within a distance of two standard deviations from the mean. In addition, $P(\mu - 3\sigma \leq y \leq \mu + 3\sigma) = P(-3 \leq z \leq 3) = 0.9973 = 99.73\%$; that is, 99.73 percent of observations from a normal population will locate within three standard deviations distance from the mean.

Exponential Distribution

An exponential distribution is featured by the following probability density function:

$$f(y) = \frac{e^{-y/\beta}}{\beta} \quad \text{for } 0 \leq y < \infty$$

with a mean and variance of

$$E(y) = \mu = \beta \quad \text{and} \quad \text{Var}(y) = \beta^2$$

The exponential distribution is often used to model the following:

1. Lifetime of some electronic components
2. Interarrival time of customers entering a service facility
3. Time between consecutive machine failures or earthquakes

Binomial Distribution

The binomial distribution is a discrete probability distribution that characterizes a binomial random variable. The binomial random variable can be used for the following situation:

1. There are n successive trials. Each trial will only have two distinct outcomes, S (success) or F (failure).
2. The probability of success $P(S) = p$, and $P(F) = 1 - P(S) = 1 - p$.
3. The result of each trial will not affect the results of any other trials.

If this situation is true, then the number of successes y out of n trials will be a binomial random variable and its probability function $p(y)$ will be

$$p(y) = \frac{n!}{y!(n-y)!} p^y (1-p)^{n-y} \quad y = 0, 1, \ldots, n$$

The mean and variance of the binomial random variable are

$$E(y) = \mu = np \quad \text{and} \quad \text{Var}(y) = \sigma^2 = np(1-p)$$

This binomial distribution is often denoted by $y \sim B(n, p)$. The following are examples of binomial random variables:

1. The number of defective parts, y, in a lot of n parts in a sequential quality inspection
2. The number of positive customer responses, y, in a survey involving n customers

Poisson Distribution

The Poisson probability distribution provides a model for the probability of occurrence of the number of rare events that happen in a unit of time, area, volume, and so on. Actually, the Poisson distribution is an extreme case of the binomial distribution, where n is very large and p is very small. That is, the probability of a rare event occurrence $p = P(S)$ is very small, but the number of trials n is very large.

In the Poisson distribution, the parameter λ ($\lambda = np$) is used. The probability function of the Poisson distribution $p(y)$ is

$$p(y) = \frac{\lambda^y e^{-\lambda}}{y!} \quad y = 0, 1, 2, \ldots$$

The mean and variance of the Poisson distribution are

$$E(y) = \lambda \quad \text{and} \quad \text{Var}(y) = \sigma^2 = \lambda$$

11.3.2 Statistical Parameter Estimation

All probability distribution models depend on population parameters, such as μ and σ^2 in the normal distribution, and p in the binomial distribution. Without these parameters, no probability distribution model can be used. In real-world applications, these population parameters are usually not available; however, statistical estimates of these population parameters can be computed based on a sample of data from the population.

The commonly used statistical estimate for μ in the normal distribution is the sample mean $\bar{y}$, where

$$\bar{y} = \frac{1}{n}\sum_{i=1}^{n} y_i$$

For a sample of n observations $y_1, y_2, \ldots, y_n$, from $y \sim N(\mu, \sigma^2)$.

The commonly used statistical estimate for σ^2 in the normal distribution is the sample variance s^2, where

$$s^2 = \frac{1}{n-1}\sum_{i=1}^{n}(y_i - \bar{y})^2$$

for a sample of n observations $y_1, y_2, \ldots, y_n$ from $y \sim N(\mu, \sigma^2)$ and $\bar{y} = \frac{1}{n}\sum_{i=1}^{n} y_i$.

The commonly used statistical estimates for p in a binomial distribution $B(n, p)$ is the sample ratio $\hat{p}$:

$$\hat{p} = \frac{y}{n}$$

where y is the actual number of successes (S) in n trials.

However, statistical estimates are only approximations of the true population parameters. When the sample size is small, there will be substantial discrepancies between population parameters and statistical estimates. As the sample size becomes larger, the discrepancies will get smaller.

11.4 Quality Measures and Six Sigma Metrics

For any product or business process, there are always performance metrics that we want to measure and improve. For example, in a loan approval process, the cycle time (the time between the application and the loan decision) is a performance metric. In Example 11.1, the oxide film thickness is a performance metric and the ideal thickness is 560 Å. Most of the actual process performance metrics are also random variables; clearly, the cycle time of each loan application is a random variable, and the oxide film thickness is also a random variable.

There are many *quality measures* that have been developed to measure the process performance with the presence of randomness. Quality measures compare the degree of randomness in the process performance and compare the degree of randomness with the process performance specification. The most commonly used process performance quality measure is the process capability index. In Six Sigma practice, many other process performance–related metrics have also been developed, such as Sigma quality level and DPMO (defects per million opportunities).

In this section, we first discuss the process capability index, and then we discuss other Six Sigma metrics.

11.4.1 Process Performance and Process Capability

Process performance is a measure of how well a process performs. It is measured by comparing the actual process performance level versus the ideal process performance level. For the oxide film building process, performance may be measured by the oxide film thickness, and its ideal performance level would be 560 Å in thickness. For most processes, the performance level is not constant. We call this variation the process variability. If the process performance can be measured by a real number, then the process variability can usually be modeled by the normal distribution, and the degree of variation can be measured by the standard deviation of that normal distribution.

If process performance level is not a constant but a random variable, then we can use the process mean and process standard deviation as key performance measures. The mean performance can be calculated by averaging a large number of performance measurements.

If processes follow the normal probability distribution, a high percentage of the process performance measurements will fall between $\pm 3\sigma$ of the process mean, where σ is the standard deviation. That is, approximately 0.27 percent of the measurements would naturally fall outside the $\pm 3\sigma$ limits and the balance of them (approximately 99.73 percent) would be within the $\pm 3\sigma$ limits.

Since the process limits extend from -3σ to $+3\sigma$, the total spread amounts to about 6σ total variation. This total spread is often used to measure the range of process variability, also called the process spread.

For any process performance measure, there are usually some performance specification limits. For example, if the oxide film thickness in a wafer is too high or too low, then the wafer will not function well. Suppose it is required that deviation from the target value of 560 Å cannot be more than 100 Å; then the specification limits would be 560 ± 100 Å, or we say that its specification spread is (460, 660), where 460 Å is the lower specification limit (LSL) and 660 Å is the upper specification limit (USL).

11.4.2 Process Capability Indices

Capability indices are simplified measures that quickly describe the relationship between the variability of a process and the spread of the specification limits.

The Capability Index C_p

The equation for the simplest capability index C_p is the ratio of the specification spread to the process spread, the latter represented by six standard deviations or 6σ.

$$C_p = \frac{\text{USL} - \text{LCL}}{6\sigma}$$

When using C_p we assume that the normal distribution is the correct model for the process.

The capability index C_p can be translated directly to the percentage or proportion of nonconforming product outside specifications, if the mean of the process performance is at the center of the specification limit.

When $C_p = 1.00$, approximately 0.27 percent of the parts are outside the specification limits (assuming that the process is centered on the midpoint between the specification limits) because the specification limits closely match the process UCL and LCL. We say this is about 2700 parts per million (ppm) nonconforming.

When $C_p = 1.33$, approximately 0.0064 percent of the parts are outside the specification limits (assuming the process is centered on the midpoint between the specification limits). We say this is about 64 ppm nonconforming. In this case, we would be looking at normal curve areas beyond $1.33 \times 3\sigma = \pm 4\sigma$ from the center.

When $C_p = 1.67$, approximately 0.000057 percent of the parts are outside the specification limits (assuming the process is centered on the midpoint between the specification limits). We say this is about 0.6 ppm nonconforming. In this case, we would be looking at normal curve areas beyond $1.67 \times 3\sigma = \pm 5\sigma$ from the center of the normal distribution.

The Capability Index C_{pk}

The major weakness in C_p is that, for many processes, the mean performance is not equal to the center of the specification limit; also many process means will drift from time to time. When that happens, the probability calculation about nonconformance will be totally wrong when we still use C_p. Therefore, one must consider where the process mean is located relative to the specification limits. The index C_{pk} is created to do exactly this.

$$C_{pk} = \text{Min}\left\{\frac{\text{USL}-\mu}{3\sigma}, \frac{\mu-\text{LSL}}{3\sigma}\right\} = \text{Min}\{C_{\text{PU}}, C_{\text{PL}}\}$$

We have the following situation. The process standard deviation is $\sigma = 0.8$ with a USL = 24, LSL = 18, and the process mean $\mu = 22$.

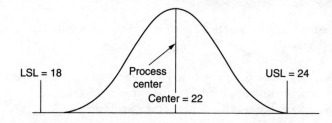

$$C_{pk} = \text{Min}\left\{\frac{24-22}{3\times 8}, \frac{22-18}{3\times 8}\right\} = \text{Min}(0.83, 1.67) = 0.83$$

It is also clear that

$$C_{\text{PU}} = 0.83 \quad \text{and} \quad C_{\text{PL}} = 1.67$$

If the process mean is exactly centered between the specification limits,

$$C_p = C_{pk} = 1.25$$

Example 11.8

For the film thickness data given in Example 11.1, LSL = 460 and USL = 660. We do not know the exact value of μ and σ; however, we can calculate that $\bar{y} = 572.02$ and $s = 24.53$. Because the sample size of this data set is fairly large ($n = 47$), we can substitute μ and σ by using $\bar{y}$ and s. Then we have

$$C_{pk} = \text{Min}\left\{\frac{660-572.02}{3\times 24.53}, \frac{572.02-460}{3\times 24.53}\right\} = \text{Min}(1.19, 1.51) = 1.19$$

MINITAB can be used to conduct a comprehensive process capability analysis. The following MINITAB output is the process capability analysis for the film thickness data.

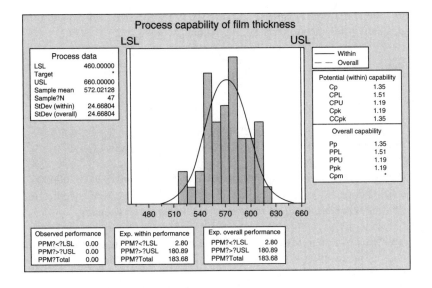

11.4.3 Sigma Quality Level (Without Mean Shift)

In 1988, the Motorola Corporation was the winner of the Malcolm Baldrige National Quality Award. Motorola bases much of its quality effort on its Six Sigma program. The goal of this program was to reduce the variation in every process to such an extent that a spread of 12σ (6σ on each side of the mean) fits within the process specification limits.

Figure 11.6 gives a graphical illustration of this Six Sigma quality. If the actual variation is measured by standard deviation σ, 6σ quality means that

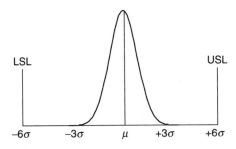

Figure 11.6 Normal Distribution and Six Sigma Quality

the total spread of the specification is six times the standard deviation on each side of the mean. For a Six Sigma quality level,

$$C_p = \frac{\text{USL} - \text{LSL}}{6\sigma} = \frac{12\sigma}{6\sigma} = 2$$

By using the normal probability distribution, it can be computed that

$P(y$ will be in specification$) = P(\text{LSL} \leq y \leq \text{USL}) = P(-6 \leq z \leq 6)$
$= 0.999999998 = 99.9999998\%$

Clearly, $P(y$ will be out of specification$) = 1 - 0.9999999998 = 0.000000002$, or 0.002 defective parts per million. Similarly, if the spread of specification is 5 times σ on each side of the mean, it is called 5 sigma quality.

Table 11.1 summarizes the relationship between C_p, Sigma quality level (without mean shift), percentage in specification, and defective ppm.

11.4.4 Sigma Quality Level (With Mean Shift)

In most actual processes, the process mean $\bar{y}$ is not usually a constant. The process mean $\bar{y}$ will shift from time to time. For example, in a manufacturing process, with the change of raw material, operator, the process may suddenly change its mean level. In a service process, with a change of server, shift, and the process mean may also change. In order to take into account this mean shift effect, Motorola allocates 1.5σ on either side of the process mean for shifting of the mean. For a Six Sigma quality level, with

Table 11.1 The Relationship Between Sigma Quality Level, Process Capability and Defective Levels

Sigma Quality Level (Without Mean Shift)	C_p	Percentage in Specification	Defective ppm
1	0.33	68.27	317,300
2	0.67	95.45	45,500
3	1.0	99.73	2,700
4	1.33	99.9937	63
5	1.67	99.999943	0.57
6	2.00	99.9999998	0.002

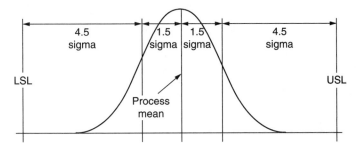

Figure 11.7 Six Sigma Quality Level with 1.5 Sigma Mean Shift

Table 11.2 The Relationship Between Sigma Quality Level, Process Capability and Defective Levels with 1.5 Sigma Mean Shift

Sigma Quality Level (With Mean Shift)	C_p	Percentage in Specification	Defective ppm
1	0.33	30.23	697,700
2	0.67	69.13	308,700
3	1.0	93.32	66,810
4	1.33	99.3790	6,210
5	1.67	99.97670	233
6	2.00	99.999660	3.4

the maximum possible mean shift of 1.5σ, the minimum distance from process mean to one of the specification limits could be as small as 4.5σ. Figure 11.7 illustrates the relationship between the mean shift and Six Sigma quality level.

Thus, even if the process mean strays as much as 1.5σ from the process center, a full 4.5σ remains. This ensures a worst-case scenario of 3.4 ppm nonconforming on each side of the distribution. With the inclusion of a 1.5σ mean shift, for the same Six Sigma quality level, the defective ppm will be much larger than that without considering the mean shift.

Table 11.2 summarizes the relationship between C_p, Six Sigma quality level (with mean shift), percentage in specification, and defective ppm.

Chapter 12

Theory of Constraints

12.1 Introduction

For all profit-earning corporations, it is natural that the goal of the corporation is to make as much profit as possible for now and in the future. Moneymaking is also a process; there is also a process management problem in running, improving, and possibly redesigning this moneymaking process. Naturally there are several questions about this moneymaking process:

1. How does this moneymaking process work?
2. What is the determining factor for the capacity of this process?
3. If we want to make more money, what is the most efficient way to improve the process?

The theory of constraints (Golratt and Cox 1986, Goldratt 1990) tries to answer these questions. Goldratt and Cox (1986) wrote a book titled *The Goal*. This book is in a novel format and describes the life of a plant manager who struggles to simultaneously manage his plant and his marriage. The term "theory of constraints" is not mentioned, but the main ideas of this theory are discussed in bits and pieces. The following terms are often mentioned in *The Goal*:

- Bottlenecks
- Throughput
- Inventory
- Return on investment
- Cash flow
- Socratic way
- Fear of change

The Goal also reminds readers that there are three basic measures used in the evaluation of the moneymaking process:

- Throughput
- Inventory
- Operational expenses

In a manufacturing plant circumstance, Goldratt and Cox think that these measures are more relevant in moneymaking than frequently used performance measures such as machine efficiency, equipment utilization, and downtime.

In *The Goal*, the following basic concepts of the theory of constraints are outlined (Goldratt and Cox 1986):

- Bottleneck resources are "resources whose capacity is equal to or less than the demand placed upon it. A non-bottleneck is any resource whose capacity is greater than the demand placed on it." If a resource presents itself as a bottleneck, then things must be done to lighten the load. Some of the appropriate steps might be to off-load material to relieve a bottleneck or to work only on the parts in the bottleneck that are needed now. Beware of lost production at a bottleneck due to poor quality or rejects.
- Balanced plants are perhaps not a good thing. Do not balance capacity with demand, but "balance the flow of product through the plant with demand from the market." The plant may be capable of generating inventories and goods at record levels, but this may jam up the plant's system. The idea is to make the flow through the bottleneck equal to market demand. One can do more with less by just producing what the market requires at the time. It is possible that the existing plant has more than enough resources to do any job, but the flow must be controlled.
- Dependent events and statistical fluctuations are important. A subsequent event depends on the one prior to it. The story of Herbie and the local scout pack describes how the slowest member of a group restrains the pace of the group (Fig. 12.1). Similarly, a bottleneck restrains the entire throughput.

Figure 12.1 The Slowest Member of a Team Sets the Speed for the Whole Team

The Goal also defines the following terms:

Throughput "The rate at which the system generates money through sales." The finished product must be sold before it generates money.

Inventory "All the money that the system has invested in purchasing things that it intends to sell." This can also be defined as solid investments or patents.

Operation Expenses "All the money that the system spends in order to turn inventory into throughput." This includes depreciation, lubricating oil, scrap, and carrying costs.

The terms *throughput*, *inventory*, and *operational expenses* can be simply defined as incoming money, money stuck inside, and money going out, respectively.

In his 1990 book *Theory of Constraints*, Goldratt provided more details about his theory; he proposed the following five-step method:

1. Identify the process constraints. A process constraint limits the firm from achieving its performance and goals. Thus constraints must be identified and prioritized for impact.
2. Decide how to exploit the process constraints. The nonconstraints in the process should be managed properly so that resources or materials are provided to feed the constraints.
3. Subordinate everything else to achieve the decisions in steps 1 and 2. Constraints may have a limit, so look for ways to reduce the effects of a constraint, or look to expand the capacities of the constraints.
4. Elevate the system's constraints. Try to eliminate the problems of the constraints. Strive to keep improving the system.
5. Go back to step 1. After the constraint has been broken, go back to step one and look for new constraints.

After Goldratt's groundbreaking books, many researchers and practitioners (for example, Goetsch and Davis 2000, Stein 1996, Woeppel 2001) applied, developed, and improved the theory of constraints.

The theory of constraints can be applied to many service processes beside that of the manufacturing sector and achieve significant gains in revenue and profitability. In the Design for Six Sigma practice, the theory of constraints can be used with process management as a process diagnosis tool to pinpoint the limitation and bottleneck of the service process under study and generate improvement ideas for the process. We first discuss some fundamental concepts of the theory of constraints in Sec. 12.2. Section 12.3

discusses the practical implementations of the theory of constraints. Because process improvement deals with many possible changes, Sec. 12.4 addresses the issues of change management.

12.2 Basic Concepts in the Theory of Constraints

In the theory of constraints, every business entity is a "moneymaking machine," or a moneymaking process. A moneymaking process is like a river; the money is like water, and the flow direction is from the customers' end to the company (business entity). Clearly, the key for this moneymaking process to work well is to "make the water flow." Figure 12.2 and Example 12.1 illustrate such an example for a restaurant.

> **Example 12.1: A Restaurant as a Moneymaking Machine**
> This moneymaking process is a restaurant. The customers are the "sources of water." The first step is marketing; if the marketing is not very good, then there will not be a sufficient flow of customers into the process (the restaurant). Fortunately, the marketing was good, so the customer arrival rate is 200 per hour. The second process step is the dining hall; the dining hall only has enough capacity to serve 100 customers per hour. The third process step is the kitchen; the kitchen has the capacity to make 200 meals per hour. Obviously, the dining hall capacity is a bottleneck, or a constraint. For this moneymaking process, the maximum flow of money cannot exceed the flow rate allowed by the bottleneck, which is 100 customers per hour. Therefore, the throughput for this process is 100 customers per hour. In order to make more money, the first thing this restaurant needs to do is to increase the capacity of the dining hall; every unit of capacity improvement will create one unit of improvement in the throughput. Before this happens, any improvement in the kitchen or in marketing will not help anything.

The theory of constraints is based on several important concepts. One is its throughput-based operation performance measures; the other is constraints and constraints management. In this section, we discuss these in detail.

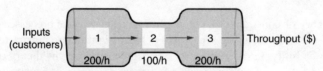

1. Marketing: Capacity 200 customers/hour
2. Dining hall: Capacity 100 customers/hour
3. Kitchen: Capacity: 200 meals/hour

Figure 12.2 A Restaurant as a Moneymaking Process

12.2.1 Throughput, Inventory, and Operating Expense

In the theory of constraints, there are three key measures for the moneymaking process: throughput, inventory, and operating expense. The definitions of these concepts are slightly different than those people usually refer to.

Throughput The money produced by the system is called throughput. By adding the time factor, throughput is also defined as the rate at which money is generated by sales. The regular definition of throughput often refers to the production output per unit time. In the theory of constraints, the production output is not a throughput unless a consumer purchases it. Throughput is also not gross revenue, because some of the revenue might be simply components purchased from suppliers and their sales simply pass through our process and do not add any value to us. In the theory of constraints, the throughput is calculated by taking gross revenue minus all totally variable expenses, such as purchased material cost, sales commissions, and any subcontract expenses.

Inventory Inventory is the money captured (locked) within the process. From the viewpoint of the theory of constraints, not only the parts, unused raw material, unsold goods, but also all of the assets (buildings, equipment, and so on) are considered as inventory.

Operating Expense Operating expense is all the money spent to turn inventory into throughput. It includes all direct or indirect payroll expenses, all supplies, and all overheads. In other words, all expenses related to time are operating expenses. In general, the operating expense is the real money you take from your pockets to produce products or services to satisfy customers.

These three measures are related to the regular financial performance measures, such as net profit, rate of return, and productivity. The detailed relationships are given as follows:

$$\text{Net profit} = \text{throughput} - \text{operating expense} \tag{12.1}$$

$$\text{Rate of return (ROI)} = \frac{\text{throughput} - \text{operating expense}}{\text{inventory}} \tag{12.2}$$

$$\text{Productivity} = \frac{\text{throughput}}{\text{inventory}} \tag{12.3}$$

Based on these relationships, Fig. 12.3 shows the relationships between throughput, inventory, operating expense, and profit.

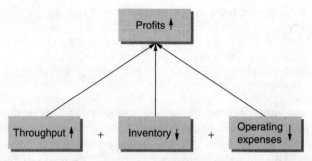

Figure 12.3 Relationships between Throughput, Inventory, Operating Expenses, and Profit

Clearly, high profitability can be achieved by increased throughput and decreased inventory and operating expense. The theory of constraints believes that the throughput is always limited by at least one constraint, as we illustrated by Example 12.1, so the throughput will not be able to increase unless the corresponding constraint is elevated, that is, the bottleneck is enlarged. Therefore, the most efficient and economical way of increasing throughput is to work on enlarging the bottleneck. Consequently, the best way to increase the profitability is to enlarge the bottleneck. Any improvement on nonbottleneck activities will not improve the throughput; therefore, it will not improve the profitability.

12.2.2 Constraints

A constraint or bottleneck is defined as any department, workstation, or operation that restricts the flow of product through the production system. Constraints management is crucial in improving process throughput and profitability. There are three types of constraints: policy, resource, and material. Each type of constraint has a different impact on the process and should be managed differently.

Policy Constraints

A policy is a rule, a measurement, or condition that dictates organizational behavior. According to Woeppel (2001), policy constraints are the most frequently encountered constraints and are the least expensive to fix. Batch size rule, resource utilization rules, and project management policy are all considered as policy constraints.

Policy constraints cannot be spotted directly, but a shortage of resources (material, machine, time, and so on) in some process steps may lead to the discovery of a policy constraint. Many times, a policy constraint is often mistaken as a resource constraint. Example 12.2 gives a very common policy constraint in the software development process.

Example 12.2: A Policy Constraint in Software Development
Many software companies suffer from long product development time and cost overruns. On the surface, many of these problems appear to be caused by the shortage of labor and resources. These bottlenecks delay the software release time to the market, so they directly reduce the throughput. However, many researchers indicate that most of these constraints are actually policy constraints. One very common policy constraint is the practice of multitasking, that is, where one programmer is involved in many software development projects simultaneously. If the programmer is involved in too many projects, then multitasking can delay the project completion time, so this practice is often a policy constraint. The following examples show how multitasking practice can delay the software project completion times.

Figure 12.4 shows three software development projects. Each has 10 modules, that is, module A through module J. Assume each module needs 2 weeks to be completed and that each project will take 20 weeks. However, in these three software development projects, there are three types of modules. Modules A, B, and C are type-1 modules and need to be written by one type of programmer (resource 1); modules D, E, F, and G are type 2 and need to be written by another type of programmer (resource 2); and modules H, I, and J are type 3 and need to be written by yet another type of programmer (resource 3). The company only has one programmer for each type, so these three projects cannot be done as described in Fig. 12.4. So the company can use the multitasking approach described in Fig. 12.5, in which each programmer (resource) is involved in all three projects simultaneously and switches between projects all the time. We can see that each software development project will take from 48 weeks up to 52 weeks to finish with the multitasking approach.

Figure 12.6 shows a nonmultitasking approach; each programmer is involved in one project at a time. We can see this approach actually reduces the software

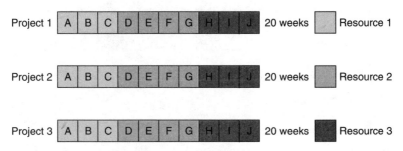

Figure 12.4 Three Software Development Projects

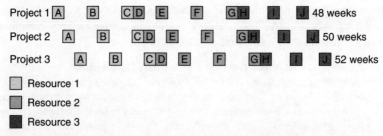

Figure 12.5 Three Software Development Projects, Multitasking

development time for all three projects, with exactly the same amount of resources (programmers). Clearly, this example shows that the multitasking practice slows down the project completion times, so it is a policy constraint.

Resource Constraints

Resource constraints are resource-related constraints. The resources are usually machines, people, skills, and market. The market constraint means that the market demand is less than the throughput of the process. According to Woepple (2001), resource constraints are less common than policy constraints. Many times, policy constraints are mistaken as resource constraints.

Material Constraints

Material constraints involve a shortage of scarce material. According to Woepple (2001), the material constraints are the least common. However, they do happen occasionally. In the year 2003, some types of steel became a scarce commodity for the automotive industry.

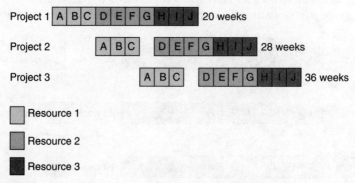

Figure 12.6 Three Software Development Projects, No Multitasking

The heart of the theory of constraints is constraints management. Constraints management is the systematic approach to identifying, managing, and loosening up the binding constraints and moving the moneymaking process to a new level. Constraints management will be discussed in Sec. 12.3.

12.3 Theory of Constraints Implementation Process

The practical implementation process of the theory of constraints is featured by the following five-step constraints management process:

1. Identify the constraint(s)
2. Decide how to exploit the identified constraint(s)
3. Subordinate everything else in the process to step 2
4. Elevate the system's constraint(s)
5. Go back to step 1

Now we describe these steps for a constraint.

12.3.1 Identify the Constraint

Identifying the constraint is very essential. The answers to the following questions will help to pinpoint the constraint:

1. Where is the work backed up? Where are the piles of work waiting?
2. Where do most problems seem to originate?
3. Are there any resources with high utilization? Which one has the highest resource utilization rate? If it is a constraint, it will never run out of work and it will always be behind.

The effects of a constraint are always blockage of the resources proceeding to the constraint resource (resulting in a queue) and starvation of resources downstream of the constraint resource, as illustrated by Fig. 12.7.

These additional questions should be answered to verify if it is really a constraint:

1. If we add another resource, would the output of the facility increase?
2. When this resource is starved or idle, will the entire production plan be thrown off?

Example 12.3: Bottleneck in a Hospital
Figure 12.8 is a flowchart that shows a simplified version of a patient's pathway through the hospital process. It is clear from this that general practitioner (GP)

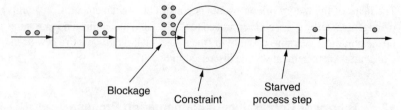

Figure 12.7 Symptoms of a Constraint

referral, appointment making, joining the waiting lists, and discharge in this mock system are all steps where large numbers of patients can be processed within a given time. The outpatient consultation and the follow-up visit are lower-volume steps where fewer patients are dealt with in the same period of time. The lowest throughput of all, however, is at the surgery stage. This is the step that constitutes the bottleneck or constraint in this fictitious patient pathway. No matter how many more patients are being dealt with at any of the other stages, the process cannot be speeded up so long as the surgery stage remains incapable of increasing its throughput.

No matter how hard clinicians and managers in this particular example try to improve throughput elsewhere in the system, they will never succeed in driving down waiting lists if the surgery stage remains incapable of processing more patients in a given time. In fact, any efforts to improve matters could actually lead to bigger waiting lists for surgery.

12.3.2 Decide How to Exploit the Identified Constraint

Once you have found where the constraint is, you should decide what to do with it. First, you may have to do more investigation on what kind of constraint it is. Is it a policy, resource, or material constraint? The following are some ideas of how to deal with the constraint:

- If it is a policy constraint, find a better way to do the job, such as was illustrated in Example 12.2.
- Increase the capacity of the constraint.
- Ensure well-trained and cross-trained employees are available to operate and maintain this constraint.

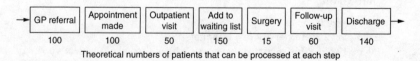

Figure 12.8 A Constraint in a Hospital

- Develop alternate routings, processing procedures, or subcontractors.
- Move inspections and tests to a position just before the constraint in order not to add additional strain to the bottleneck.

In the constraint or bottleneck, every hour of wasted time, whether it is downtime or idle time, is a lost hour of the whole production system. So every technique should be used to reduce every form of wasted time. Employing well-trained people to run the bottleneck is very essential. According to a story by Woepple (2001), in one of his consulting projects, he found a bottleneck process step. Unfortunately, this most important step in the whole production system was run by the lowest-paid people and the employee turnover in this step was really high, which made the problem even worse. When the pay structure was changed and the people in this position became the highest paid, the employee turnover was greatly reduced and the throughput was greatly increased.

12.3.3 Subordinate Everything Else in the Process to the Constraint

If you cannot increase the capacity of the bottleneck, you have to live with it. Your throughput will not be greater than the capacity of the bottleneck. The best you can hope for is that the throughput is exactly equal to the bottleneck capacity. In this case, the plan will be to achieve the production pace equal to, and not less than, the bottleneck capacity. An important precondition to achieve this is the smooth production flow from the beginning of the process to the end, with exactly the rate of the bottleneck capacity. Any fluctuation of the flow rate will cause more blockage and starving, thus reducing the throughput.

The following are some basic rules in the theory of constraints (Goldratt and Cox 1986, Goldratt 1990):

1. An hour lost at a bottleneck is an hour lost for the entire system.
2. An hour saved at a nonbottleneck is a mirage.
3. Do not balance the capacity; balance the flow.
4. The level of resource utilization of a nonbottleneck is not determined by its own potential but by some other constraint in the system.
5. Bottlenecks govern both throughput and inventory in the system.
6. Priorities can be set only by examining the system's constraints. Lead times are the result of a schedule.

Goldratt also proposed a production control strategy based on the theory of constraints. He called it the *drum-buffer-rope (DBR)* strategy (Goldratt and Cox 1986, Goldratt 1990, Mabin and Balderstone 2000). This strategy is based on the following ideas:

- Drum-buffer-rope (DBR) is a production control technique used to implement the theory of constraints.
- If a system has a bottleneck, its production rate controls the pace of the system. Its beat drives the system, hence the name *drum* for this control point.
- A buffer is placed in front of the bottleneck to protect the bottleneck from fluctuations and variations in the feeding rate to the bottleneck. The buffer size is measured in the standard time required for the bottleneck to process all items in the buffer.
- The buffer is connected to the raw material dispatching point via a feedback loop called the *rope*. The dispatching point will release only that amount that will keep the buffer inventory build up.

Figure 12.9 shows how this drum-buffer-rope strategy works. In this drum-buffer-rope production control system, the drum is the pace maker because the optimized production rate is exactly equal to the bottleneck capacity, so the bottleneck capacity is the pace of the production. For all the process steps before and after the bottleneck, it is also ideal to set the paces of production rates equal to that of the bottleneck in order to avoid blockages of resources, clogs of the process steps, as well as starved process steps. The buffer right before the bottleneck is the buffer inventory that is designed to offset the possible fluctuations in the flow. If the upstream process steps before the bottleneck produce too much, the buffer inventory will hold the surplus; if the upstream process steps produce too little, this buffer inventory will be used to feed the bottleneck in order to keep the bottleneck busy all the time. The inventory level in the buffer is used as the control signal. If the buffer inventory level becomes too high, then the rope will feed this information to the beginning step of the process; it will ask the incoming flow rate (the upstream production rate) to be reduced. On the other hand, if the buffer inventory level becomes too low, the rope will feed this information to the beginning of the process; it will ask the incoming flow rate (the upstream production rate) to be increased.

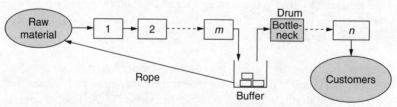

Figure 12.9 Drum-Buffer-Rope Production Control System

The theory of constraints also believes that the balanced capacity may not be a good approach for maintaining high throughput. The most important issue is to balance flow. This is illustrated by Example 12.4.

Example 12.4: Capacity Balance versus Flow Balance
Table 12.1 describes a sequential production process with five workstations. Specifically, the incoming job is first worked on station 1, then on each of stations 2 to 4, and finally on station 5. The second column in Table 12.1 lists the current task completion times for each workstation. Clearly, workstation 4 takes the longest time to complete its work, so it is the slowest workstation and the bottleneck of the whole production flow. A proposed improvement plan is improvement option 1, which balances the capacity of each workstation to be 53 seconds, in terms of work completion time. This is the balanced capacity approach. However, from the viewpoint of the theory of constraints, option 1 is not the desirable approach. Because there will always be fluctuations in the production that will affect the actual capacity of the process steps, if an upstream process is clogged, it will immediately become a bottleneck for the whole production system, as illustrated in Fig. 12.10. Improvement option 2 is a better approach, because it gives downstream process steps progressively bigger capacities so the clogs are less likely to form (Fig. 12.10).

12.3.4 Elevate the System's Constraint

After identification of the constraint, efforts should be made to increase the capacity of the constraint. For a policy constraint, a low cost or even a no-cost fix is possible. For a resource or material constraint, additional investment is usually needed. Elevation of the system constraint will usually increase the throughput. However, when the old constraint (the weakest link) is gone,

Table 12.1 Workstation Task Completion Times

Workstation	Work Completion Time (Seconds)	Improvement Option 1	Improvement Option 2
1	45	53	59
2	40	53	56
3	60	53	53
4	70	53	50
5	50	53	47
Total time	265	265	265

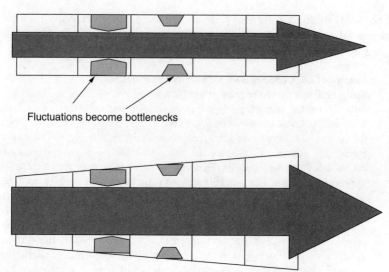

Fluctuations become bottlenecks

Progressively larger capacity downstream will make clogs unlikely

Figure 12.10 Balanced Capacity versus Unbalanced Capacity

there will always be another part of the process (second weakest link) that becomes a constraint. Example 12.5 shows this process.

Example 12.5: Constraints in a Restaurant
At the opening of a new restaurant, marketing is not good; there are only 50 customers per hour entering the restaurant. The dining room capacity is 100, and the kitchen capacity is 200 meals per hour. Clearly marketing is the constraint (Fig. 12.11). After a marketing campaign, more customers are attracted to the restaurant, at a rate of 250 customers per hour. But the dining hall can only hold 100 customers per hour, so the dining hall becomes a new constraint (Fig. 12.12).

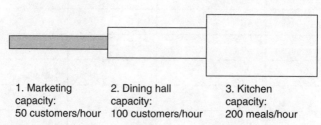

1. Marketing capacity: 50 customers/hour
2. Dining hall capacity: 100 customers/hour
3. Kitchen capacity: 200 meals/hour

Figure 12.11 Restaurant Grand Opening

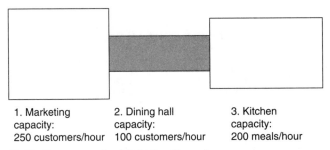

1. Marketing capacity: 250 customers/hour
2. Dining hall capacity: 100 customers/hour
3. Kitchen capacity: 200 meals/hour

Figure 12.12 Restaurant after Marketing Campaign

After that, the restaurant struggles to enlarge the capacity of the dining hall, by increasing the dining area, adding more dining tables, and increasing the table turnover rate. Then the capacity of the dining hall beomes 250 customers per hour. However, the kitchen can only cook 200 meals per hour, so the kitchen becomes a new bottleneck (Fig. 12.13). Now restaurant management will have to work on this third constraint.

12.4 Change Management

No matter which aspect of the DFSS deployment you are involved in, whether it is applying the theory of constraints, DFSS process design, or DFSS service product design, changes are inevitable and will shake many guarded and old paradigms. People's reaction to change varies from denial to pioneering and passes through many stages. On this venue, the objective of a DFSS team leader, such as a Black Belt, is to develop alliances for his or her efforts as the team progresses through the process. We depict the different stages of change in Fig. 12.14. The stages are linked by what is called *frustration curves*. We suggest that the DFSS team leader draw such

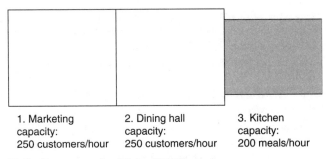

1. Marketing capacity: 250 customers/hour
2. Dining hall capacity: 250 customers/hour
3. Kitchen capacity: 200 meals/hour

Figure 12.13 Restaurant after Dining Hall Expansion

428 Chapter Twelve

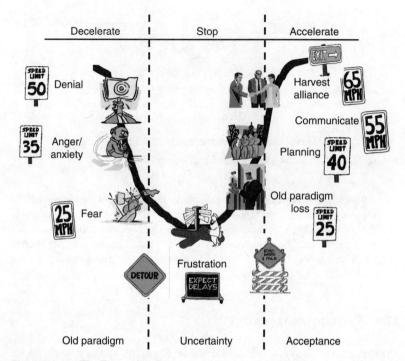

Figure 12.14 The Frustration Curve

a curve periodically for each team member and use some or all of the strategies listed below to move his or her team members to the positive side, the "recommitting" phase.

There are several strategies to use to deal with change. To help decelerate (reconcile), the Black Belt needs to listen with empathy, acknowledge difficulties, and define what is over and what is not. To help stop the old paradigm and reorient the team to the DFSS paradigm, the DFSS team leader should encourage redefinition, utilize management to provide structure and strength, rebuild a sense of identity, gain a sense of control and influence, and encourage opportunities for creativity. To help recommit (accelerate) the team in the new paradigm, the team leader should reinforce the new beginning, provide clear purpose, develop a detailed plan, be consistent in the spirit of Six Sigma, and celebrate success.

A successful change management should have the following elements:

1. *Grassroots participation:* You need to get support from all the people involved and affected.
2. *Upper-level management support:* Upper-level management involvement in change is definitely needed. Without upper-level management support, any change will be impossible. However, it is advised that micromanagement from the upper level should be avoided.
3. *Collaboration of team members:* Change requires joint efforts involving many people. The more people involved in all aspects of change, the better.
4. *Implementation of change in a stepwise fashion:* Because change implementation can be very draining and tiring, people can get very stressed and tired after some changes are implemented; people need to take a rest. However, you also need to guard against reversal of change.
5. *Measurement of the work:* You need to develop measurable metrics for the change, such as a quantitative measure or "change gate," that is, the exit criteria for a change stage. In this way, you will really know what type and how much progress you have made.
6. *Development of strategies:* You need several strategies. One is the overall strategy. The others are special strategies for every change stage. It is advised that several alternative strategies be prepared, so in case the first strategy does not work, you still have other alternatives available.
7. *Recognition of the fact that changes are often political:* If you only prepared the technical change plan and business change plan, you will almost certainly fail. Changes heavily involve people; you need to work on the people part of change and do your homework.
8. *Cultural sensitivity:* You need to be aware that cultural factors affect a person's ability to change, and you need to know how to deal with it. There are three levels of culture that must be addressed: culture of the country, culture of the organization, and culture of the department. You need to understand all three in order to facilitate the change.

REFERENCES

Aaker, D. A., *Building Strong Brand*, New York: The Free Press, 1996.
Aaker, D., and Joachimsthaler, E., *Brand Leadership*, New York: The Free Press, 2000.
Aaker, D., *Managing Brand Equity*, New York: The Free Press, 1991.
Akao Y., *Quality Function Deployment: Integrating Customer Requirements into Product Design*, Portland, OR: Productivity Press, 1990.
Allen, J., Robinson, C., and Stewart, D. (eds.), *Lean Manufacturing: A Plant Floor Guide*, Dearborn, MI: Society of Manufacturing Engineers, 2001.
Altshuller, G. S., *Creativity as Exact Science*, New York: Gordon & Breach, 1988.
Altshuller, G. S., "On the Theory of Solving Inventive Problems," *Design Methods and Theories*, vol. 24, no. 2, 1990, pp. 1216–1222.
Anderson, E., Rornell, C., and Lehmann, D., "Customer satisfaction, Market Share and Profitability: Findings from Sweden," *Journal of Marketing*, July 1994, pp. 53–66.
Anjard, R. P., "Process Mapping: A Valuable Tool for Construction Management and Other Professionals," *MCB University Press*, vol. 16, no. 3/4, 1998, pp 79–81.
Arnold, D. *The Handbook of Brand Management*, Reading, MA: Addison-Wesley, 1992.
Augustine, N. R., *Augustine's Laws*, New York: Viking, 1983.
Banks, J., Carson, J. S., Nelson, B. L., and Nicol, D. M., *Discrete-Event System Simulation*, 3rd ed., Englewood Cliffs, NJ: Prentice-Hall, 2001.
Batra, R., Lehmann, D., and Singh, D., The Brand Personality Component of Brand Goodwill: Some Antecedents and Consequences, *Brand Equity and Advertising: Adverting's Role in Building Strong Brands*, Aaker, D., and Biel, A., (eds.), Lawrence Erlbaum Hillsdale, NJ: Associates Publishers, 1993, pp. 83–96.
Berry, M., and Linoff, G. S., *Master Data Mining*, New York: Willey, 2000.
Biel, A. L., "Converting Brand Image into Equity," *Brand Equity and Advertising: Advertising's Role in Building Strong Brands*, Aaker, D., and Biel, A. (eds.), Hillsdale, NJ: Lawrence Erlbaum Associates Publishers, 1993, pp. 83–96.
Black, J. T., and Hunter, S. L., *Lean Manufacturing Systems and Cell Design*, Dearborn, MI: Society of Manufacturing Engineers, 2003.
Bothe, D. R., *Measuring Process Capabilities*, New York: McGraw-Hill, 1997.
Bremer, M., "Value Stream Mapping," *Rainmakers*, vol. 1, issue 8, May 28, 2002, www.imakenews.com/rainmakers/e_article000046701.cfm.
Bryant, J., "Customer Oriented Value Engineering," *Value World*, vol. 9, no. 1, 1986, p. 7.
Bussey, L. E., *The Economic Analysis of Industrial Projects*. Englewood Cliffs, NJ: Prentice-Hall, 1998.
Buzzel, R. D., and Gale, B. T., *The PIMS Principles, Linking Strategy to Performance*, New York: The Free Press, 1987.
Caulkin, S., "Chaos Inc.," *Across the Board*, July 1995, pp. 32–36.
Chowdhury, S., *Design for Six Sigma*, Chicago, IL: Dearborn Trade, a Kaplan Professional Company, 2002.
Clausing, D. P., *Total Quality Development: A Step by Step Guide to World-Class Concurrent Engineering*, New York: ASME Press, 1994.
Claxton, G., *Hare Brain, Tortoise Mind*, London: Fourth Estate, 1997.
Cohen, L., "Quality Function Deployment and Application Perspective from Digital Equipment Corporation," *National Productivity Review*, vol. 7, no. 3, Summer 1988, pp. 197–208.
Cohen, L., *Quality Function Deployment: How to Make QFD Work for You*, Reading, MA: Addison-Wesley, 1995.
Davis, S. M., *Brand Asset Management*, San Francisco: Jossey-Bass Inc., 2000.
de Brentani, U., "The New Product Process in Financial Services: Strategy for Success," *International Journal of Bank Marketing*, vol. 11, no. 3, 1993, pp. 15–22.

Deming, E., *Out of the Crisis*, Cambridge, MA: Massachusetts Institute of Technology, Center for Advanced Engineering Study, 1982.

Dixon, J. R., *Design Engineering: Inventiveness, Analysis, and Decision Making*, New York: McGraw-Hill, 1966.

Edelstein, H. A., *Introduction to Data Mining and Knowledge Discovery*, 3rd ed., Potomac, MD: Two Crows Corporation, 1999.

Fallon, C., *Value Analysis: Being a Revision of Value Analysis to Improve Productivity*, 2nd ed., North Brook, IL: Society of American Value Engineers, 1980.

Farquhar, P. H., "Managing Brand Equity," *Marketing Research*, vol. 3, 1989, pp. 24–33.

Fredriksson, B., *Holistic Systems Engineering in Product Development*, The Saab-Scania Griffin, Saab-Scania, AB, Linkoping, Sweden, November 1994.

Fuglseth, A., and Gronhaug, K., "IT-Enabled Redesign of Complex and Dynamic Business Processes; the Case of Bank Credit Evaluation," *OMEGA International Journal of Management Science*. vol. 25, 1997, pp. 93–106.

Galbraith, J. R., *Designing Complex Organizations*, Reading, MA: Addison-Wesley, 1973.

Gale, B., *Managing Customer Value*, New York: The Free Press, 1994.

Garvin, D. A., *Managing Quality: The Strategic and Competitive Edge*, Free Press, 1988.

George, M. L., *Lean Six Sigma for Service: How to Use Lean Speed and Six Sigma Quality to Improve Services and Transactions*, New York: McGraw-Hill, 2003.

Goetsch, D. L., and Davis, S. B., *Quality Management, Introduction to Total Quality Management for Production, Processing, and Services*, 3rd ed., Upper Saddle River, NJ: Prentice-Hall, 2000.

Goldratt, E., and Cox, J., *The Goal: A Process of Ongoing Improvement*, revised edition, Croton-on-Hudson, NY: North River Press, 1986.

Goldratt, E., *Theory of Constraints*, Great Barrington, MA: North River Press, 1990.

Harrell, C., and Tumay, K., *Simulation Made Easy: A Manager's Guide*, Norcross, GA: Industrial Engineering and Management Press, 1995.

Harris, R. L., "A Guide to Value Consciousness," *SAVE International Conference*, Southfield, MI, 1968.

Harry, M. J., *The Vision of 6-Sigma: A Roadmap for Breakthrough*, Phoenix, AZ: Sigma Publishing Company, 1994.

Harry, M. J., "Six Sigma: A Breakthrough Strategy for Profitability," *Quality Progress*, May 1998, pp. 60–64.

Hauser, J. R., and Clausing, D., "The House of Quality," *Harvard Business Review*, vol. 66, no. 3, May–June 1988, pp. 63–73.

Jacobson, R., and Aaker D., "The Strategic Role of Product Quality," *Journal of Marketing*, vol. 51, 1987, pp. 31–44.

Kapferer, J. N., *Strategic Brand Management*, 2nd ed., London: McGraw-Hill, 1997.

Kaufman, J., *Executive Overview*, Houston, TX: Cooper Industries, 1981.

Kaufman, J. J., *Value Engineering for the Practitioner*, 2nd ed., Raleigh, NC: North Carolina State University Press, College of Education, 1989.

Keller, G., et al. *Sap R/3 Business Blueprint: Understanding the Business Process Reference Model*, Upper Saddle River, NJ: Prentice-Hall, 1999.

Keller, K. L., "Conceptualizing, Measuring, and managing Customer-Based Brand Equity," *Journal of Marketing*, vol. 57, 1993, pp. 1–22.

Kim, S. J., Suh, N. P., and Kim, S. G., "Design of Software System Based on Axiomatic Design," *Annals of the CIRP*, vol. 40/1, 1991, pp. 165–170.

Kochan, N., (ed.), *The World's Greatest Brands*, New York: New York University Press, 1997.

Koltler, P. H., *Marketing Management: Analysis, Planning and Control*, 8th ed., Englewood Cliff, NJ: Prentice-Hall, 1991.

Kota, S., "Conceptual Design of Mechanisms Using Kinematics Building Blocks—A Computational Approach," Final Report of the NSF Design Engineering Program, Grant #DDM 9103008, October 1994.

Leigh, A., and Maynard, M., *ACE Teams—Creating Star Performance In Business*, Oxford: Butterworth-Heinnemann, 1993.

Levine, M., *A Branded World*, Hoboken, NJ: John Wiley and Sons, 2003.

Lientz, B. P., and Rea, K. P., *Breakthrough IT Change Management*, Burlington, MA: Elsevier Butterworth-Heinemann, 2004.

Liker, J. K., *The Toyota Way*, New York: McGraw-Hill, 2004.

Mabin, V. J., and Balderstone, S. J., *The World of the Theory of Constraints*, Boca Raton, FL: St. Lucie Press, 2000.

Magrab, E. B., *Integrated Product and Process Design and Development*, Boca Raton, FL: CRC Press LLC, 1997.

Mann, D. L., *Hands-on Systematic Innovation*, CREAX Press, 2002.

Mann D. L., *Hands-on Systematic Innovation for Business and Management*, Clevedon, UK: IFR Press, 2004.

Mann, D. L., and Domb, E., "40 Inventive (Business) Principles with Examples," *TRIZ Journal*, September 1999.

Marsh, D., Waters, F., and Mann, D., "Using TRIZ to Resolve Educational Delivery Conflicts Inherent to Expelled Students in Pennsylvania," *Triz-Journal*, November 2002, http://www.triz-journal.com.

Martin, D., and Martin, R., *TeamThink: Using the Sports Connection to Develop, Motivate, and Manage a Winning Business Team*, New York: Dutton, 1993.

Maskell, B. H., *Performance Measurement for World Class Manufacturing*, Productivity Press, 1991.

Mejabi, O., *Private Communications*, 2003.

Miles, B. L., "Design for Assembly: A Key Element within Design for Manufacture," *Proceedings of the Institution of Mechanical Engineeres, Part D, Journal of Automobile Engineering*, no. 203, 1989, pp. 29–38.

Miles, L. D., *Techniques for Value Analysis and value Engineering*, New York: McGraw-Hill, 1961.

Miles, L. D., *Techniques of Value Analysis and Engineering*, 2nd ed., New York: McGraw-Hill, 1972.

Monden, Y., *The Toyota Management System: Linking the Seven Key Functional Areas*, Portland, OR: Productivity Press, 1993.

Mostow, J., "Toward Better Models of the Design Process," *The AI Magazine*, 1985, pp. 44–57.

Mudge, A., *Value Engineering: A Systematic Approach*, New York: McGraw-Hill, 1971.

Murphy, J., *Brand Valuation: Establishing a True and Fair Value*, London: Hutchinson Business Books, 1990.

Nordlund, M., Tate, D., and Suh, N. P., "Growth of axiomatic design through Industrial Practice," *3rd CIRP Workshop on Design and implementation of Intelligent Manufacturing Systems*, Tokyo, Japan, June 19–21, 1996, pp. 77–84.

O'Brien, B. C., *Perspective*, Park, R. J. (ed.) (unpublished).

Oakland, J. S., *Total Quality Management: The Route to Improving Performance*. 2nd ed., Oxford: Butterworth-Heineman, 1994.

Ohno, T., *Toyota Production System: Beyond Large Scale Production*, Portland, OR: Productivity Press, 1990.

Pande, P. S., Neuman, R. P., and Cavanagh, R. R., *The Six Sigma Way: How GE, Motorola, and Other Top Companies are Honing Their Performance*, New York: McGraw-Hill, 2000.

Park, R. J., *Value Engineering, A Plan for Invention*, Boca Raton, FL: Saint Lucie Press, 1999.

Park, R. J., *Value Engineering*, Birmingham, MI: R.J. Park & Associates, Inc., 1992.

Peppard, J., and Rowland, P., *The Essence of Business Process Re-engineering*, Hemel Hempstead: Prentice-Hall Europe, 1995.

Peter, L. J., *The Peter Pyramid*, London: Allen & Unwin, 1986.

Peters, T., *In Search of Excellence*, New York HarperCollins, November 1982.

Peters, T., *The Circle of Innovation*, London: Hodder & Stoughton, 1997.

Pugh, S., *Creating Innovative Products Using Total Design*, edited by Clausing, D. and Andrade, R. Reading, MA: Addison-Wesley, 1996.

Pugh, S., *Total Design: Integrated Methods for Successful Product Engineering*, Reading, MA: Addison-Wesley, 1991.

Rae, K., "TRIZ and Software—40 Principle Analogies," *Triz-Journal*, Pt. 1, September 2001a, http://www.triz-journal.com.

Rae, K., "TRIZ and Software—40 Principle Analogies," *Triz-Journal*, Pt. 2, November 2001*b*, http://www.triz-journal.com.

Ramaswamy, R., *Design and Management of Service Processes*, Reading, MA: Addison-Wesley, 1996.

Rantanen, K., "Altshuler's Methodology in Solving Inventive Problems," *ICED-88*, Budapest, August 23–25, 1988.

Rea, L. M., and Parker, R. A., *Designing and Conducting Survey Research: A Comprehensive Guide*, San Francisco, CA: Jossey-Bass, 1992.

Reeve D., *Value Engineering Analysis of the Oakland County Youth Assistance Program*, Master Thesis, Rochester Hills, MI: Oakland University, 1975.

Reis, A., *Positioning: The Battle for Your Mind*, New York: McGraw-Hill, 1981.

Retseptor, G., "40 Inventive Principles in Quality Management," *Triz-Journal*, March 2003, http://www.triz-journal.com.

Roberts, W., *Leadership Secrets of Attila The Hun*, London: Transworld, 1989.

Rother, M., and Shook, J., *Learning to See: Value-Stream Mapping to Create Value and Eliminate Muda*, Brookline MA: The Lean Enterprise Institute, Inc., 2003.

Schmenner, R. W., *Plant and Service Tours in Operations Management*, 4th ed., New York: McMillan, 1994.

Sherden, W. A., *Market Ownership*, New York: American Management Association, 1994.

Shigeru, M., *Management for Quality Improvement, The Seven New QC Tools*, Cambridge, MA: Productivity Press, 1988.

Shingo, S., *The Toyota Production System from an Industrial Engineering Viewpoint*, Portland, OR: Productivity Press, 1989.

Stein, R. E., "Re-Engineering the Manufacturing System: Applying the Theory of Constraints," New York: Marcel Dekker, 1996.

Suh, N. P., *The Principles of Design*, New York: Oxford University Press, 1990.

Suh, N. P., "Impact of Axiomatic Design," *3rd CIRP Workshop on Design and the Implementation of Intelligent Manufacturing Systems*, Tokyo, Japan, June, 19–22, 1996, pp. 8–17.

Suh, N., *Axiomatic Design: Advances and Applications*, Oxford: Oxford University Press, 2001.

Taguchi, G., Chowdhury, S., and Taguchi S., *Robust Engineering*, New York: McGraw-Hill, 2000.

Taguchi, G., *Taguchi Methods,* vol. 1, *Research and Development*, Japan Standard Association, Livonia, MI: ASI Press, 1994.

Taguchi, G., *Taguchi on Robust Technology Development*, New York: ASME Press, 1993.

Taguchi, G., and Wu, Y., *Introduction to Off-line Quality Control*, Magaya, Japan: Central Japan Quality Control Association, 1979. (Available from American Supplier Institute, Livonia, MI.)

Taguchi, G., Elsayed, E., and Hsiang, T., *Quality Engineering in Production Systems*, New York: McGraw-Hill, 1989.

Taguchi, G., *Introduction to Quality Engineering*, White Plains, NY: UNIPUB/Kraus International Publications, 1986.

Tauber, F. M., "Brand Leverage: Strategy for Growth in a Cost Controlled World," *Journal of Advertising Research*, vol. 28, (August-September), 1988, pp. 26–30.

Tsourikov, V. M., "Inventive Machine: Second Generation," *Artificial Intelligence & Society*, Spring or Verlag, no. 7, 1993, pp. 62–77.

Wasserman, M., "Sales Market Value," in *SAVE Proceedings*, Park R. J. (ed.), Southfield, MI, 1977.

Watkins, T., *Economics of the Brand*, London: McGraw-Hill, 1986.

Western Electric Co. Inc., *Statistical Quality Control Handbook*, New York: Western Electric Co., 1956.

Woeppel, M. J., *Manufacturer's Guide to Implementing the Theory of Constraints*, Boca Raton, FL: CRC Press LLC, 2001.

Womack, J. P., and Jones, D. T., *Lean Thinking*, New York: The Free Press, 2003.

Womack, J. P., Jones, D. T., and Roos, D., *The Machine That Changed the World*, New York: Rawson Associates, 1990.

Yang, K., "Improving Automotive Dimensional Quality by Using Principal Component Analysis," *Quality and Reliability Engineering International*, vol. 12, 1996, pp. 401–409.

Yang, K., and El_Haik, B., *Design for Six Sigma: A Roadmap for Product Development*, New York: McGraw-Hill, 2003.

Younker, D. L., *Value Engineering: Analysis and Methodology*, New York: Marcel Dekker, 2003.

Zhang, J., Chai, K., and Tan, K., "40 Inventive Principles with Applications in Service Operations Management," *Triz-Journal*, December 2003, http://www.triz-journal.com.

Zlotin, B., Zusman, A., Kaplan, L., Visnepolschi, S., Proseanic V., and Malkin S. A., "TRIZ Beyond Technology: The Theory and Practice of Applying TRIZ to Non-technical Areas," *Triz-Journal* January 2001.

INDEX

A

Aaker's brand identity model, 193–196, 199
Abandonment rate
　in service factory processes, 330
　in telephone service processes, 342
Accuracy in purchasing and supply
　　processes, 353
Achievement goals in job plans, 177–178
Actions in TRIZ, 233–234
　continuity of useful action principle,
　　269–270
　partial or excessive actions principle, 266
　periodic action principle, 269
　preliminary action principle, 262–263
　preliminary antiaction principle, 262
Activities in FAST diagrams, 154
Actual costs in cost visibility analysis, 136
Administrative costs in cost visibility
　　analysis, 137
Affinity diagrams, 105
Allowances in cost visibility analysis, 137
Alternatives
　in DFSS service process design, 45
　in DFSS service product design, 38–39
　in value engineering, 131–132
Analysis
　brand strategy, 212–216
　cost visibility, 136–137
　in information phase, 150–154
　for processes
　　logistics and distribution, 348–349
　　manufacturing, 325–326
　　office, 328–329
　　product development, 315–316
　　professional service, 340–341
　　project shop, 345–346
　　purchasing and supply, 353
　　pure service shop, 335–336
　　retail service store, 338–339
　　service factory, 331–332
　　telephone service, 343–344
　　transportation, 350–351
　in TRIZ
　　analysis diagrams, 233–234
　　contradiction, 245

　　functional, 231–236, 244–245
　　S-curve, 245
　　value, 56, 94
Analyze phase
　in DFSS service process design, 44–45
　in DFSS service product design, 36–39
Another dimension principle, 267–268
Another sense principle, 274–275
Antiaction principle, 262
Antiweight principle, 261–262
Apparent solution level in inventions,
　　227–228
Arithmetic means, 396–397
Arrows in IDEF0 process mapping,
　　358–359
Assembly lines, 321–323
Assisting functions in TRIZ, 232
Associations with brands
　in brand equity, 206
　description, 201–202
　product-related, 194–196
Asymmetry principle, 259–260
Attitudes
　Likert scale for, 68
　in plan development, 178–179
Attributes in quality function
　　deployment, 123
Automobile dealership construction case
　　study, 180–181
Autonomy for creativity, 310
Average queue length
　in checkout lines, 337
　in service factory processes, 330

B

Balanced plants, 414
Basic functions
　in FAST diagrams, 153
　for information phase, 143
　in TRIZ, 231
Batch flow shops, 320–321
Beforehand cushioning principle, 263
Behavior, buying, 190–192
Benchmarking in quality function
　　deployment, 110–111, 113

437

Index

Benefits
  in brands, 199
  in value, 50–51
Best costs for functions, 163
Best value method, 50
Bias in surveys, 59
Binomial distributions, 403–404
Black Belt members, 15
Blank columns and rows in quality function deployment, 112–113
Blessing in disguise principle, 270–271
Blocking, process, 305
Boarding time in transportation processes, 350
Book center case study, 122–126
Bottleneck resources, 414
Box plots, 395–396
Boxes in IDEF0 process mapping, 358–359
Brainstorming
  in creative phase, 165
  in design analyze phase, 45
  in process design, 379
  in process excellence, 311
  in quality function deployment, 123
  in value engineering, 131–132
Brands
  associations
    in brand equity, 206
    description, 201–202
    product-related, 194–196
  buying behavior with, 190–192
  definition, 187–188
  development of, 187–190, 206–207
    communication and marketing in, 222–223
    evaluation in, 223–225
    implementation, 222–223
    key factors in, 207–210
    process overview, 210–212
    strategy analysis, 212–216
    strategy development, 217–222
  equity, 203–206
  identity of
    brand as organization, 196
    brand as person, 196–198
    brand as product, 194–196
    in brand development, 207–208
    brand persona, 202–203
    Davis's brand image model, 200–203
    features, 192–193
    models, 193–194
    value proposition in, 198–200
  loyalty, 204
  name awareness, 204
  perceived quality, 205–206
Budgets for surveys, 63
Buffer inventories in lean operation, 372
Buffers in DBR strategy, 424
Burden overhead in cost visibility analysis, 136
Business case in quality function deployment, 113
Business contradiction matrices, 290–297
Business inventive principles, 255–283
Business requirements in define phase, 30–34
Business strategy changes with brands, 196
Buying behavior, 190–192

C

Calm atmosphere principle, 280
Capability in quality measures, 406
Capability indices, 407–409
Carrying capacity utilization, 348
Case studies
  process management, 380–391
  quality function deployment, 122–126
  value engineering
    automobile dealership construction, 180–181
    engineering department organization analysis, 181–185
Cause-and-effect diagrams
  in process design, 45
  in process diagnosis, 378
Cellular manufacturing
  in lean operation, 372–373
  in manufacturing process, 318–320
Central tendency, measures of, 396–397
Champion members, 15
Change management, 427–429
Channel strategy, 222
Cheap short-living objects principle, 274
Checklist questions for surveys, 66
Checkout errors, 337
Checkout lines, 337
Chronological order for survey questions, 70
Classification in TRIZ, 230, 245–246
Closed-ended survey questions, 65–67

Cluster sampling, 77
Collaboration in change management, 429
Color changes principle, 277
Commercial costs in cost visibility analysis, 137
Commitment in process excellence, 309
Communication
  in brand development
    implementation, 222–223
    promotion, 209
  in brand strategy development, 221–222
  in process excellence, 311
Comparisons, paired, 170–172
Competitor analysis
  in brand development, 210
  in brand strategy analysis, 216
  customer value analysis, 94
  in quality function deployment, 110–111
Complexity of surveys, 59
Composite structures principle, 281–283
Concepts
  in FAST diagrams, 153
  in TRIZ problem-solving process, 246
Conflicts in quality function deployment, 113, 121
Constraint theory, 413–416
  basic concepts, 416
  change management, 427–429
  implementation process, 421–427
  material constraints, 420–421
  policy constraints, 418–420
  resource constraints, 420
  throughput, inventory, and operating expenses in, 417–418
Constraints
  elevating, 425–427
  exploiting, 422–423
  identifying, 421–422
  for processes, 305–307
  subordination for, 423–425
Continuity of useful action principle, 269–270
Continuous random variables, 400
Contracts in brand positioning, 218–219
Contradictions in TRIZ, 227
  analysis, 245
  business contradiction matrices, 290–297
  definition, 230
  eliminating, 246–254
  physical, 239–240

  table of inventive principles, 284–289
  technical, 238–239
Controllable factors in brand development, 207
Convenience factors
  in service organizations, 7
  for surveys, 58
  in value, 51
Conventional solution level in inventions, 227–228
Coordination, product-brand, 220–221
Copying principle, 273–274
Core processes, 300–301
Correcting functions in TRIZ, 232
Cost-function relationships, 158–164
Costs
  in information development
    cost visibility analysis, 134–137
    sources, 137–138
  in processes
    logistics and distribution, 347–348
    manufacturing, 324
    office, 328
    product design and development, 314–315
    project shop, 345
    purchasing and supply, 352
    pure service shop, 335
    service factory, 331
    transportation, 350
  of surveys, 58–59
  and value, 52
Country of origin in product brands, 195
Creative design in DFSS, 26–27
Creative phase in value engineering job plans, 164–166
Creativity
  in functions, 145–149
  in process excellence, 309–310
Credibility in brand positioning, 219
Critical-to-delivery (CTDs) characteristics, 101
Critical-to-quality (CTQs) characteristics, 101
Critical-to-satisfaction (CTS) metrics, 31–34
Cultural sensitivity in change management, 429
Culture in brand development, 210
Current state value stream mapping, 364–369
Curvature principle, 265

Cushioning principle, 263
Customer-oriented design, 25–26
Customer service quality and interaction in processes, 3, 8, 10–11
  logistics and distribution, 348
  office, 328
  professional service, 340
  pure service shop, 334
  retail service store, 337–338
  service factory, 330
  telephone service, 343
  transportation, 350
Customer surveys, 57
  administering, 72–73
  instrument design, 64
    length, 71–72
    question order, 70–71
    question types, 65–69
    question wording, 69–70
  interviewer selection and training, 64
  sample size determination for, 78–82
  sampling methods in, 73–78
  stages of, 60–65
  types of, 57–60
  use of, 61–62
Customers
  in brand strategy, 213–214
  desirability indexes, 108
  needs and requirements
    in DFSS design phase, 30–34
    in value creation, 56
    in quality function deployment
      attributes, 106–107, 123
      competitive assessments, 110–111
      intent, 106
  value for
    in business excellence, 21
    competitive analysis, 94
    deployment, 94–99
    maps, 89–93
Cycle time in value stream mapping, 367

D

Data analysis of surveys, 65
Data collection
  in DFSS analyze phase, 45
  in process diagnosis, 378
Data-driven management, 393
Davis's brand image model, 200–203
DBR (drum-buffer-rope) strategy, 423–424
Decision-making in process excellence, 310–311
Declined stage in technical system evolution, 243
Defects in processes
  office, 327
  pure service shop, 333
  service factory, 330
Define, Measure, Analyze, Design, and Verify (DMADV) procedure, 20
Define phase
  in DFSS service process design, 42–43
  in DFSS service product design, 30–34
Delivery
  in logistics and distribution processes, 347
  in project shop processes, 344–345
  in service, 3, 8–9, 11, 23
Demand-driven production systems, 373–374
Demographics in brand development, 209–210
Dependent events, 414
Dependent functions in FAST diagrams, 154
Deploying value, 56
Descriptive statistics
  graphical, 394–396
  numerical, 396–399
Design
  for Lean Six Sigma, 20–21
  in process management, 302–303
  in service, 23
  for value maximization, 55–56
Design for Six Sigma. *See* DFSS (Design for Six Sigma)
Design phase
  in DFSS service process design, 45–46
  in DFSS service product design, 40–41
Design process, 378–379
DFSS (Design for Six Sigma), 18–20
  customer-oriented design in, 25–26
  in service industry, 27–30
  service process design phases, 42–43
    analyze, 44–45
    define, 42–43
    design, 45–46
    measure, 43–44
    verify, 46
  service product design phases, 30
    analyze, 36–39

define, 30–34
design, 40–41
measure, 34–36
verify, 41–42
system design and creative design in, 26–27
Taguchi method and fire prevention philosophy in, 27
Diagnosis, process, 302, 378
Diagnostics in quality function deployment, 119
Diagrams
FAST, 150–154
establishing, 157–158
symbols and graphs in, 154–157
functional analysis, 233–234
tree, 105
Differentiation of brands
in brand positioning, 218
failures in, 195
Direct involvement, 60
Direct measurement, 60
Directions in FAST diagrams, 154–155
Discarding principle, 277–278
Discipline in process excellence, 309
Discovery level in inventions, 229
Discrete random variables, 399–400
Distribution
in brand development, 209
logistics and distribution processes, 5, 346–349
Distributions, statistical
binomial, 403–404
expected value, variance, and standard deviation, 400–401
exponential, 403
frequency, 394–395
normal, 401–403
Poisson, 404
probability, 399–405
statistical parameter estimation, 404–405
DMADV (Define, Measure, Analyze, Design, and Verify) procedure, 20
DMAIC strategy, 18–19
Documentation of processes
elements of, 307–308
process mapping, 377–378
Dot plots, 394–395
Drop-off points in transportation processes, 349–350

Drum-buffer-rope (DBR) strategy, 423–424
Duplicated survey elements, 74
Dynamic interactions in processes, 305
Dynamic routing, 316–317
Dynamics of value, 53–55
Dynamics principle, 266

E

Economic factors
in service organizations, 7
in value, 51
Efficiency
in job shop processes, 317
lean operation techniques for, 369
in product design and development processes, 314
80/20 percent relationships, 170
Emotional benefits in brands, 199
Encounters, service, 11
Engineering department organization analysis case study, 181–185
Enriched atmosphere principle, 279–280
Equipment for processes, 306
Equity, brand, 203–206
Errors in processes
office, 327
professional service, 339
project shop, 345
pure service shop, 333
retail service store, 337
service factory, 330
telephone service, 343
Esteem value, 54
Evaluation
in brand development, 223–225
function, 150–154
project, 17
in TRIZ, 231
in value engineering, 132, 166–172
Events
dependent, 414
rare, 404
Evolution in TRIZ
of technical systems, 240–243
trends in, 230
Excellence
ideality as, 237–238
in processes, 309–311
roles in, 21

442 Index

Excessive actions principle, 266
Exchange value, 54
Exit time in transportation processes, 350
Expected value, 400–404
Expenses in constraint theory, 415, 417–418
Exponential distributions, 403
Extensions of brands, 196
Eye-openers in quality function deployment, 113, 121–122

F

Facility design, 8
Factory service process, 4
Factory within a factory concept, 318
Fair-value zones in customer value maps, 91
FAST (functional analysis system technique) diagrams, 150–154
   establishing, 157–158
   symbols and graphs in, 154–157
Feedback
   in project shop processes, 345
   in technical system evolution, 242
   in TRIZ, 271–272
Fields in TRIZ, 233, 236
Final results in TRIZ, 230, 245
Financial resources for processes, 306
Fire prevention philosophy, 27
Fit in brand positioning, 220
Fixed burden in cost visibility analysis, 136
Fixed costs in cost visibility analysis, 136
Flexibility
   in manufacturing processes, 324
   in office processes, 328
Flow shops, 320–321
Flowcharts
   in process mapping, 356–357
   Six Sigma, 17
Flows
   in cellular manufacturing, 319
   in lean operation, 370–372
   in value stream mapping, 364
Fluidity principle, 275
Foreign elements in surveys, 74
Freedom for creativity, 310
Freight costs in cost visibility analysis, 137
Frequency distributions, 394–395
Frustration curves, 427–428
Full rollout
   in DFSS service process design, 46
   in DFSS service product design, 42

Functional analysis
   in information phase, 150–154
   in TRIZ, 230–236, 244–245
Functional analysis system technique (FAST) diagrams, 150–154
   establishing, 157–158
   symbols and graphs in, 154–157
Functional benefits in brands, 199
Functional factors
   in service organizations, 6–7
   in value, 50
Functionality in TRIZ, 230
Functions
   for creativity, 145–149
   defining, 139–142
   in information phase, 138–150
   processes in, 149–150
   in product brands, 194–195
   of products, 52
   in TRIZ
      modeling, 231–236, 244–245
      resources for, 237
      statements, 232–233
   in value engineering
      evaluating, 131
      identifying, 130–131
      types of, 142–145
Funnel pattern for survey questions, 70
Future state value stream mapping, 375–377

G

Goals
   in quality function deployment, 114
   of surveys, 61
   in value engineering job plans, 177–178
Government regulations, 306–307
Graphical descriptive statistics, 394–396
Graphs in FAST, 154–157
Grassroots participation in change management, 429
Green Belt members, 15
Group technology in cellular manufacturing, 318
Growth stage in technical system evolution, 242
Guidelines for processes, 307

H

Handover
   in DFSS service process design, 46
   in DFSS service product design, 42

Hard brand associations, 206
Harmful functions in TRIZ, 232
Hierarchy of needs, 54–55
High-order functions in FAST, 150
High performance capability in DFSS
    design, 40
High-value zones in customer value maps, 91
Highest-order functions in FAST diagrams, 152
Histograms, 394–395
Holes principle, 276
Homogeneity principle, 277
House of quality, 105–106
How Much in quality function
    deployment, 110
Hows in quality function deployment, 107
    correlation, 109–110
    identification, 114–117
    importance calculations, 117
    quality characteristics, 124

I

Ideal final result (IFR) in TRIZ
    calculating, 245
    in ideality, 238
    as limit, 230
Ideality in TRIZ
    calculating, 245
    elements of, 237–238
IDEF0 process mapping, 358–359
Identification tasks in process mapping, 377
Identify, Design, Optimize, and Verify
    (IDOV) procedure, 20
Identifying constraints, 421–422
Identity
    of brands
        brand as organization, 196
        brand as person, 196–198
        brand as product, 194–196
        in brand development, 207–208
        brand persona, 202–203
        Davis's brand image model, 200–203
        features, 192–193
        models, 193–194
        value proposition in, 198–200
    in TRIZ, 230
IDOV (Identify, Design, Optimize, and
    Verify) procedure, 20
IFR (ideal final result) in TRIZ
    calculating, 245
    in ideality, 238
    as limit, 230

Image in brand strategy self-analysis, 215
Image models, brand, 200–203
Implementation
    in brand development, 222–223
    in constraint theory, 421–427
    in process management, 303, 379–380
    of surveys, 65
    in value engineering job plans, 132, 176–179
Importance in quality function deployment
    calculations, 117–119
    ratings, 108–109
Improvement in inventions, 228
In-person interviews, 58–59
In-person surveys, 73
Incremental costs in cost visibility
    analysis, 136
Independent functions in FAST diagrams, 154
Infancy stage in technical system
    evolution, 242
Information
    surveys for, 62–63
    in TRIZ, 237
Information flow
    in purchasing and supply processes, 353
    in value stream mapping, 365
Information phase in value engineering job
    plans, 133
    cost-function relationship in, 158–164
    FAST diagram construction in, 154–158
    function analysis and evaluation in,
        150–154
    function determination in, 138–150
    information development in, 134–138
Instrument design, survey, 64
    length, 71–72
    question order, 70–71
    question types, 65–69
    question wording, 69–70
Intermediary principle, 272–273
Interval-scale variables, sample size for,
    81–82
Interval scales for survey questions, 67
Interviewers
    bias in, 59
    selecting and training, 64
Invention outside technology level,
    228–229
Inventive principles in TRIZ
    business, 255–283
    contradiction table of, 284–289
    regular, 246–254

Inventories
  in constraint theory, 415, 417–418
  in lean operation, 372
  in processes
    manufacturing, 324
    office, 327
    purchasing and supply, 353
Inventory questions for surveys, 66
Inverted funnel pattern for survey questions, 71

J
Job plans in value engineering, 132
  creative phase, 164–166
  evaluation phase, 166–172
  implementation phase, 176–179
  information phase, 133
    cost-function relationship in, 158–164
    FAST diagram construction in, 154–158
    function analysis and evaluation in, 150–154
    function determination in, 138–150
    information development in, 134–138
  planning phase, 172–174
  reporting phase, 174–176
Job shops, 316–318

K
Kano model of quality, 111–112
Key customer identification
  in DFSS service process design, 43
  in DFSS service product design, 30
Knowledge
  in process design, 379
  in process excellence, 310
  in TRIZ, 237

L
Labor in cost visibility analysis, 136
Layout in process mapping, 377
Lead time
  in processes
    manufacturing, 323
    office, 327
    product design and development, 314
    purchasing and supply, 353
  in value stream mapping, 367
Leading survey questions, 69
Lean operation principles, 362–363
  in DFSS service process design, 45

one-piece flow in, 370–372
pull-based production in, 373–374
techniques, 369–375
value stream mapping
  current state, 364–369
  future state, 375–377
waste elimination, 362–364
work cells in, 372–373
Lean Six Sigma design, 20–21
Lemons into lemonade principle, 270–271
Length
  questionnaire, 71–72
  queue
    in checkout lines, 337
    in service factory processes, 330
Levels of measurement for survey questions, 67
Liabilities in value, 51
Lifecycle costs in product design and development, 314–315
Likert scales, 68
Line flow shops, 321–323
Local quality principle, 258–259
Logistics and distribution processes, 5, 346–349
Lowest-order functions in FAST diagrams, 153
Loyalty, brand, 204

M
Machine village process, 370
Mail-out surveys
  administering, 72
  advantages and disadvantages of, 58
Main basic functions in TRIZ, 231
Maintenance in process management, 303, 380
Management support in change management, 429
Manufacturing processes, 316
  analysis, 325–326
  batch flow shop, 320–321
  cellular manufacturing, 318–320
  job shop, 316–318
  line flow shop, 321–323
  performance metrics, 323–324
Mapping processes, 302, 355–356
  analyzing, 44
  in case study, 381

developing, 43
flowcharts in, 356–357
IDEF0, 358–359
tasks in, 377–378
value stream, 359–362
  analyzing, 44, 378
  current state, 364–369
  future state, 375–377
Maps, customer value, 89–93
Market-perceived profiles
  price, 88–89
  quality, 84–88
Market position, 216
Market segment, 218
Marketing
  in brand development implementation, 222–223
  in brand strategy development, 221–222
Maslow, Abraham, 54
Master Black Belt members, 15
Materials
  as constraints, 420–421
  in cost visibility analysis, 136
  in purchasing and supply processes, 352
  in value stream mapping, 365
Matrices
  business contradiction, 290–297
  in quality function deployment, 107–109, 114–117
Maturity stage in technical system evolution, 242–243
Maximizing value, 55–56
Means, 396–397
Measure phase
  in DFSS service process design, 43–44
  in DFSS service product design, 34–36
Measurement levels for survey questions, 67
Measurement of performance. *See* Performance metrics
Measures of central tendency, 396–397
Measures of relative standing, 398–399
Measures of variation, 397–398
Medians
  determining, 397
  in histograms, 395
  in percentiles, 399
Merging principle, 260
Methods and tools in Six Sigma, 17
Metrics. *See* Performance metrics

Missing elements in surveys, 74
Models
  brand identity, 193–196, 199
  brand image, 200–203
  functional, 244–245
  quality, 111–112
Moneymaking machines, 416
Monitoring processes, 380
Motivation and needs hierarchy, 54–55
Muda, 363
Multiple-choice survey questions, 65–66
Multipurpose survey questions, 69

N
Name awareness for brands, 204
Needs hierarchy, 54–55
Nested doll principle, 261
Net profit in constraint theory, 417
Nominal scales for survey questions, 67
Nonbasic but beneficial functions in TRIZ, 231–232
Nonprobability survey sampling, 75–78
Nonproduct-related characteristics in brand personality, 198
Normal distributions, 401–403
Numerical descriptive statistics, 396
  measures of central tendency, 396–397
  measures of relative standing, 398–399
  measures of variation, 397–398

O
Objectives
  in FAST diagrams, 153–154
  in quality function deployment, 113
  realistic, 110
  of surveys, 61
Objects in TRIZ, 232–233
Obsolete material inventory, 353
Office processes, 4, 326–329
On-time arrivals in transportation processes, 350
On-time receipts in purchasing processes, 352
One-piece flow in lean operation, 370–372
One-shop service organizations, 2
Open-ended survey questions, 68–69
Operation expenses in constraint theory, 415, 417–418
Operation items in quality function deployment, 124, 126
Opinions, Likert scale for, 68

Optimizing processes, 301–302
Order of survey questions, 70–71
Order processing, 353
Ordinal scales for survey questions, 67
Organization, brands as, 196
Organizational infrastructure, 15–16
Other way around principle, 264–265
Outlining task in process mapping, 377
Overflow calls in telephone service processes, 342
Overhead in cost visibility analysis, 136

P

Paced assembly lines, 322
Paired comparisons, 170–172
Paradigm shift principle, 279
Parameter changes principle, 278
Parameter estimation, statistical, 404–405
Pareto voting, 169–170
Partial or excessive actions principle, 266
PATH (promise, acceptance, trust, and hope) for brand names, 189
Path functions in FAST diagrams, 154
PDLS (process description languages), 307–308
Perceived quality
 brands, 205–206
 and value, 52–53
Percentiles
 in histograms, 395
 in measures of relative standing, 399
Perceptions with brands
 in buying decisions, 191
 in competitor analysis, 216
Performance
 in cost-function relationships, 160
 in DFSS service process design, 44
 in DFSS service product design, 40
 in product brands, 195
 and value, 52
Performance metrics
 for batch flow shops, 321
 for cellular manufacturing, 320
 for change management, 429
 for DFSS service product design, 34–36
 for line flow shops, 322–323
 for processes, 308
  job shop, 318
  logistics and distribution, 347–348
  manufacturing, 323–324

 office, 327–328
 product design and development, 314
 professional service, 339
 project shop, 345
 purchasing and supply, 352–353
 pure service shop, 333–335
 retail service store, 337
 service factory, 330–331
 telephone service, 342
 transportation, 350
 for quality, 406
 for value, 51–53
Periodic action principle, 269
Persona, brand, 202–203
Personalities, brand, 197–198
Personnel for processes, 306
Persons, brands as, 196–198
Physical contradictions in TRIZ, 239–240
Pickup points in transportation processes, 349–350
Pilot implementations for processes, 379
Pilot tests
 in DFSS service process design, 46
 in DFSS service product design, 41
Placement in brand development, 209
Planning in value engineering, 132, 172–174, 178–179
Planning matrices in quality function deployment, 109
Plans, job. *See* Job plans in value engineering
Poisson distributions, 404
Policy constraints, 418–420
Politics in change management, 429
Population of data
 for probability distributions, 399
 for surveys, 63–64, 73–75
Positioning in brand strategy development, 217–220
Pregnancy stage in technical system evolution, 241
Preliminary action principle, 262–263
Preliminary antiaction principle, 262
Premium freight expenditures, 352
Pretesting surveys, 64
Price
 in brand development, 208–209, 222
 in brand personality, 198
 and value, 52
Price profiles, market-perceived, 88–89

Primary path functions in FAST
    diagrams, 154
Privacy of surveys, 58–59
Probability distributions, 399–400
    binomial, 403–404
    expected value, variance, and standard
        deviation, 400–401
    exponential, 403
    normal, 401–403
    Poisson, 404
    statistical parameter estimation, 404–405
Probability sampling, 75–76
Problems in TRIZ
    classification, 230, 245–246
    definition, 230, 243–245
    problem-solving process, 243–246
Process description languages (PDLS),
    307–308
Processes and process management
    capability of
        indices, 407–409
        quality measures, 406
    case study, 380–391
    control in
        in DFSS service process design, 46
        in DFSS service product design, 42
    core, 300–301
    design in, 378–379
        in case study, 387–391
        in DFSS service process design, 43–45
        in services, 8
    diagnosis in, 378
        in case study, 381–387
        in DFSS service process design, 44–45
    documentation, 307–308
    elements of, 304–305
    excellence in, 309–311
    features of, 303–304
    implementation, 379–380
    introduction, 299–303
    lean operation principles. *See* Lean
        operation principles
    logistics and distribution, 346–349
    maintaining, 380
    manufacturing and production, 316–326
    mapping. *See* Mapping processes
    office and transaction, 326–329
    optimizing, 301–302
    performance metrics for, 308
    product design and development, 312–316
    professional service, 339–341
    project shop, 344–346
    purchasing and supply, 352–355
    pure service shop, 332–336
    quality measures of, 406
    resources and constraints for, 305–307
    retail service store, 336–339
    service factory, 329–332
    supporting, 301
    telephone service, 341–344
    transportation, 349–351
Product-brand coordination, 220–221
Product-related associations, 194–196
Product scope in brands, 194
Production batch size in value stream
    mapping, 367
Production lead time, 323
Production lines, 321–323
Production processes, 316–326
Productivity in constraint theory, 417
Products
    in brand development, 208
    in brand strategy self-analysis, 215
    brands as, 193–196
    design, 8, 312–316
    in services, 9
Professional service processes, 5,
    339–341
Profitability in service organizations, 6
Profits
    in constraint theory, 417
    in cost visibility analysis, 137
Project charters
    in DFSS service process design, 42–43
    in DFSS service product design, 30
Project scope
    in FAST diagrams, 152
    in information development, 138
Project shop processes, 5, 344–346
Projects
    evaluation, 17
    execution, 16
    flowcharts, 17
    management, 17
    in quality function deployment, 113–114
    selection, 16
    team members, 15
Promise, acceptance, trust, and hope (PATH)
    for brand names, 189
Promotion in brand development, 209

Proportions, sample size for, 78–81
Psychological factors
  in service organizations, 6–7
  in value, 50–51
Public relations in brand development, 223
Pull-based production, 373–374
Purchasing and supply processes, 5, 352–355
Pure service shop processes, 4, 332–336
Purpose of surveys, 61
Push-based production, 374

Q

QFD. *See* Quality function deployment (QFD)
Quality
  perceived, 205–206
  in processes
    manufacturing, 323
    professional service, 340
    project shop, 345
    pure service shop, 334
    retail service store, 337–338
    service factory, 330
    telephone service, 343
    transportation, 350
  product brands, 195
  and value, 52–53
Quality function deployment (QFD), 101–103
  analysis in, 112–113
  benefits, assumptions, and realities in, 103–104
  case study, 122–126
  in DFSS, 19
  example, 113–122
  history, 103
  Kano model of quality, 111–112
  methodology overview, 104–111
  summary, 126–128
  for value deployment, 56
Quality levels
  without mean shifts, 409–410
  with mean shifts, 410–411
Quality measures, 405–406
  process capability indices, 407–409
  process performance and capability, 406
Quality profiles, market-perceived, 84–88
Questionnaires
  length of, 71–72
  in quality function deployment, 125

Questions in surveys
  order, 70–71
  types, 65–69
  wording, 69–70
Queue length
  in checkout lines, 337
  in service factory processes, 330
Quick setup time reductions, 375

R

Random flow patterns, 319
Random sampling, 76
Random variables, 399–405
Ranges, 397–398
Ranking in paired comparisons, 171
Rapid development stage in technical system evolution, 242
Rare events, Poisson distributions for, 404
Rate of return (ROI) in constraint theory, 417
Rating questions for surveys, 66
Ratings in quality function deployment, 108–109
Realistic objectives, 110
Recovering principle, 277–278
Redesign process, 378–379
Redials in telephone service processes, 342
Refining
  in DFSS service process design, 46
  in DFSS service product design, 41
Regulations for processes, 306–307
Relationship matrices, 107–108, 114–117
Relative change principle, 279
Relative standing, measures of, 398–399
Remove tension principle, 263–264
Reneges in telephone service processes, 342
Reporting phase in value engineering job plans, 132, 174–176
Reports from surveys, 65
Requirements
  in DFSS service product design, 30–34
  in value creation, 56
Resonance principle, 268–269
Resources
  bottleneck, 414
  as constraints, 420
  for processes, 305–307
    mapping, 377
    professional service, 340
    project shop, 345
    pure service shop, 335
  in TRIZ, 230, 236–237

Response rates of surveys, 58–59
Restaurant service
  analyze phase in, 36–39
  customer needs for, 31–34
Results
  of surveys, 61–62
  in TRIZ, 230, 245
Retail service store processes, 4, 336–339
Return on brand development, 223
Revenue in service organizations, 6
Revisions in value stream mapping, 360
Rework
  in professional service processes, 340
  in project shop processes, 345
Roadblocks in value engineering job plans, 174
Robustness
  in DFSS service product design, 40
  in product design and development processes, 314
ROI (rate of return) in constraint theory, 417
Roles in business excellence, 21
Rollout
  in DFSS service process design, 46
  in DFSS service product design, 42
Roofs in quality function deployment, 109–110
Rope in DBR strategy, 424
Routing in job shop processes, 316–317

S
S-curves in TRIZ, 240–243, 245
Samples of data, 399
Sampling for surveys
  methods for
    cluster, 77
    frames, 63–64, 73–75
    nonprobability, 75–78
    probability, 75–76
    random, 76
    systematic, 76–77
  selection process, 64
  size, 64, 78–82
Scales for survey questions, 66–68
Schedules for surveys, 63
Science in inventions, 228–229
Scope
  in FAST diagrams, 152
  in information development, 138
  in product brands, 194

Screening techniques in value engineering job plans, 166–172
Secondary functions for information phase, 143
Secondary research, 60
Secondary useful functions, 231–232
Segmentation principle, 257–258
Selection
  survey interviewers, 64
  survey samples, 64
  in value engineering job plans, 169–172
Self-analysis in brand strategy, 214–216
Self-expressive benefits in brands, 199
Self-service principle, 273
Serial flow patterns, 319
Service
  DFSS for. *See* DFSS (Design for Six Sigma)
  Six Sigma for, 21–23
  system design and creative design in, 26–27
Service costs
  in pure service shop processes, 335
  in service factory processes, 331
Service factors in value, 51
Service factory processes, 329–332
Service industry applications of TRIZ, 254–255
Service organizations, 1–6
  Six Sigma in. *See* Six Sigma overview
  success factors for, 6–12
Service processes, 3
Service products, 2–3
  in DFSS. *See* DFSS (Design for Six Sigma)
  value creation in, 47–48
  value maximization in, 55–56
Service time in processes
  professional service, 340
  pure service shop, 334
  service factory, 330
  telephone service, 342
Shipment damage and loss, 347
Shopping carts, 338
Short-living objects principle, 274
Shortages in purchasing and supply processes, 352
Sidewalk surveys, 78
Significance in quality function deployment, 113
Simulation in process diagnosis, 378
Single-piece flow, 370–372

Six Sigma overview, 12
  fundamental beliefs, 13–14
  methods and tools, 17
  organizational infrastructure, 15–16
  projects in
    evaluation, 17
    execution, 16
    flowcharts, 17
    management, 17
    selection, 16
    team members, 15
  summary, 18
  training, 16
Six Sigma standard, 14
Size, sample, 64, 78–82
Skills in process excellence, 310
Skipping principle, 270
Small invention inside paradigm level in inventions, 228
Snowball sampling, 78
Soft brand associations, 206
Solution generation in TRIZ, 230, 246
Space resources in TRIZ, 236
Specifications in FAST diagrams, 153–154
Spread, process, 406–407
Stakeholders, communication with, 311
Standard deviation, 398, 400–404
Standard normal distributions, 402
Starving, process, 305
Statements in TRIZ, 232–233
Statistics
  basics, 393–394
  descriptive
    graphical, 394–396
    numerical, 396–399
  quality measures, 405–406
    process capability indices, 407–409
    process performance and capability, 406
    quality levels with mean shifts, 410–411
    quality levels without mean shifts, 409–410
  random variables and probability distributions, 399–400
    binomial, 403–404
    expected value, variance, and standard deviation, 400–401
    exponential, 403
    normal, 401–403
    Poisson, 404
  statistical parameter estimation, 404–405

Stepwise implementation in change management, 429
Strategy analysis in brand development, 212–216
Strategy development
  for brands, 217–222
  in change management, 429
Stratified random sampling, 77
Strengths
  in brand strategy competitor analysis, 216
  in quality function deployment, 121
Stress in surveys, 59
Subjects in TRIZ, 232–233
Substance resources in TRIZ, 236
Substantial invention inside technology level in inventions, 228
Success factors for service organization, 6–12
Superior performance in DFSS service product design, 40
Supply and purchasing processes, 5, 352–355
Supporting functions
  in FAST diagrams, 154
  in TRIZ, 232
Supporting processes, 301
Surveys
  customer. See Customer surveys
  in quality function deployment, 125
Sustainability in brand positioning, 219–220
Symbols
  in FAST, 154–157
  in flowcharts, 356–357
System design, 26–27
Systematic survey sampling, 76–77

T
Taguchi method, 27
Taking out principle, 258
Target market segment in brand positioning, 218
Targets in quality function deployment, 110
Team members
  in change management, 429
  in Six Sigma organizations, 15–16
Technical competitive assessments, 111
Technical contradictions, 238–239, 246–254
Technical importance ratings, 108
Technical systems, evolution of, 240–243
Techniques for equipment, 306
Technological evolution analysis, 230

Technology in inventions, 228–229
Telephone service processes, 5, 341–344
Telephone surveys, 59
  administering, 72–73
  advantages and disadvantages of, 59
Theory of inventive problem solving (TIPS).
  See TRIZ (theory of inventive problem solving)
Thin and flexible principle, 275–276
Threatening survey questions, 69
Throughput
  in constraint theory, 415, 417–418
  in processes
    manufacturing, 324
    office, 327–328
    pure service shop, 334
    service factory, 331
Time pressure of surveys, 58
Time resources in TRIZ, 236
Time to complete in project shop processes, 345
Time to market in brand development, 209
Time waiting for assistance in retail service store processes, 337
Tools
  in Six Sigma, 17
  in TRIZ, 230, 245–246
Total costs in cost visibility analysis, 137
Training
  equipment users, 306
  in process excellence, 310
  Six Sigma, 16
  survey interviewers, 64
Transaction costs in office processes, 328
Transaction processes, 326–329
Transportation processes, 5, 349–351
Transportation time, 350
Travel costs, 350
Tree diagrams, 105
Tree pattern for survey questions, 71
TRIZ (theory of inventive problem solving)
  contradictions in, 227
    analysis of, 245
    business contradiction matrices, 290–297
    definition, 230
    eliminating, 246–254
    physical, 239–240
    table of inventive principles, 284–289
    technical, 238–239
  in DFSS, 19
  elements in, 230–231

function modeling and functional analysis in, 231–236
ideality in, 237–238
inventive principles in
  business, 255–283
  contradiction table of, 284–289
  regular, 246–254
problem-solving process in, 243–246
resources in, 236–237
S-curves and evolution of technical systems in, 240–243
service industry applications of, 254–255
for value deployment, 56
Turn lemons into lemonade principle, 270–271

U

Uncontrollable factors in brand development, 209–210
Uniqueness in brand positioning, 219
Units of analysis in survey sampling, 73–75
Universality principle, 260–261
Unnecessary costs, 129
Unpaced assembly lines, 322
Unspoken wants, 111
Upper-level management support in change management, 429
Use value, 54
Users in product brands, 195
Uses in product brands, 195

V

Validation
  in DFSS service process design, 46
  in DFSS service product design, 42
Validation plans for processes, 379
Value
  analyzing, 56
  in brand positioning, 219
  creating, 47–48
  deploying, 56
  elements of, 48–51
  maximizing, 55–56
  metrics for, 51–53
  versatility and dynamics of, 53–55
Value councils, 179
Value engineering, 129–132
  case studies
    automobile dealership construction, 180–181
    engineering department organization analysis, 181–185

Value engineering (*Cont.*)
  evaluation, planning, reporting, implementation in, 132
  functions in
    evaluating, 131
    identifying, 130–131
    types of, 142–145
  job plans in. *See* Job plans in value engineering
Value management, 83–84
  customer value in
    competitive analysis, 94
    deployment, 94–99
    maps, 89–93
    market-perceived profiles
      price, 88–89
      quality, 84–88
Value of systems in cost-function relationships, 160
Value proposition in brands, 198–200
Value stream mapping, 359–362
  analysis of
    in DFSS service process design, 44
    in process diagnosis, 378
  current state, 364–369
  future state, 375–377
Variability, process, 406
Variable burden in cost visibility analysis, 137
Variables
  expressed in proportions, 78–81
  random, 399–405
Variance, 398, 400–404
Variation measures, 397–398
Vehicle usage
  in logistics and distribution processes, 348
  in transportation processes, 350

Verbs in TRIZ, 232–233
Verify phase
  in DFSS service process design, 46
  in DFSS service product design, 41–42
Versatility of value, 53–55
Virtual cells, 319
Vision development in brand strategy analysis, 212–213
Voice of the customer (VOC), 25, 101
Voting, Pareto, 169–170

W
Waiting time
  in pure service shop processes, 333–334
  in service factory processes, 330
  in telephone service processes, 342
Waste in processes, 300, 362–364
Weak columns and rows in quality function deployment, 112–113
Weighting in paired comparisons, 171
Whats in quality function deployment
  determining, 123
  identifying, 114–115
  obtaining, 106–107
  weak, 119
Work cells in lean operation, 372–373
Work-in-process inventories
  in lean operation, 372
  in manufacturing process, 324
  in office processes, 327
Working populations for survey sampling, 74

Y
Yaesu Book Center case study, 122–126